Lecture Notes in Artificial Intelligence 4676

Edited by J. G. Carbonell and J. Siekmann

Subseries of Lecture Notes in Computer Science

T0224493

Matthias Klusch   Koen Hindriks
Mike P. Papazoglou   Leon Sterling (Eds.)

Cooperative Information Agents XI

11th International Workshop, CIA 2007
Delft, The Netherlands, September 19-21, 2007
Proceedings

 Springer

Series Editors

Jaime G. Carbonell, Carnegie Mellon University, Pittsburgh, PA, USA
Jörg Siekmann, University of Saarland, Saarbrücken, Germany

Volume Editors

Matthias Klusch
German Research Center for Artificial Intelligence
Stuhlsatzenhausweg 3, 66121 Saarbruecken, Germany
E-mail: klusch@dfki.de

Koen Hindriks
TU Delft, ICT, Man-Machine-Interaction
Delft, The Netherlands
E-mail: k.v.hindriks@tudelft.nl

Mike P. Papazoglou
Tilburg University, INFOLAB
PO Box 90153, 5000 LE Tilburg, The Netherlands
E-mail: mikep@uvt.nl

Leon Sterling
The University of Melbourne
442 Auburn Road, Hawthorn, Victoria 3122, Australia
E-mail: leon@csse.unimelb.edu.au

Library of Congress Control Number: 2007934752

CR Subject Classification (1998): I.2.11, I.2, H.4, H.3.3, H.2.8, C.2.4, H.5

LNCS Sublibrary: SL 7 – Artificial Intelligence

ISSN   0302-9743
ISBN-10   3-540-75118-1 Springer Berlin Heidelberg New York
ISBN-13   978-3-540-75118-2 Springer Berlin Heidelberg New York

Springer is a part of Springer Science+Business Media

springer.com

© Springer-Verlag Berlin Heidelberg 2007
Printed in Germany

Typesetting: Camera-ready by author, data conversion by Scientific Publishing Services, Chennai, India
Printed on acid-free paper   SPIN: 12161805   06/3180   5 4 3 2 1 0

Preface

These are the proceedings of the 11th International Workshop on Cooperative Information Agents (CIA 2007), held at the Delft University of Technology, The Netherlands, September 19–21, 2007.

In today's world of ubiquitously connected heterogeneous information systems and computing devices, the intelligent coordination and provision of relevant added-value information at any time, anywhere is of key importance to a variety of applications. This challenge is envisioned to be coped with by means of appropriate intelligent and cooperative information agents.

An information agent is a computational software entity that has access to one or multiple heterogeneous and geographically dispersed data and information sources. It pro-actively searches for and maintains information on behalf of its human users, or other agents preferably just in time. In other words, it is managing and overcoming the difficulties associated with information overload in open, pervasive information and service landscapes.

Each component of a modern cooperative information system is represented by an appropriate intelligent information agent capable of resolving system and semantic heterogeneities in a given context on demand. Cooperative information agents are supposed to accomplish both individual and shared joint goals depending on the actual user preferences in line with given or deduced limits of time, budget and resources available. One major challenge of developing agent-based intelligent information systems in open environments like the Internet and the Web is to balance the autonomy of networked data, information, and knowledge sources with the potential payoffs of leveraging them by the use of cooperative and intelligent information agents.

The objective of the international workshop series on cooperative information agents (CIA), since its establishment in 1997, is to provide a small but very distinguished interdisciplinary forum for researchers, programmers, and managers to become informed about, present, and discuss the latest high-quality results in research and development of agent-based intelligent and cooperative information systems and applications for the Internet and the Web. Each event of the series offers regular and invited talks of excellence that are given by renown experts in the field, a selected set of system demonstrations, and honors innovative research and development of information agents by means of best a paper award and system innovation award giving. The proceedings of the series are regularly published as volumes of the *Lecture Notes in Artificial Intelligence* (LNAI) series of Springer.

In keeping with its tradition, this year's workshop featured a sequence of regular and invited talks of excellence given by leading researchers covering a broad area of topics of interest. CIA 2007 featured 4 invited and 20 regular papers selected from 38 submissions. The result of the peer-review of all contributions is included in this volume that is, we believe, again rich in interesting, inspiring, and advanced work on the research and development of intelligent information agents worldwide. All workshop proceedings have been published by Springer as *Lecture Notes in Artificial Intelligence* volumes: 1202 (1997), 1435 (1998), 1652 (1999), 1860 (2000), 2182 (2001), 2446 (2002), 2782 (2003), 3191 (2004), 3550 (2005), 4149 (2006).

This year the *CIA System Innovation Award* and the *CIA Best Paper Award* were sponsored by Whitestein Technologies AG, Switzerland, and the CIA workshop series, respectively. There was also some financial support available to a limited number of students as (co-)authors of accepted papers to present their work at the CIA 2007 workshop; these grants were sponsored by the IEEE FIPA standard committee, and the Belgian-Dutch Association for Distributed AI (BN-VKI).

The CIA 2007 workshop was organized in cooperation with the Association for Computing Machinery (ACM), in particular the ACM special interest groups SIG-Web, SIG-KDD, SIG-CHI and SIG-Art. We are very grateful and indebted to our sponsors for their financial support, which made this event in its convenient form only possible. The sponsors of CIA 2007 were:

ICT Research Center, TU Delft, The Netherlands

Whitestein Technologies, Switzerland

IEEE Computer Society Standards Organisation Committee on Intelligent and Physical Agents (FIPA)

Belgian-Dutch Association for Distributed AI (BNVKI)

Dutch research school for Information and Knowledge Systems (SIKS)

We are particularly thankful to the authors and invited speakers for contributing their latest results in relevant areas to this workshop, as well as to all members of the Program Committee, and the external reviewers for their critical reviews of submissions. Finally, a particularly cordial thanks goes to the local organization team from TU Delft for providing us with a traditionally comfortable and all-inclusive location, and a very nice social program in the beautiful, typically Dutch city of Delft.

September 2007

Matthias Klusch
Koen Hindriks
Mike Papazoglou
Leon Sterling

Organization

Co-chairs

Matthias Klusch   DFKI, Germany, *Chair*
Koen Hindriks   TU Delft, The Netherlands
Mike P. Papazoglou   University Tilburg, The Netherlands
Leon Sterling   University Melbourne, Australia

Program Committee

Wolfgang Benn   TU Chemnitz, Germany
Sonia Bergamaschi   University of Modena, Italy
Abraham Bernstein   University of Zurich, Switzerland
Monique Calisti   Whitestein Technologies, Switzerland
Boi Faltings   EPF Lausanne, Switzerland
Rune Gustavsson   TH Blekinge, Sweden
Heikki Helin   TeliaSonera, Finland/Sweden
Michael Huhns   University of South Carolina, USA
Catholijn Jonker   TU Delft, The Netherlands
Hillol Kargupta   UMBC, USA
Uwe Keller   DERI Innsbruck, Austria
Ryszard Kowalczyk   Swinburne University, Australia
Manolis Koubarakis   TU Crete, Greece
Sarit Kraus   Bar-Ilan University, Israel
Daniel Kudenko   University York, UK
Alain Leger   France Telecom, France
Victor Lesser   University Massachusetts, USA
Jiming Liu   Hong Kong Baptist University, China
Stefano Lodi   University of Bologna, Italy
Werner Nutt   FU Bozen-Bolzano, Italy
Sascha Ossowski   University Rey Juan Carlos, Spain
Terry Payne   University of Southampton, UK
Adrian Pearce   University of Melbourne, Australia
Alun Preece   University of Aberdeen, UK
Cartic Ramakrishnan   Wright State University, USA
Jeffrey Rosenschein   Hebrew University, Israel
Amit Sheth   Wright State University, USA
Steffen Staab   University of Koblenz, Germany
Katia Sycara   Carnegie Mellon University, USA

Valentina Tamma	University Liverpool, UK
Jan Treur	VU Amsterdam, The Netherlands
Rainer Unland	University Duisburg-Essen, Germany
Gottfried Vossen	University Muenster, Germany
Gerhard Weiss	SCCH, Austria
Steve Willmott	UPC Barcelona, Spain

External Reviewers

Vincent Louis
Mohan Baruwal Chhetri
Jakub Brzostowski
Eduardo Rodrigues Gomes
Ingo Mueller
Meenakshi Nagarajan
Lutz Neugebauer
Matt Perry
Frank Seifert
Kuldar Taveter
Marco Wiering
Steven Willmott

Table of Contents

Invited Contributions

Information Search and Processing

Applications

Rational Cooperation

Interaction and Cooperation

Trust

Managing Sensors and Information Sources Using Semantic Matchmaking and Argumentation

Alun Preece

University of Aberdeen, Computing Science, Aberdeen, UK
apreece@csd.abdn.ac.uk

Abstract. Effective deployment and utilisation of limited and constrained intelligence resources — including sensors and other sources — is seen as a key issue in modern multinational coalition operations. In this talk, I will examine the application of semantic matchmaking and argumentation technologies to the management of these resources. I will show how ontologies and reasoning can be used to assign sensors and sources to meet the needs of missions, and show how argumentation can support the process of gathering and reasoning about uncertain evidence obtained from sensor probes.

1 Introduction

Effective deployment and utilisation of limited and constrained intelligence, surveillance and reconnaisance (ISR) resources is seen as a key issue in modern network-centric joint-forces operations. For example, the 2004 report *JP 2-01 Joint and National Intelligence Support to Military Operations* states the problem in the following terms: "ISR resources are typically in high demand and requirements usually exceed platform capabilities and inventory. ... The foremost challenge of collection management is to maximize the effectiveness of limited collection resources within the time constraints imposed by operational requirements."[1]

Our work focusses upon the application of Virtual Organisation technologies to manage coalition resources. In the past we have shown an agent-based VOs can manage the deployment and utilisation of network resources in a variety of domains, including e-business, e-science, and e-response [1,4]. Two distinguishing features of our work are (1) the use of semantically-rich representations of user requirements and resource capabilities, to support matchmaking using ontologies and reasoning, and (2) the use of argumentation to support negotiation over scarce resources, decisions about which resources to use, and the combining of evidence from information-providing resources (e.g. sensors).

In this talk, I will examine the application of (1) and (2) to the management of ISR resources in the context of coalition operations. The first part of the talk describes an ontology-based approach to the problem of assigning sensors and sources to meet the needs of missions. The second part then looks at how argumentation and subjective logic can facilitate the process of gathering uncertain evidence through a series of sen-

---

[1] http://www.dtic.mil/doctrine/jel/new_pubs/jp2_01print.pdf, pages III–10–11, accessed April 27, 2007.

M. Klusch et al. (Eds.): CIA 2007, LNAI 4676, pp. 1–4, 2007.
© Springer-Verlag Berlin Heidelberg 2007

sor probes, and combining that evidence into a set of arguments in support of, and in opposition to, a particular decision.

2 Semantic Matchmaking of Sensors and Missions

The assignment of ISR assets to multiple competing missions can be seen as a process comprising two main steps: (1) assessing the fitness for purpose of alternative ISR means to accomplish a mission, and (2) allocating available assets to the missions. Our work draws upon current military doctrine, specifically the Missions and Means Framework (MMF) [5] which provides a model for explicitly specifying a *mission* and quantitatively evaluating the utility of alternative warfighting solutions: the *means*.

Figure 1 shows how missions map to ISR means. Starting from the top left the diagram sketches the analysis of a mission as a top-down process that breaks a mission into a collection of operations (e.g. search-and-rescue), each of which is broken down further into a collection of distinct tasks having specific capability requirements (e.g. wide-area surveillance). On the right hand side, the diagram depicts the analysis of capabilities as a bottom-up process that builds up from elementary components (e.g. electro-optincal/infrered (EO/IR) camera) into systems (e.g. camera turret), and from systems up into platforms equipped with or carrying those systems (e.g. an unmanned aerial vehicle (UAV)).

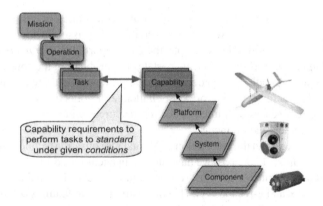

Fig. 1. Overview of the Mission and Means Framework (MMF)

In particular, we propose the use of ontologies to support the following activities:

- specifying the requirements of a mission;
- specifying the capabilities provided by ISR assets (sensors, platforms and other sources of intelligence, such as human beings);
- comparing — be a process of automated reasoning —the specification of a mission against the specification of available assets to either decide whether there is a solution (a single asset or combination of assets) that satisfies the requirements of a mission, or alternatively providing a ranking of solutions according to their relative degree of utility to the mission.

Although one can envisage a single ontology supporting the entire sensor-mission matchmaking process, actually we adhere to the Semantic Web vision of multiple inter-linking ontologies covering different aspects of the domain. First, we provide an ontology based on the MMF, which is basically a collection of concepts and properties that are essential to reason about the process of analyzing a mission and attaching the means required to accomplish it (mission, task, capability, or asset). Then we provide a second ontology that refines some of the generic concepts in the MMF ontology so as to represent the ISR-specific concepts that constitute our particular application domain. This second ontology comprises several areas frequently organized as taxonomies, such as a classification of sensors (acoustic, optical, chemical, radar) and information sources, a classification of platforms (air, sea, ground and underwater platforms), a classification of mission types, or a classification of capabilities. As noted in the previous section, there are existing ontologies covering at least part of each of these domains.

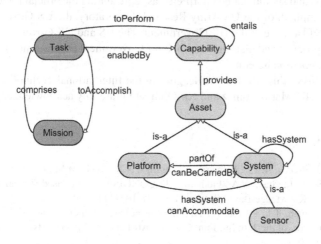

Fig. 2. Main ontological concepts and their relationships

Figure 2 shows a high level view of the main concepts and relationships that support our semantic matchmaking approach. On the left hand side, we find the concepts related to the mission: a mission comprises several tasks that need to be accomplished. On the right hand side we find the concepts related to the means: a sensor is a system that can be carried by or constitutes part of a platform; inversely, a platform can accommodate or have one or more systems, and both platforms and systems are assets; an asset provides one or more capabilities; a capability can entail a number of more elementary capabilities, and is required to perform certain type of tasks and inversely, a task is enabled by a number of capabilities.

The talk will give further details of the ontology and its application.

3   Arguing About Evidence in Partially Observable Domains

In this talk, we also show how argumentation can be used to manage the process of gathering and reasoning about evidence from sensors and sources [2]. Because such sources are fallible, and the military domain typically involves environments that are only partially observable, we needed to devise a novel framework for argumentation in domains containing uncertainty [3]. The concept of argument schemes is built into the framework, allowing for a rich set of primitives to be utilised in the argumentation process. We have also attempted to cater for other important concepts in argument such as accrual of arguments, defaults, and burden of proof. While the lowest levels of the framework are general enough to be applied to almost any area in which argument is used, the higher levels are aimed at evidential reasoning, incorporating abstract models of sensors and the notion of obtaining information from the environment.

Acknowledgments. This research was sponsored by the US Army Research Laboratory and the UK Ministry of Defence and was accomplished under Agreement Number W911NF-06-3-0001. The views and conclusions contained in this document are those of the author(s) and should not be interpreted as representing the official policies, either expressed or implied, of the US Army Research Laboratory, the US Government, the UK Ministry of Defence or the UK Government. The US and UK Governments are authorized to reproduce and distribute reprints for Government purposes notwithstanding any copyright notation hereon.

Contributions to this talk from colleagues in the International Technology Alliance project, especially Mario Gomez and Nir Oren, are gratefully acknowledged.

References

1. Norman, T.J., Preece, A.D., Chalmers, S., Jennings, N.R., Luck, M.M., Dang, V., Nguyen, T., Deora, V., Shao, J., Gray, W.A., Fiddian, N.J.: CONOISE: Agent-based formation of virtual organisations. Knowledge-Based Systems 17(2–4), 103–111 (2004)
2. Oren, N., Norman, T.J., Preece, A.: Argumentation based contract monitoring in uncertain domains. In: Proc. of the 20th Int. Joint Conf. on Artificial Intelligence, Hyderabad, India, pp. 1434–1439 (2007)
3. Oren, N., Norman, T.J., Preece, A.: Subjective logic and arguing with evidence. Artificial Intelligence Journal (page to appear)
4. Preece, A., Chalmers, S., McKenzie, C.: A reusable commitment management service using semantic web technology. Knowledge-Based Systems 20(2), 143–151 (2007)
5. Sheehan, J.H., Deitz, P.H., Bray, B.E., Harris, B.A., Wong, A.B.H.: The military missions and means framework. In: Proceedings of the Interservice/Industry Training and Simulation and Education Conference, pp. 655–663 (2003)

Towards a Delegation Framework for Aerial Robotic Mission Scenarios

P. Doherty[1] and John-Jules Ch. Meyer[2]

[1] Dept of Computer and Information Science,
Linköping University, Sweden
patdo@ida.liu.se
[2] Dept of Information and Computing Sciences,
Utrecht University, The Netherlands
jj@cs.uu.nl

Abstract. The concept of delegation is central to an understanding of the interactions between agents in cooperative agent problem-solving contexts. In fact, the concept of delegation offers a means for studying the formal connections between mixed-initiative problem-solving, adjustable autonomy and cooperative agent goal achievement. In this paper, we present an exploratory study of the delegation concept grounded in the context of a relatively complex multi-platform Unmanned Aerial Vehicle (UAV) catastrophe assistance scenario, where UAVs must cooperatively scan a geographic region for injured persons. We first present the scenario as a case study, showing how it is instantiated with actual UAV platforms and what a real mission implies in terms of pragmatics. We then take a step back and present a formal theory of delegation based on the use of 2APL and KARO. We then return to the scenario and use the new theory of delegation to formally specify many of the communicative interactions related to delegation used in achieving the goal of cooperative UAV scanning. The development of theory and its empirical evaluation is integrated from the start in order to ensure that the gap between this evolving theory of delegation and its actual use remains closely synchronized as the research progresses. The results presented here may be considered a first iteration of the theory and ideas.

1 Background

The use of Unmanned Aerial Vehicles (UAVs) which can operate autonomously in dynamic and complex operational environments is becoming increasingly more common. While the application domains in which they are currently used are still predominantly military in nature, we can expect to see widespread usage in the civil and commercial sectors in the future as guidelines and regulations are developed by aeronautics authorities for insertion of UAVs in civil airspace.

One particularly important application domain where UAVs could be of great help in the future is in the area of catastrophe assistance. Such scenarios include natural disasters such as earthquakes or tsunamis or man-made disasters caused by terrorist activity. In such cases, civil authorities often require a means of

M. Klusch et al. (Eds.): CIA 2007, LNAI 4676, pp. 5–26, 2007.
© Springer-Verlag Berlin Heidelberg 2007

acquiring an awareness of any situation at hand in real-time and the ability to monitor the progression of events in catastrophe situations. Unmanned aerial vehicles offer an ideal platform for acquiring the necessary situation awareness to proceed with rescue and relief in many such situations. It is also often the case that there is no alternative in acquiring the necessary information because one would like to avoid placing emergency services personal in the line of danger as much as possible.

For a number of years, The Autonomous Unmanned Aerial Vehicle Technologies Lab (AutUAVTech Lab) at Linköping University, Sweden, has pursued a long term research endeavour related to the development of future aviation systems in the form of autonomous unmanned aerial vehicles [7,8]. The focus has been on both high autonomy (AI related functionality), low level autonomy (traditional control and avionics systems), and their integration in distributed software architectural frameworks [9] which support robust autonomous operation in complex operational environments such as those one would face in catastrophe situations.

More recently, our research has moved from single platform scenarios to multi-platform scenarios where a combination of UAV platforms with different capabilities are used together with human operators in a mixed-initiative context with adjustable platform autonomy. The application domain we have chosen to pursue is emergency services assistance. Such scenarios require a great deal of cooperation among the UAV platforms and between the UAV platforms and human operators. We require a principled means of developing such cooperative frameworks and believe that formal specification is a requirement that will help control this complexity and also ensure the development of safe and robust multi-platform system interaction frameworks.

One of the key conceptual ideas that tie together mixed initiative problem solving, adjustable autonomy and cooperative behavior is that of delegation. In the first iteration of formal specifications we have chosen to focus on the development of a formal theory of delegation and to test it empirically in the context of one or more parts of a complex emergency services rescue scenario.The methodology we are using to develop the required technology is to iterate on formal specification and empirical evaluation in parallel in order to ensure that the gap between theory and practice does not become insurmountable.

For a number of years, the Intelligent Systems Group at Utrecht University, Holland, has pursued a long term research project with the goal of developing formal specifications of multi-agent systems and the development of 2APL/3APL [12,6], an agent programming language with a formal semantics. Recently, the two research groups have begun a collaborative endeavor which focuses on the development of formal specifications of cooperative behavior and instantiation of such a framework in the UAV domain. One of the starting points is a formal specification of delegation and its instantiation in an emergency services scenario.

As a basis for a formal theory of delegation, our starting point is recent work by Castelfranchi and Falcone [4,10]. This is a general human-inspired theory of (several forms of) delegation. It appears that this offers a very good starting point for our purposes. In particular their notions of *strong* or *strict* delegation with *open* and *closed* variants are appropriate in our setting.

We also require a formal logical framework to tighten up some of the insights provided by Castelfranchi and Falcone and to extend and provide new additions required for our robotics-inspired theory of delegation. For this we have chosen the KARO formalism [13] which is an amalgam of dynamic logic and epistemic/doxastic logic.

The UAV platforms currently used in our experimentation are rotor-based systems which have been extended with an avionics system for low-level autonomy and a complex CORBA-based distributed software architecture for high-level autonomy and integration with avionics. In order to integrate cooperative functionality with the existing system, we have chosen to use JADE [2] as a basis for this and ACL as the communication language between UAVs. The delegation framework will be embedded in this extension.

1.1 Paper Structure

The article will be structured as follows. In section 2, a catastrophe assistance scenario will be described which involves the use of 2 or more UAVs. In section 3, we briefly describe the actual UAV platforms used in our flight tests. In section 4, we describe the existing software architecture used with our platforms and a new addition which wraps a JADE layer around the legacy system. In section 7, the concept of strong delegation is introduced and formalized using the KARO formalism and 2APL. In section 8, we return to the scenario and use the new theory of delegation to formally specify many of the communicative interactions related to delegation used in the cooperative UAV scanning scenario. Section 9 offers some conclusions.

This article should be viewed as tentative and as a first iteration of work that will be energetically pursued in the future. It is exploratory in nature and although we do hope that much of the work reported will remain intact, it may be subject to some change and modification in the future.

2 An Emergency Service Scenario

On December 26, 2004, a devastating earthquake of high magnitude occured off the west coast off Sumatra. This resulted in a tsunami which hit the coasts of India, Sri Lanka, Thailand, Indonesia and many other islands. Both the earthquake and the tsunami caused great devastation. During the initial stages of the catastrophe, there was a great deal of confusion and chaos in setting into motion rescue operations in such wide geographic areas. The problem was exacerbated by shortage of manpower, supplies and machinery. Highest priorities in the initial stages of the disaster were search for survivors in many isolated areas where road systems had become inaccessible and providing relief in the form of delivery of food, water and medical supplies.

Let's assume for a particular geographic area, one had a shortage of trained helicopter and fixed-wing pilots and/or a shortage of helicopters and other aircraft.
Let's also assume that one did have access to a fleet of autonomous unmanned
helicopter systems with ground operation facilities. How could such a resource
be used in the real-life scenario described?

A pre-requisite for the successful operation of this fleet would be the existence
of a multi-agent (UAV platforms, ground operators, etc.) software infrastructure
for assisting emergency services in such a catastrophe situation. At the very least,
one would require the system to allow mixed initiative interaction with multiple
platforms and ground operators in a robust, safe and dependable manner. As
far as the individual platforms are concerned, one would require a number of
different capabilities, not necessarily shared by each individual platform, but by
the fleet in total. These capabilities would include:

* the ability to scan and search for salient entities such as injured humans,
  building structures or vehicles;
* the ability to monitor or surveil these salient points of interest and continually collect and communicate information back to ground operators and
  other platforms to keep them situationally aware of current conditions;
* the ability to deliver supplies or resources to these salient points of interest if
  required. For example, identified injured persons should immediately receive
  a relief package containing food, medical and water supplies.

Although quite an ambitious set of capabilities, several of them have already been
achieved to some extent using our experimental helicopter platforms, although
one has a long way to go in terms of an integrated, safe and robust system of
systems.

Realistically, each of the capabilities listed above would have to be used in a
cooperative manner among the UAV platforms in use. This would imply a high-
level communication infrastructure among the platforms and between the platforms and ground operators. The speech-act based FIPA ACL language would
be one candidate for such a communication infrastructure. At some level of abstraction, each of the UAV platforms and ground operators should be viewed
as agents in a multi-agent system. A suitable candidate for such an abstraction
would be JADE (Java Agent DEvelopment framework) or some such similar
framework.

As will become clear shortly, delegation of tasks among the platforms and
ground operators will be essential in pulling off such complex multi-agent cooperative scenarios. To be more specific in terms of the scenario, we can assume
there are two separate legs or parts to the emergency relief scenario in the context
sketched previously.

Leg I. In the first part of the scenario, it is essential that for specific geographic
areas, the UAV platforms should cooperatively scan large regions in an attempt to identify injured persons. The result of such a cooperative scan
would be a saliency map pinpointing potential victims, their geographical
coordinates and sensory output such as high resolution photos and thermal

images of potential victims. The resulting saliency map would be generated as the output of such a cooperative UAV mission and could be used directly by emergency services or passed on to other UAVs as a basis for additional tasks.

Leg II. In the second part of the scenario, the saliency map generated in Leg I would be used as a basis for generating a logistics plan for several of the UAVS with the appropriate capabilities to deliver food, water and medical supplies to the injured identified in Leg I. This of course would also be done in a cooperative manner among the platforms.

It is worth remarking that Leg I has actually been done using two of our experimental UAV platforms and Leg II has been executed in simulation. In these cases, a fully integrated delegation based cooperative framework was not yet in place, but the experiments have provided much data as to what would be required. This is in fact very much part of what this paper is about.

3   The Physical UAV Platform

The AutUAVTech Lab has developed a number of UAV platforms which include both fixed-wing and rotor-based systems. These include both micro- and mini-sized platforms. For the purposes of this paper, it is instructive to focus on one of the more complex rotor-based platforms that has been extended with a JADE level and has been used in our initial multi-UAV platform experimentation.

The WITAS[1] UAV platform [9] is a slightly modified Yamaha RMAX helicopter (Fig. 1). It has a total length of 3.6 m (including main rotor) and is powered by a 21 hp two-stroke engine with a maximum takeoff weight of 95 kg. The helicopter has a built-in attitude sensor (YAS) and an attitude control system (YACS).

The hardware platform (box on the left side of the RMAX in Fig. 1) which has been developed and integrated with the WITAS RMAX contains three PC104 embedded computers. The primary flight control (PFC) system runs on a PIII (700Mhz), and includes a wireless Ethernet bridge, a RTK GPS receiver, and several additional sensors including a barometric altitude sensor. The PFC is connected to the YAS (attitude sensors) and YACS (attitude controller), an image processing computer and a computer for deliberative capabilities. The image processing (IPC) system runs on the second PC104 embedded computer (PIII 700MHz), and includes a thermal camera and a color CCD camera mounted on a pan/tilt unit, a video transmitter and a recorder (miniDV). The deliberative/reactive (D/R, DRC) system runs on the third PC104 embedded computer (Pentium-M 1.4GHz) and executes all high-end autonomous functionality. Network communication between computers is physically realized with serial line RS232C and Ethernet. Ethernet is mainly used for CORBA applications (see below), remote login and file transfer, while serial lines are used for hard real-time networking.

---

[1] WITAS is an acronym for the Wallenberg Information Technology and Autonomous Systems Lab which hosted a long term UAV research project (1997-2005).

Fig. 1. The WITAS RMAX Helicopter

4   The Software Architecture

A hybrid deliberative/reactive software architecture has been developed for the UAV. Conceptually, it is a layered, hierarchical system with deliberative, reactive and control components. Fig. 2 presents the functional layer structure of the architecture and emphasizes its reactive-concentric nature.

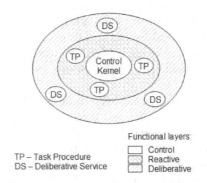

TP – Task Procedure
DS – Deliberative Service

Functional layers:
Control
Reactive
Deliberative

Fig. 2. Functional structure of the architecture

The Control Kernel is a mixed continuous/discrete system and realizes all flight control modes such as hovering, trajectory following, take-off and landing. Its components contain continuous control laws and mode switching is realized using event-driven hierarchical concurrent state machines (HCSMs). An extended version of HCSMs is presented in [14]. HCSMs can be represented as state transition diagrams and are similar to statecharts [11].

The high-level part of the system has reduced timing requirements and is responsible for coordinating the execution of reactive Task Procedures (TPs) at the reactive layer and for implementation of high-level deliberative capabilities.

A TP is a high-level procedural execution component which provides a computational mechanism for achieving different robotic behaviors by using both deliberative and control components from the control and deliberative layers in a highly distributed and concurrent manner. The control and sensing components of the system are accessible for TPs through a Helicopter Server which in turn uses an interface provided by the Control Kernel. A TP can initiate one of the autonomous control flight modes available in the UAV (e.g. take off, vision-based landing, hovering, dynamic path following or reactive flight modes for interception and tracking). Additionally, TPs can control the payload of the UAV platform which currently consists of video and thermal cameras mounted on a pan-tilt unit in addition to a stereo camera system. TPs can receive data delivered by the PFC and IPC computers i.e. helicopter state and camera system state (including image processing results), respectively.

The software implementation of the High-Level system is based on CORBA (Common Object Request Broker Architecture), which is often used as middleware for object-based distributed systems. It enables different objects or components to communicate with each other regardless of the programming languages in which they are written, their location on different processors, or the operating systems they are running in. A component can act as a client, a server, or as both. The functional interfaces to components are specified via the use of IDL (Interface Definition Language). The majority of the functionalities which are part of the architecture can be viewed as CORBA objects or collections of objects, where the communication infrastructure is provided by CORBA facilities and other services such as real-time and standard event channels.

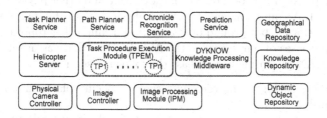

Fig. 3. Some deliberative, reactive and control services

This architectural choice provides us with an ideal development environment and versatile run-time system with built-in scalability, modularity, software relocatability on various hardware configurations, performance (real-time event channels and schedulers), and support for plug-and-play software modules. Fig. 3 presents some (not all) of the high-level services used in the WITAS UAV system. Those services run on the D/R computer and interact with the control system through the Helicopter Server. The Helicopter Server on one side uses CORBA to be accessible by TPs or other components of the system; on the other side it communicates through shared memory with the HCSM based interface running in the real-time part of the DRC software.

Many of the D/R services should be accessible as services to FIPA compliant agents which can be delegated and used by other FIPA compliant agents to achieve mission goals cooperatively. In order to do this, the existing architecture has been extended with an additional conceptual layer.

5 JADE Extension to the Existing Architecture

The CORBA-based distributed architecture described in section 4 has recently been extended with what is conceptually an additional outer layer in order to leverage the functionality of JADE [2]. "JADE (Java Agent Development Framework) is a software environment to build agent systems for the management of networked information resources in compliance with the FIPA specifications for interoperable multi-agent systems." [1]. Since both the 2APL platform and 2APL agents are FIPA compliant and the format of the messages that are communicated between 2APL agents is based on FIPA standards which use speech acts as the communicative paradigm, the choice of JADE provides an excellent foundation for developing and integrating our delegation framework with our existing UAV platforms and 2APL.

The new outer layer is two-tiered and consists of a collections of FIPA compliant agents that have a number of services associated with them. The lower tier of the outer layer consists of an ACL/CORBA Gateway agent. The Gateway agent serves as an intermediary between the CORBA legacy architecture, which contains a number of services implemented as CORBA servers or objects. The FIPA compliant agents in the upper tier of the outer layer serve as intermediaries between the physical platform on which they reside and other physical platforms or external FIPA compliant agents. Other physical platforms may include not only UAV platforms, but ground operator stations and other robotic platforms.

For the purposes of our scenario, the ACL/CORBA Gateway agent associated with our RMAX platforms has at least the following services:

* TakeOff() – An agent may request that a specific platform takes off autonomously.
* Land() – An agent may request that a specific platform lands autonomously
* FlyTo() – An agent may request that a specific platform flies to a designated waypoint.

The Gateway agent for an RMAX or other UAV platform may or may not have some of the following services which are associated with cooperative scanning:

* CoopScan() – An agent with this service has the capability of coordinating a cooperative scanning mission among different UAV platforms.
* Scan() – An agent with this service has the capability of participating in a (cooperative) scanning mission upon request.
* CScan() – An agent with this service can actually determine the specific flight plans for UAV platforms participating in a cooperative scanning mission.
* FlyScan() – An agent with this service has the ability to fly a specific flight plan generated by an agent with the CScan capability.

* GetScan() – An agent with this service can access an agent with a CScan service and request its output for specific UAV platforms participating in the cooperative scanning mission.

Each of these services is also parameterized. The degree of instantiation of parameters for a specific service will in fact reflect a degree of delegation also. The fewer parameters provided, the more open the delegation of a service is.

The upper tier consists of

* an Interface Agent – The interface agent is the official communication interface used by all external platforms or agents to communicate with a specific UAV platform. It is through this agent that all delegation and information requests made.
* a Resource Agent – The Resource agent is an internal information agent that is capable of responding to queries pertaining to both the static and dynamic resources associated with a particular UAV platform. This would include physical sensors, high-level services such as planners, low-level services such as flight mode capabilities, fuel and battery levels, etc. The agent always has an up-to-date dynamic description of available and locked resources, etc., in addition to static resource descriptions. Static resource descriptions provide information about potential resource capabilities while dynamic resource descriptions provide information about actual resource capabilities.
* and additional virtual service agents specific to composite service functionality associated with specific UAV platforms.

An example of a virtual service agent used in our scenario would be a Coop-Scanning Agent. Any platform containing such an agent has the ability to put together a cooperative scanning mission. A CoopScan service is associated with such agent and is registered in the JADE platform Directory Facilitator which provides yellow page lookup of specific services and specific agents that can provide these services.

6   Leg I: Cooperative Scanning

In the remainder of the paper, we will focus on Leg I of this ambitious mission scenario and use it as a vehicle for describing how a theory of delegation would be used.

6.1   The Generic Setup

Let's assume that at any one time, k different UAV platforms may be in the general vicinity of an area that needs to be scanned and n $(n < k)$ platforms have the potential to participate in a scan of the area. Of those n platforms that have the potential in theory, only m $(1 \leq m \leq n)$ platforms actually can participate in an upcoming mission during a particular time window which is assumed. The $k - n$ platforms can not participate due to absence of static resources. They simply are not capble of participating in such missions. The $n - m$ platforms can not participate due to absence of dynamic resources. In theory, they have the necessary static resources such as cameras, but lack one or more of the necessary resources because they are being used or locked relative to some other task. For instance, these platofrms may lack fuel or battery power, or may be scheduled to participate in other missions which overlap with the time window for the upcoming scan.

It is assumed that the region to be scanned can be specified as a convex polygonal region. In order to put together a cooperative scanning mission, the following must be the case:

* At least one platform must be in the vicinity with a CoopScan() service and this service must be available for use during the specific mission time window associated with a cooperative scan mission. This platform has the role of coordinating the use of platforms involved in a cooperative scan mission.
* At least one platform must be in the vicinity with a CScan() service and this service must be available for use during the specific mission time window associated with a cooperative scan mission. The platform with the CoopScan service will use this platform for actually determining the sub-scan paths that each of the participating platforms will fly in order to jointly scan a polygonal region together.
* At least one platform must be in the vicinity with a Scan() service and this service must be available for use during the specific mission time window associated with a cooperative scan mission. Hopefully there is more than one platform offering this service since such platforms will do the real scanning work. The platform with the CoopScan service will coordinate activity between these platforms and the platform containing the CScan service.
* Each of the platforms with a Scan service, also has a FlyScan() service. This service is used to actually fly the sub-scan path associated with this platform and generated for the platform by the platform with the CScan() service.

6.2   Basic Communication and Delegation Processes

The communication and delegation processes involved in a cooperative mission scenario using RMAX platforms with the specifications described previously should serve as a case study for developing the delegation framework we have in mind. In the ideal situation, we can first use the delegation framework as a descriptive tool for these interactions and study how well it fits an actual mission scenario. After some refinement via the use of different case studies, we should expect the delegation framework to function in a prescriptive manner as a means for formally specifying the *right* way to delegate tasks and reason about them in future missions. The long term goal is to provide a formal specification framework for delegation in cooperative mission scenarios.

Let us analyze the current scenario in this context. In broad outline, the following communication and delegation processes are involved:

1. A ground operator (GOP) station receives a request for a convex polygonal region R1 to be scanned during a temporal interval I_1 bounded below by time t_1 and above by time t_2, $I_1 = [t_1, t_2]$. The mission goal is to provide a scan of this region within the specified interval.

2. The GOP Agent (Delegator) would request a call for proposals (cfp) to any agent that could provide a CoopScan service within the mission time window I_1. This is an open form of delegation with some constraints.

3. Suppose a delegation agreement is made between the GOP Agent and the Coop1 Agent , who has the CoopScan service available within the mission time window I_1 and can commit to offering it for use and does so after checking it has the appropriate static and dynamic resources.

4. At this point, the main task has been openly delegated and the GOP agent may also choose to monitor the task execution process or not. The GOP may choose to constrain how the scan is done to a greater or lesser extent by providing additional constraints at this time. For example, it may limit the types of platforms used, or for those that are used, may provide max/min altitudes and velocities. If it does not do this, then the Coop1 Agent will have a more open delegation and consequently, more autonomy.

5. One has a number of choices as to when a delegation agreement has been made between the GOP Agent and the Coop1 Agent. For instance, is the agreement made based only on the fact that the Coop1 Agent has the Coop-Scan service available within the mission time window I_1, or should the agreement only be made after the Coop1 Agent mediates a complete mission as described in the steps below?

6. The Coop1 Agent must now put together a cooperative scanning mission and make sure it is completed within the mission time window I_1.

7. The Coop1 Agent now requests a cfp for any agents having a CScan service that is accessible during I_1 and chooses such an agent. Let's call this agent the CScan1 Agent. A delegation agreement is made between Coop1 Agent and CScan1 Agent.

8. The Coop1 Agent now requests a cfp for any agents having a Scan service that is accessible during I_1. Let's suppose that two RMAX agents, RMAX1 Agent and RMAX2 Agent can provide the scanning service in some parts of region R1 during I_1. Two new delegation agreements are set up between Coop1 Agent and RMAX1 Agent and RMAX2 Agent, respectively.

9. The Coop1 Agent now provides the CScan1 Agent with the appropriate information about the participating platforms RMAX1, and RMAX2, in the form of camera apertures, and max and min velocities, etc. It then requests that the CScan1 Agent do its job of computing subregion-scan plans for region R1.

10. Note that there are several recursive delegation agreements which are active and time is also progressing. There are any number of ways where these delegations could fail due to contingencies or temporal constraint violations. Let's assume that the CScan1 Agent is successful though and has generated subplans for each of RMAX1 and RMAX2.

11. Under the assumption that the respective delegation agreements between Coop1 Agent and RMAX1 and RMAX2 Agents concerning the Scan service have not been violated, the Coop1 Agent initiates new delegation agreements with RMAX1 and RMAX2 relative to the FlyTo service to fly to the initial starting points for the cooperative scan.

12. Upon completion of these tasks, repectively, the Coop1 Agent will initiate two new delegation agreements respectively relative to the FlyScan services of RMAX1 and RMAX2. It is during this agreement that the flight plans generated by the CScan1 Agent are passed on to RMAX1 and RMAX2. This is also a form of open delegation because only the waypoint sequences, max and min velocities and max and min altitudes are sent to RMAX1 and RMAX2 along with a scan pattern type. It is up to RMAX1 and RMAX2 to fly these patterns appropriately using their respective motion planners and obstacle avoidance capabilities.

13. Upon completion of these tasks, the appropriate sensor data or saliency maps regarding injured persons is provided to the Coop1 Agent, who may then provide it to the original GOP Agent as part of the initial delegation agreement. After this, the cooperative scan mission is considered accomplished.

Many of the details of these processes have been left out and there are many forms and degrees of open and closed delegation going on here. In particular, the degree of parameterization of services reflects a wide spectrum of degrees of open delegation. The more parameters provided the less autonomy the delegated agent has in achieving the task. Another issue of extreme importance is providing a means of adapting to failures in tasks and consequently, delegation agreements. In real-life scenarios such as those considered here, there are always many chances that tasks may not succeed or may only partially succeed. One requires mechanisms to cancel delegation of activities and also repair them through new delegation agreements if possible. An additional factor which makes this type of mission extremely complex is taking into account both static and dynamic resources required in addition to temporal and other constraints which may be violated due to contingencies out of control of the participating platforms.

In section 7, we will introduce our first attempt at formalizing a notion of strong delegation and then relate it to the case study just described in some detail.

7   Strong Delegation

In this section we briefly review Falcone & Castelfranchi's notion of strong/strict delegation [4,10].

Falcone & Castelfranchi's approach to delegation builds on a BDI model of agents, that is, agents having beliefs, goals, intentions, plans. Falcone & Castelfranchi do not give a formal semantics of their operators, but in order to get a better understanding we will use KARO logic ([13]) as the underlying framework. This will aid us to consider how to program a multi-agent system with delegation capability.

7.1 KARO Logic

The KARO formalism is an amalgam of dynamic logic and epistemic / doxastic logic, augmented with several additional (modal) operators in order to deal with the motivational aspects of agents. KARO has operators for belief (Bel_A), expressing that "agent A believes that"[2], and action ($[A : \alpha]$, "after A's performance of α it holds that"). The belief operator Bel_A is assumed to satisfy the usual (positive and negative) introspection properties $Bel_A\phi \to Bel_ABel_A\phi$ and $\neg Bel_A\phi \to Bel_A\neg Bel_A\phi$ (cf. [13,15]). Furthermore, there are additional operators for ability ($Able_A$), desires (Des_A) and commitment (Com_A):

* $Able_A\alpha$ expressing that agent A is able (has the capability) to perform action α
* $Des_A\phi$ expressing that agent A desires (a state described by formula) ϕ
* $Com_A(\alpha)$ expressing that A is committed to action ('plan') α (i.e. "having put it onto its agenda"). (This operator is very close to Cohen & Levesque's notion of $INTEND_2$ ("intention_to_do") [5]).

Furthermore, the following syntactic abbreviations serving as auxiliary operators are used:

* (dual) $\langle A : \alpha \rangle\varphi = \neg[A : \alpha]\neg\varphi$, expressing that the agent has the opportunity to perform α resulting in a state where φ holds.
* (opportunity) $Opp_A\alpha = \langle A : \alpha \rangle$tt, i.e., an agent has the opportunity to do an action iff there is a successor state.
* (practical possibility) $PracPoss_A(\alpha, \varphi) = Able_A\alpha \wedge Opp_A\alpha \wedge \langle A : \alpha \rangle\varphi$, i.e., an agent has the practical possibility to do an action with result φ iff it is both able and has the opportunity to do that action and the result of actually doing that action leads to a state where φ holds;
* (can) $Can_A(\alpha, \varphi) = Bel_A PracPoss_A(\alpha, \varphi)$, i.e., an agent can do an action with a certain result iff it believes (it is 'aware') it has the practical possibilty to do so;
* (realisability) $\Diamond_A\varphi = \exists a_1, \ldots, a_n PracPoss_A(a_1; \ldots; a_n, \varphi)$, i.e., a state property φ is realisable by agent A iff there is a finite sequence of atomic actions of which the agent has the practical possibility to perform it with the result φ, i.e., if there is a 'plan' consisting of (a sequence of) atomic actions of which the agent has the practical possibility to do them with φ as a result.
* (goal) $Goal_A\varphi = \neg\varphi \wedge Des_A\varphi \wedge \Diamond_A\varphi$, i.e., a goal is a formula that is not (yet) satisfied, but desired and realisable.
* (intend) $Int_A(\alpha, \varphi) = Can_A(\alpha, \varphi) \wedge Bel_A Goal_A\varphi$, i.e., an agent (possibly) intends an action with a certain result iff the agent can do the action with that result and it moreover believes that this result is one of its goals.

---

[2] In fact, we use a slightly simplified form of KARO which originally also contains a separate notion of knowledge.

In order to formalize Falcone & Castelfranchi's notion of delegation we also need three additional operators:[3]

* $Dep(A, B, \tau)$, expressing that A is dependent on B for the performance of task τ
* $MB_{AB}\phi$, expressing that ϕ is mutual (or common) belief amongst agents A and B
* $SC(B, A, \tau)$, expressing that B is socially committed to A to achieve τ for A, or that there is a *contract* between A and B concerning task τ.

We will not provide the formal semantics of this logic here, but refer to [13] for details.[4]

7.2   Falcone and Castelfranchi's Notion of Strong Delegation

First, we define the notion of a task as a pair consisting of a goal and a plan for that goal, or rather, a plan and the goal associated with that plan. Paraphrasing Falcone & Castelfranchi into KARO terms we consider a notion of strong/strict delegation represented by a speech act S-Delegate(A, B, τ) of A delegating a task $\tau = (\alpha, \phi)$ to B, specified as follows:

S-Delegate(A, B, τ), where $\tau = (\alpha, \phi)$
Preconditions:

(1) $Goal_A(\phi)$
(2) $Bel_A Can_B(\tau)$ (Note that this implies $Bel_A Bel_B(Can_B(\tau))$)
(3) $Bel_A(Dep(A, B, \alpha))$

Postconditions:

(1) $Goal_B(\phi)$ and $Bel_B Goal_B(\phi)$
(2) $Com_B(\alpha)$.
(3) $Bel_B Goal_A(\phi)$
(4) $Can_B(\tau)$ (and hence $Bel_B Can_B(\tau)$, and by (1) also $Int_B(\tau)$)
(5) $MB_{AB}(\text{"the statements above"} \wedge SC(B, A, \tau))$

---

[3] Here we deviate slightly from Falcone & Castelfranchi's presentation: instead of a dependency operator they employ a preference operator. However, the informal reading of this is based on a dependency relation, cf. [10], footnote 7). We take this relation as primitive here.

[4] Of course, this semantics should be extended to cater for e.g. the dependence and social commitment operators. The easiest way to do this is to use semantical functions associated with these operators yielding the extensions of the dependency and social commitment relation, respectively, per world (or, more appropriately in the KARO framework, per model-state pair). The mutual belief operator can be given semantics along the usual lines of common knowledge/belief operators such as given e.g. in [15,16].

Informally speaking this expresses the following: the preconditions of the S-delegation act of A delegating task τ to B are that (1) ϕ is a goal of delegator A (2) A believes that B can (is able to) perform the task τ (which implies that A believes that B himself believes that he can do the task!) (3) A believes that with respect to the task τ he is dependent on B.

The postconditions of the delegation act mean: (1) B has ϕ as his goal and is aware of this (2) he is committed to the task (3) B believes that A has the goal ϕ (4) B can do the task τ (and hence believes it can do it, and furthermore it holds that B intends to do the task, which was a separate condition in F&C's set-up), and (5) there is a mutual belief between A and B that all preconditions and other postconditions mentioned hold, as well as that there is a contract between A and B, i.e. B is socially committed to A to achieve τ for A. In this situation we will call agent A the delegator and B the contractor.

Typically a social commitment (contract) between two agents induces obligations to the partners involved, depending on how the task is specified in the delegation action. This dimension has to be added in order to consider how the contract affects the autonomy of the agents, in particular the contractor's autonomy. We consider a few relevant forms of delegation specification below.

7.3   Closed vs. Open Delegation

Falco & Castelfranchi furthermore discuss the following variants of task specification:

* closed delegation: the task is completely specified: both goal and plan should be adhered to.
* open delegation: the task is not completely specified: either only the goal has to be adhered to while the plan may be chosen by the contractor, or the specified plan contains 'abstract' actions that need further elaboration (a 'sub-plan') to be dealt with by the contractor.

So in open delegation the contractor may have some freedom to perform the delegated task, and thus it provides a large degree of flexibility in multi-agent planning, and allows for truly distributed planning. We will see its usefulness in our scenario shortly.

The specification of the delegation act in the previous subsection was in fact based on closed delegation. In case of open delegation α in the postconditions can be replaced by an α', and τ by $\tau' = (\alpha', \phi)$. Note that the third clause, viz. $Can_B(\tau')$, now implies that α' is indeed believed to be an alternative for achieving ϕ, since it implies that $Bel_B[\alpha']\phi$.)

7.4   Strong Delegation in Agent Programming

When devising a system like the one we have in mind for our scenario, we need programming concepts that support delegation and in particular the open variant of delegation. In the setting of an agent programming language such as 2APL [6], we may use plan generation rules to establish a contract between two agents. Very briefly, a 2APL agent had a belief base, a goal base, a plan base, a set of capabilities (basic actions it can perform), and sets of rules to change its bases: PG rules, PR rules and PC rules. PG-rules have the form $\gamma \leftarrow \beta \mid \pi$, meaning that if the agent has goal γ and belief β then it may generate plan π and put it in its plan base. PR rules can be used to repair plans if execution of the plan fails: they are of the form $\pi \leftarrow \beta \mid \pi'$, meaning that if π is the current plan (which is failing), and the agent believes β then it may revise π into π'. PC-rules are rules for defining macros and recursive computations. (We will not specify them here).

The act of strong delegation can now be programmed in 2APL by providing the delegator with a rule

$$\phi \leftarrow Can_B(\tau) \wedge Dep(A, B, \tau) \mid SDelegate(A, B, \tau)$$

(where $\tau = (\alpha, \phi)$, which means that the delegation act may be generated by delegator A exactly when the preconditions that we described earlier are met. The action $SDelegate(A, B, \tau)$ is a communication action requesting to adapt the goal and belief bases of B according to the KARO specification given earlier, and should thus, when successful (depending under additional assumptions such as that there is a some authority or trust relation between A and B), result in a state where contractor B has ϕ in its goal base, $Goal_A(\phi)$, $Can_B(\tau)$ and $MB('contract')$ in its belief base, and plan α in its plan base. That is to say, in the case of a closed delegation specification. If the specification is an open delegation, it instead will have an alternative plan α in its plan base and a belief $Can_B(\alpha', \phi)$ in its belief base. It is very important to note that in the case of such a concrete setting of an agent programmed in a language such as 2APL, we may provide the Can-predicate with a more concrete interpretation: $Can_B(\alpha, \phi)$ is true if (either ϕ is in its goal base and α is in its plan base already, or) B has a PG-rule of the form $\phi \leftarrow \beta \mid \alpha'$ for some β that follows from B's belief base, and the agent has the resources available for executing plan α. This would be a concrete interpretation of the condition that the agent has the ability as well as the opportunity to execute the plan!

7.5 Further Issues

Regarding delegation of tasks to contractors in the framework described so far, there are a number of further issues that are interesting and also necessary for putting it to practice. The first has to do with transfer of delegation: are agents always allowed to transfer delegation further on to other agents? The answer is dependent on the application, and this should be specified as a parameter in the delegation speech act (and resulting contract). Another important question is what happens if the delegated task fails. Is the contractor allowed to repair the task (by plan repair, in 2APL: using PR-rules) on its own, or must this be reported back (and ask permission to repair the plan) to the delegator first? This has to do to what degree the delegation is specified as open. If it is open in the sense that the contractor gets a goal and may devise a plan himself, then it stands to reason that he may also repair/revise the plan if it fails. If one wants that the contractor reports back and asks for PR permission, this can be handled in 2APL by a PR rule of the form

$$\pi \leftarrow \beta \mid send(A, request, permission\_to\_repair(\pi))$$

that is triggered by an event signaling the failure of plan π. Delegator A will have a rule for handling this message, and the contractor B will need a (PC-)rule that handles the reply from the delegator, resulting in the contractor proceeding with the plan repair if the permission has been granted. Thus, in this case all plan repair rules are split into two rules, one to ask permission and the other to perform the plan repair in case permission is granted.

8   Integrating Delegation Theory with UAV Systems

We now consider how the UAV cooperative scanning mission described in some detail in section 6, could be specified more succinctly using the theory of strong delegation presented in section 7.

8.1   Strong Delegation and the Cooperative Scanning Scenario

The ground operation agent, GOP has responsibility for scanning region R_1 for injured persons and this mission should be completed during the interval $I_1 = [10:00, 10:45]$. So, $Goal_{GOP}(Scan(R_1, constraints_1))$, where $constraints_1$ is a property list containing the following parameters:

$TimeInterval, PatternType, EntityIdent, MaxAlt, MinAlt,$
$MaxVelocity, MinVelocity$

Here is an example of an instantiated list for this mission:

$(TimeInterval : [10 : 00, 10 : 45], PatternType : Rectangular(width : 5m),$
$EntityIdent : (humanbodies), Platform : (Rotor), MaxAlt : 300m, MinAlt :$
$150m, Maxvelocity : 8m/s, MinVelocity : 4m/s)$

In this case the temporal duration of the mission is specified, the type of pattern that should be flown is specified as rectangular, the type of entities one should try to identify and classify is stated as human bodies, the type of platform that may participate is restricted to rotor-based and both the altitude and velocity constraints are specified.

GOP first does a yellow page lookup for agents that have services for coordinating scanning missions (in the case of FIPA agents, the directory facilitator is used). Suppose the agent, Coop1 is found to have this capability via its registered service, $CoopScan(region, constraints)$. Let

$$\tau_1 = (\alpha_1, \phi_1) = (CoopScan(R_1, constraints_1), Scan(R_1, constraints_1)).$$

Then,

$$Bel_{GOP}Can_{Coop_1}(\tau_1) \text{ and } Bel_{GOP}(Dep(GOP, Coop1, \alpha_1))$$

At this point, GOP will send a speech act to acquire a delegation agreement with Coop1: $S\text{-}Delegate(\text{GOP}, \text{Coop1}, \tau_1)$. If the contract is successful then, upon acknowledgment, the following holds:

$Goal_{Coop1}(\phi_1)$ and $Bel_{Coop1}Goal_{Coop1}(\phi_1)$ and $Com_{Coop1}(\alpha_1)$ and $Can_{Coop1}(\tau_1)$ and $MB_{GOP,Coop1}(SC(\text{Coop1}, \text{GOP}))$

in addition to a number of other mutual beliefs and implications as specified in section 7. These interactions basically cover steps **1-5** in section 6.2.

Coop1 now has to acquire agreements with at least one platform with CScanning capability and one or more platforms with Scanning capability. This is again done via a yellow page lookup. Suppose Coop1 finds that agent CScan1 has CScanning capability via its registered service, $CScanExec(region, constraints)$ service. Let

$$\tau_2 = (\alpha_2, \phi_2) = (CScanExec(\text{R}_1, \text{constraints}_{cs1}), CScan(\text{R}_1, \text{constraints}_{cs1}).$$

At this point, Coop1 will send a speech act to acquire a delegation agreement with CScan1 with similar types of pre- and postconditions:

7. $S\text{-}Delegate(\text{Coop1}, \text{CScan1}, \tau_2)$.

Note that at this point, Coop1 has not yet provided an instantiation of the constraints, constraints_{cs1} that CScan1 requires to execute its single action plan $CScanExec()$. It first has to negotiate a delegation agreement with the actual platforms that will participate in the cooperative scanning mission.[5]

Coop1 again initiates a yellow page lookup. Suppose Coop1 finds that agents RMAX1 and RMAX2 meet the constraints specified in constraints_1, and that they have scanning capability via the registered service, $FlyScan(region, constraints)$. Let

$$\tau_3 = (\alpha_3, \phi_3) = (FlyScan(\text{R}_1, \text{constraints}_{rm1}), Scan(\text{R}_1, \text{constraints}_{rm1}))$$

and

$$\tau_4 = (\alpha_4, \phi_4) = (FlyScan(\text{R}_1, \text{constraints}_{rm2}), Scan(\text{R}_1, \text{constraints}_{rm2}))$$

8. At this point, Coop1 will send speech acts to each platform to acquire delegation agreements:
   a. $S\text{-}Delegate(\text{Coop1}, \text{RMAX1}, \tau_3)$,
   b. $S\text{-}Delegate(\text{Coop1}, \text{RMAX2}, \tau_4)$.

Note that at this point, delegation agreements are setup for RMAX1 and RMAX2 as participants in the scanning mission, but these agents do not yet know the specific region they will scan, just that it will be in region R1. In the same vein, Coop1 does not yet have the constraints, constraints_{rm1} and constraints_{rm2} associated with RMAX1 and RMAX2, so only partially instantiated lists are used as a result of this delegation agreement.

---

[5] This is similar in vein to a constraint programming approach to agent coordination, such as the one using the GrAPL language studied in [3].

9.–10. As a result of **8**, Coop1 can now provide the constraints, constraints$_{rm1}$ and constraints$_{rm2}$ to CScan1 by incorporating the necessary parameters into the constraint list constraints$_{cs1}$ required by the CScan1 in order to execute the *CScanExec()* single action plan with these two platoforms as input. These constraints include the camera apertures for each of the RMAXs, the max velocities allowed and the current positions, respectively.

11. Coop1 now has to acquire delegation agreements with RMAX1 and RMAX2 to fly to specific waypoint coordinates (assigned in the respective constraint lists), so they are positioned appropriately when the scanning mission is initiated. The following speech acts are sent by Coop1 to ensure this:

11a. *S-Delegate*(Coop1, RMAX1, τ_5), where $\tau_5 = (\alpha_5, \phi_5)$ and the goal, $\phi_5 = FlyTo(\text{constraints}_{rm1_1})$ and the *plan*,
$\alpha_5 = FlyTo(\text{constraints}_{rm1_1})$.

11b. *S-Delegate*(Coop1, RMAX2, τ_6), where $\tau_6 = (\alpha_6, \phi_6)$ and the goal, $\phi_6 = FlyTo(\text{constraints}_{rm2_1})$ and the *plan*,
$\alpha_6 = FlyTo(\text{constraints}_{rm2_1})$

If there is a green light for the mission to continue, the RMAX's are positioned and ready to fly, then one final delegation agreement is made between Coop1 and the RMAX platforms regarding the actual execution of the individual scanning patterns generated by CScan1 (and assigned in the respective constraint lists). The following speech acts are sent by Coop1 to ensure this:

12. The Coop1 Agent is now ready to delegate the specific FlyScan goals with specific coordinates generated by the CScan Agent to each of the RMAXs:

12a. *S-Delegate*(Coop1, RMAX1, τ_7), where $\tau_7 = (\alpha_7, \phi_7)$ and the goal, $\phi_7 = Scan(\text{R}_2, \text{constraints}_{rm1_2})$ and the *plan*,
$\alpha_7 = FlyScan(\text{R}_2, \text{constraints}_{rm1_2})$.

12b. *S-Delegate*(Coop1, RMAX2, τ_8), where $\tau_8 = (\alpha_8, \phi_8)$ and the goal, $\phi_8 = Scan(\text{R}_3, \text{constraints}_{rm2_2})$ and the *plan*,
$\alpha_8 = FlyScan(\text{R}_3, \text{constraints}_{rm2_2})$

The final result of this mission would be the generation of a saliency map based on the scans of both the RMAX's.

8.2 Open and Closed Delegation

As an example of the distinction between open and closed delegation and the rich spectrum of choices in between, we now consider the GOP's choices as to how to delegate a scanning goal to Coop1. Each alternative below gravitates towards a more progressively open delegation.

Recall, the strong delegation example at the beginning of section 8:

Option A: *S-Delegate*(GOP, Coop1, τ_1), where $\tau_1 = (\alpha_1, \phi_1)$ and $\phi_1 = Scan(\text{R}_1, \text{constraints}_1)$ and $\alpha_1 = CoopScan(\text{R}_1, \text{constraints}_1)$, where the the constraints, constraints$_1$ used in τ_1 are completely instantiated:

($TimeInterval$: $[10 : 00, 10 : 45]$, $PatternType$: $Rectangular(width : 5m)$, $EntityIdent$: $(humanbodies)$, $Platform$: $(Rotor)$, $MaxAlt$: $300m$, $MinAlt$: $150m$, $Maxvelocity$: $8m/s$, $MinVelocity$: $4m/s$)

GOP also asserts that the single action plan Coop1 should use to achieve the scanning goal is $CoopScan(region, constraints)$ with its arguments completely specified. From the perspective of GOP this is a completely closed delegation. Both the goal and the means of achieving it are completely specified. Coop1 appears to have limited flexibility and thus little autonomy as regards the achievement of the goal. Note though, that from another perspective, there is no guarantee that the single plan action is deterministic or without internal choice. In fact, we know that there is a great deal of recursive delegation occurring in this single action plan.

The second option is no longer closed delegation, but is closer to closed delegation than open delegation:

Option B: $S\text{-}Delegate(\mathsf{GOP}, \mathsf{Coop1}, \tau_B)$, where $\tau_B = (\alpha_B, \phi_B)$ and $\phi_B = Scan(\mathsf{R_1}, \mathsf{constraints_B})$ and $\alpha_B = CoopScan(\mathsf{R_1}, \mathsf{constraints_B})$,

where constraints$_B$ is a semi-instantiated list:

($TimeInterval$: $[10{:}00, ??]$, $PatternType$:??, $EntityIdent$: $(humanbodies)$, $Platform$: $(??)$, $MaxAlt$:??, $MinAlt$: $150m$, $Maxvelocity$: $8m/s$, $MinVelocity$: $4m/s$)

In this case, there are no constraints as to when the mission can end, the scan pattern type may be chosen arbitrarily, any type of platform can be used, and there is no constraint on maximum altitude. The semi-instantiated list of constraints implies more choice for Coop1 and consequently more autonomy for Coop1 in how it achieves the goal.

The 3rd and final option is a form of highly open delegation:

Option C: $S\text{-}Delegate(\mathsf{GOP}, \mathsf{Coop1}, \tau_C)$, where $\tau_C = (\alpha_C, \phi_C)$ and $\phi_C = Scan(\mathsf{R_1}, \mathsf{constraints_C})$ and $\alpha_C = \emptyset$.

In this case, it is completely up to Coop1 how it will achieve the scanning goal and the constraints may be more or less specified. Minimally specified constraints of course results in the most open form of delegation possible in this context.

9   Conclusion

We introduced a realistic UAV case study as a vehicle for exploring the type and nature of cooperative missions one would come across when developing complex autonomous UAV systems and higher-order deliberative capability for such systems. In parallel, we have developed and proposed a theory of delegation formalized using KARO and 2APL. We then analyzed the communicative interactions pertaining to delegation from the case study, using the new theory of delegation. There is a very good correspondence between the real-life requirements of the cooperative UAV scanning mission and the theoretical constructs proposed. This leads us to believe we are on the right track in terms of linking theory with application. The JADE layer extension described has been implemented and tested in a multi-platform cooperative scanning mission successfully flown at an emergency services testing facility in Revinge in southern Sweden. Use of the S-delegation speech acts were part of the mission and provided us with a succinct framework for executing such missions. Although the full KARO/2APL reasoning and programming system is not yet part of the deployed UAV system architecture, our initial experimentation provides convincing evidence that we are on the right track in this respect also.

Future work will include integration of the KARO/2APL reasoning system with the existing JADE layer of the UAV software architecture and additional experimentation with other cooperative missions. We will also focus on integration with task planner and execution monitor pairs in order to deal with failures in cooperative plans due to unforeseen contingencies and their dynamic repair. The delegation theory will be extended accordingly.

Acknowledgments

This work is supported in part by the National Aeronautics Research Program NFFP04 S4203 and the Swedish Foundation for Strategic Research (SSF) Strategic Research Center MOVIII.

We thank Mehdi Dastani for useful discussions concerning 2APL. We also thank David Landen and Tommy Persson for JADE related discussions and software development associated with the cooperative scanning scenario.

References

1. Bellifemine, F., Bergenti, F., Caire, G., Poggi, A.: JADE – a java agent development framework. In: Bordini, R.H., Dastani, M., Dix, J., Seghrouchni, A. (eds.) Multi-Agent Programming - Languages, Platforms and Applications, Springer, Heidelberg (2005)
2. Bellifemine, F., Caire, G., Greenwood, D.: Developing Multi-Agent Systems with JADE. John Wiley and Sons, Ltd, West Sussex, England (2007)
3. de Boer, F.S., de Vries, W., Meyer, J.-J.C., van Eijk, R.M., van der Hoek, W.: Process Algebra and Constraint Programming for Modelling Interactions in MAS, AAECC (Applicable Algebra in Engineering, Communication and Computing), pp. 113–150. Springer, Heidelberg (2005)
4. Castelfranchi, C., Falcone, R.: Toward a Theory of Delegation for Agent-Based Systems, in: Robot. Autonom. Syst. 24, 141–157 (1998)
5. Cohen, P.R., Levesque, H.J.: Intention is Choice with Commitment. Artificial Intelligence 42(3), 213–261 (1990)
6. Dastani, M., Ch, J.-J, Meyer.: A Practical Agent Programming Language, in Proc. AAMAS07 Workshop on Programming Multi-Agent Systems (ProMAS (M. Dastani, A. El Fallah Seghrouchni, A. Ricci & M. Winikoff, eds.), Honolulu, Hawaii, 2007, pp. 72-87 (2007)

7. Doherty, P.: Advanced research with autonomous unmanned aerial vehicles. In: Proceedings on the 9th International Conference on Principles of Knowledge Representation and Reasoning (2004)
8. Doherty, P.: Knowledge representation and unmanned aerial vehicles. In: Proceedings of the IEEE Conference on Intelligent Agent Technolology (IAT 2005), IEEE Computer Society Press, Los Alamitos (2005)
9. Doherty, P., Haslum, P., Heintz, F., Merz, T., Persson, T., Wingman, B.: A Distributed Architecture for Autonomous Unmanned Aerial Vehicle Experimentation. In: Proceedings of the International Symposium on Distributed Autonomous Robotic Systems, pp. 221–230 (2004)
10. Falcone, R., Castelfranchi, C.: The Human in the Loop of a Delegated Agent: The Theory of Adjustable Social Autonomy. IEEE Transactions on Systems, Man and Cybernetics–Part A: Systems and Humans 31(5), 406–418 (2001)
11. Harel, D.: Statecharts: A visual formalism for complex systems. Science of Computer Programming 8(3), 231–274 (1987)
12. Hindriks, K.V., de Boer, F.S., van der Hoek, W., Meyer, J.-J.Ch.: Agent Programming in 3APL. Int. J. of Autonomous Agents and Multi-Agent Systems 2(4), 357–401 (1999)
13. van der Hoek, W., van Linder, B., Meyer, J.-J.Ch.: An Integrated Modal Approach to Rational Agents. In: Wooldridge, M., Rao, A. (eds.) Foundations of Rational Agency. Applied Logic Series, vol. 14, pp. 133–168. Kluwer, Dordrecht (1998)
14. Merz, T., Rudol, P., Wzorek, M.: Control System Framework for Autonomous Robots Based on Extended State Machines. In: Proceedings of the International Conference on Autonomic and Autonomous Systems (2006)
15. Meyer, J.-J.Ch., van der Hoek, W.: Epistemic Logic for AI and Computer Science, Cambridge Tracts in Theoretical Computer Science, vol. 41. Cambridge University Press, Cambridge (1995)
16. Meyer, J.-J.Ch., Veltman, F.: Intelligent Agents and Common Sense Reasoning. In: Blackburn, P., van Benthem, J.F.A.K., Wolter, F. (eds.) Handbook of Modal Logic, vol. 18, pp. 991–1029. Elsevier, Amsterdam (2007)

Analysis of Negotiation Dynamics

Koen Hindriks, Catholijn M. Jonker, and Dmytro Tykhonov

Delft University of Technology, Man-Machine Interaction, Mekelweg 4,
2628CD Delft, The Netherlands
{K.Hindriks,C.Jonker,D.Tykhonov}@tudelft.nl

Abstract. The process of reaching an agreement in a bilateral negotiation to a large extent determines that agreement. The tactics of proposing an offer and the perception of offers made by the other party determine how both parties engage each other and, as a consequence, the kind of agreement they will establish. It thus is important to gain a better understanding of the tactics and potential other factors that play a role in shaping that process. A negotiation, however, is typically judged by the efficiency of the outcome. The process of reaching an outcome has received less attention in literature and the analysis of the negotiation process is typically not as rigorous nor is it based on formal tools. Here we present an outline of a formal toolbox to analyze and study the dynamics of negotiation based on an analysis of the types of moves parties to a negotiation can make while exchanging offers. This toolbox can be used to study both the performance of human negotiators as well as automated negotiation systems.

1 Introduction

Negotiation is an interpersonal decision-making process necessary whenever we cannot achieve our objectives single-handedly [10]. Parties to a negotiation need each other to obtain an outcome which is beneficial to both and is an improvement over the current state of affairs for either party. Both parties need to believe this is the case before they will engage in a negotiation. Although by engaging in a negotiation one party signals to the other party that there is potential for such gain on its side, it may still leave the other party with little more knowledge than that this is so. Research shows that the more one knows about the other party the more effective the exchange of information and offers [9]. Furthermore, humans usually do have some understanding of the domain of negotiation to guide their actions, and, as has been argued, a machine provided with domain knowledge may also benefit from such domain knowledge [3].

It is well-known that many factors influence the performance and outcome of humans in a negotiation, ranging from the general mindset towards negotiation to particular emotions and perception of fairness. As emphasized in socio-psychological and business management literature on negotiation, viewing negotiation as a *joint problem-solving task* is a more productive mindset than viewing negotiation as a *competition* in which one party wins and the other looses [4, 9, 10]. Whereas the latter

M. Klusch et al. (Eds.): CIA 2007, LNAI 4676, pp. 27–35, 2007.
© Springer-Verlag Berlin Heidelberg 2007

mindset typically induces hard-bargaining tactics and rules out disclosure of relevant information to an opponent, the former leads to joint exploration of possible agreements and induces both parties to team up and search for trade-offs to find a win-win outcome. Different mindsets lead to different negotiation strategies. A similar distinction between hard- and soft-bargaining tactics has also been discussed in the automated negotiation system literature where the distinction has been referred to as either a *boulware* or a *conceder* tactics [2].

Emotions and perception of fairness may also determine the outcome of a negotiation. People may have strong feelings about the "rightness" of a proposed agreement. Such feelings may not always be productive to reach a jointly beneficial and efficient agreement. It has been suggested in the literature to take such emotions into account but at the same time to try to control them during negotiation and rationally assess the benefits of any proposals on the table [4, 10].

Apart from the factors mentioned above that influence the dynamics of negotiation, many other psychological biases have been identified in the literature that influence the outcome of a negotiation, including among others partisan perceptions, overconfidence, endowment effects, and reactive devaluation [8, 10].

In order to gain a better understanding of the negotiation dynamics and the factors that influence the negotiation process it is crucial to not only mathematically evaluate the efficiency of negotiation outcomes but also to look at the pattern of offer exchanges, what Raiffa [8] calls the *negotiation dance*. In the remainder we present part of a formal toolbox to analyze patterns in offer exchanges and present some initial findings in the literature.

2 Towards a Formal Toolbox for Negotiation Dynamics Analysis

The insights of which factors influence the negotiation process as well as outcome as described in the previous section were gained by means of experiments performed e.g. by psychologists, and social scientists. More recently, the development of automated negotiation software has provided a basis to experiment and collect data about the negotiation process through human-computer interaction [1, 7]. Here we introduce part of a toolbox that allows formal analysis of the negotiation dynamics in experiments with humans as well as with machines.

Our interest is in analyzing, classifying and in precisely characterizing aspects of the negotiation dynamics that influence the final agreement of a negotiation. The main interest thus is in proposing concepts and metrics that relate these factors to specific aspects of the negotiation dynamics and to thus also gain a better understanding of the final outcome of a negotiation.

The key concept in the analysis toolbox that we propose is that of various categories or classes of negotiation actions, including in particular the offers made by each party. A proposed offer can be classified based on the utility it provides to the proposing party ("Self") as well as to the other party ("Other"). The possible classifications are visualized in Figure 1.

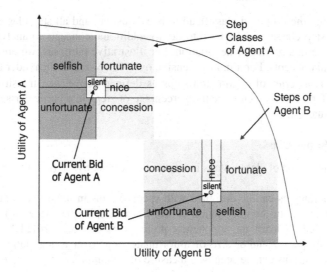

Fig. 1. Visualization of step classes in common outcome space

We distinguish six negotiation step classes, which are formally defined below. Before formally defining the concepts below, some additional notation is introduced. $U_S(b)$ denotes the utility of "Self" with respect to bid b. Similarly, $U_O(b)$ denotes the utility of "Other" with respect to b. We use $\Delta_a(b, b') = U_a(b')-U_a(b)$, $a \in \{S,O\}$, to denote the utility difference of two bids b and b' in the utility space of agent a. We also write $\Delta_a(s)$ to denote $\Delta_a(b, b')$ for a step $s = b \rightarrow b'$. Here we present a precise definition of the classes of negotiation steps proposed in [1] extended as discussed above. These step categories define the core of the step-wise analysis method [5].

Definition of Step Classes:

Let $s = b_S \rightarrow b'_S$ be a step in the bidding by Self (the definition for Other is completely symmetric). Then the negotiation step s taken by Self is classified as a:

- *Fortunate Step*, denoted by (S+, O+), iff:
  $\Delta_S(s) > 0$, and $\Delta_O(s) > 0$.
- *Selfish Step*, denoted by (S+, O≤), iff:
  $\Delta_S(s) > 0$, and $\Delta_O(s) \leq 0$.
- *Concession Step*, denoted by (S-, O≥), iff:
  $\Delta_S(s) < 0$, and $\Delta_O(s) \geq 0$.
- *Unfortunate Step*, denoted by (S≤, O-), iff:
  $\Delta_S(s) \leq 0$, and $\Delta_O(s) < 0$.
- *Nice Step*, denoted by (S=, O+), iff:
  $\Delta_S(s) = 0$, and $\Delta_O(s) > 0$.
- *Silent Step*, denoted by (S=, O=), iff:
  $\Delta_S(s) = 0$, and $\Delta_O(s) = 0$.

Observe that the proposed classification is exhaustive, and all step classes are disjoint. These step classes can be used to define additional concepts to analyze the negotiation dance in a particular negotiation. For illustrative purposes, we present just a few additional concepts. For a more extensive overview we refer the reader to [5].

A trace t is a series of negotiation steps as defined above, i.e., transitions $b \rightarrow b'$ with b, b' offers. For a given trace the percentage of steps in a particular step class is defined as usual.

Definition. % per Class

The percentage $\%_c(t)$ of class c steps in a trace t is defined by: $\%_c(t) = \#t_c / \#t$.

Negotiation strategies can be designed with specific aims in mind that should be observable as patterns in the negotiation dance. For example, the success of a strategy that is supposed to learn its opponent's preferences can be verified by checking whether the frequency and/or size of unfortunate steps over a negotiation trace decreases. Such patterns can be seen as a measure of *adaptability of a party to its opponent*. Another useful measure of the *sensitivity to the opponent's preferences* can be defined by comparing the percentage of *fortunate, nice and concession steps* that increase the opponent's utility to the percentage of *selfish, unfortunate and silent steps* that decrease it. Intuitively, an agent that only performs steps that increase its opponent's utility can be said to be (very) sensitive to the needs of its opponent.

Definition. Sensitivity to Opponent Preferences

The measure for sensitivity of agent a to its opponent's preferences is defined for a given trace t by:

$$\text{sensitivity}_a(t) = \frac{\%_{Fortunate}(t_a) + \%_{Nice}(t_a) + \%_{Concession}(t_a)}{\%_{Selfish}(t_a) + \%_{Unfortunate}(t_a) + \%_{Silent}(t_a)}$$

In case no selfish, unfortunate or silent steps are made we stipulate that *sensitivity(a,t)=∞*. If *sensitivity$_a$(t)<1*, then an agent is more or less insensitive to opponent preferences; if *sensitivity$_a$(t)>1*, then an agent is more or less sensitive to the opponent's preferences, with complete sensitivity for *sensitivity(a,t)=∞*. Typically, this sensitivity measure varies with different domains and different opponents and averages over more than one trace need to be computed. Note that the notion of sensitivity is *asymmetric*: one agent may be sensitive to the other's preferences, but not vice-versa.

3   Experimental Results

In this section, we present some experimental findings to illustrate the usefulness of our analysis toolbox.

Bosse and Jonker [1] performed two experiments with human subjects. The negotiation dances produced were analyzed with the step analysis method, although silent steps and nice steps were not considered as special cases of the concession step.

In the first experiment eighteen subjects participated and consisted of AI students in The Netherlands (12 males and 6 females). Their age varied between 19 and 27 years. The participants had to negotiate against each other (refered to as HH negotiations) and were motivated by the challenge to obtain the highest utility. Furthermore, they were challenged to outperform the computer in the corresponding Computer-Computer negotiation process (CC) they were also allowed to perform. The computer agent used the ABMP negotiation strategy, see [6]. In the HH process, one person is assigned the role of the buyer, and the other one is assigned the role of the seller. In the CC process, a computer buyer negotiates with a computer seller, both using the profile of the corresponding human negotiator. By keeping the negotiation profile stable over the two processes, it is guaranteed that the utility spaces remains the same, and that the resulting traces are thus comparable.

In the second experiment 76 subjects (43 males and 33 females) participated. The experiment took place during an introductory course for family members of AI students. Most of the participants (about 75%) were parents of the students, their age varying between 45 and 55 years. The other 25% were brothers and sisters of the students, their age varying between 17 and 24 years. Almost all of the participants did not have any background in AI. Education and occupation were a fair representation of the general population in The Netherlands.

The participants formed 38 teams of two persons, and each team was assigned to a computer. Each team was told that they could negotiate as a team against the computer. This deviation was necessary for that occasion, due to a lack of available computers. Each group participated in two negotiation processes: a Human-Computer (HC) process and a CC process. In the HC process, all teams played the buyer role, and use their own personal profile. Computer roles used the ABMP strategy of [6]. In the CC process, a computer buyer uses the profile of the human team. By keeping the negotiation profile stable over the two processes, it is guaranteed that the utility spaces remain the same, and that the resulting traces are thus comparable.

One trend observed in both experiments, is that the Nash distance and the EPP distance (both measures for fairness of the negotiation) were significantly shorter in the CC traces than in the HH traces, see Table 1, and Table 3 shows that they were shorter in the CC traces than in the HC traces. Furthermore, these distances seem to be shorter in the HH traces than in the HC traces. Thus, the CC negotiations turned out to have the "fairest" outcome, followed by the HH traces. The outcomes of the HC traces were the least balanced. This can be seen in the first two cells in Table 3, where the mean (human) buyer utility (0.89) was much higher than the mean (computer) seller utility (0.72). This is an important finding, because when the same negotiation spaces are explored by two computer negotiators, the buyer utility hardly drops (0.87), whilst the seller utility increases significantly (0.83).

Table 1. Performance in Experiment 1

	Buyer Utility	Seller Utility	Pareto Distance	Nash Distance	EPP Distance	Number of rounds
HH traces	0.87	0.80	0.05	0.22	0.16	7.00
CC traces	0.88	0.89	0.03	0.12	0.06	8.00
t-value	0.376	2.807	-0.786	-3.988	-3.463	1.540
p-value	0.717	0.023	0.455	0.004	0.009	0.146

Table 2. Steps made in Experiment 1

	Fortunate (S+ O+)	Concession (S- O+)	Selfish (S+ O-)	Unfortunate (S- O-)
HH, buyer	3 (6.52%)	36 (78.26%)	2 (4.35%)	5 (10.87%)
HH, seller	5 (11.36%)	32 (72.73%)	4 (9.09%)	3 (6.82%)
CC, buyer	0 (0%)	58 (89.23%)	0 (0 %)	7 (10.77 %)
CC, seller	0 (0 %)	48 (82.76%)	0 (0 %)	10 (17.24 %)

Table 3. Performance in Experiment 2

	Buyer Utility	Seller Utility	Pareto Distance	Nash Distance	EPP Distance	Number of rounds
HC traces	0.89	0.72	0.05	0.30	0.23	8.84
CC traces	0.87	0.83	0.06	0.17	0.10	8.91
t-value	-1.729	3.684	0.309	-5.161	-6.228	0.066
p-value	0.092	0.001	0.759	0.000	0.000	0.948

Table 4. Steps made in Experiment 2

	Fortunate (S+ O+)	Concession (S- O+)	Selfish (S+ O-)	Unfortunate (S- O-)
HC, buyer	23 (7.62%)	232 (76.82%)	17 (5.63%)	30 (9.93%)
HC, seller	2 (0.68%)	251 (85.37%)	0 (0 %)	41 (13.95%)
CC, buyer	0 (0 %)	287 (94.41 %)	0 (0 %)	17 (5.59 %)
CC, seller	0 (0 %)	267 (90.51 %)	0 (0 %)	28 (9.49 %)

Apparently the ABMP strategy used by the computer seller is not robust to being exploited by a human buyer. This observation is supported by the data in Tables 2 and 4. In both situations, the computers made more unfortunate steps than the humans. In addition, the computer sellers made more unfortunate steps than the computer buyers.

A last important finding concerns the diverse bidding behaviour by humans. As shown in Table 2, human negotiators sometimes make steps that improve the utility for both parties. Of course, doing this has the risk of making selfish steps. In its current state, the ABMP agent hardly makes these kinds of steps. Nevertheless, in some cases the unpredictable human behavior actually resulted in better results.

The extended step-wise analysis technique has been applied to a number of negotiation strategies for software agents [5]. A tournament with the strategies was set up and run. The following strategies have been studied: The ABMP strategy [6], a concession oriented strategy, which computes bids to offer next without taking domain or opponent knowledge into account, the Trade-off strategy is based on similarity criteria [3], and exploits domain knowledge to stay close to the Pareto Frontier. The

Random Walker strategy serves as a "baseline" strategy. It randomly jumps through the negotiation space, and can be run with or without a break-off point.

A full analysis was made of the type of steps made on AMPO vs City domain [9]. Here we show some examples of negotiation dances typical for the negotiation strategies we selected for the tournament (see figure 2).

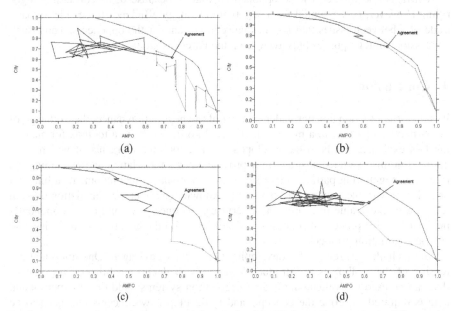

(a)   (b)

(c)   (d)

Fig. 2. Dynamics of negotiation process for: a) Random Walker (City) vs Trade-Off strategy (AMPO), b) Trade-Off (City) vs Trade-Off strategy (AMPO), c) Trade-Off (City) vs ABMP strategy (AMPO), d) Random Walker (City) vs ABMP strategy (AMPO)

The Trade-off strategy uses bids sent by its opponent to estimate what is the best possible trade-off between issues using similarity criteria. It assumes rationality of the opponent's strategy, which means that values of the issues in the opponent's bid represent its preferences. The Random-Walker strategy (fig. 2a), however, does not match this assumption and, as a result, the Trade-Off is rather strongly affected by it and performs multiple unfortunate steps.

Figure 1b shows that the Trade-Off strategy performs well against itself because both opponent's use the same assumption of rationality of the negotiation strategies. In addition, good predictability[1] of the issues produces only few unfortunate steps. Here unfortunate step is a result of a mismatch in the weights of the similarity functions and actual weights of the opponent's issues cause wrong trade-offs between issues and results in unfortunate steps.

ABMP strategy has very similar negotiation paths in all experiment due to its concession strategy and high dimensionality (10 issues) of the negotiation space. ABMP

---

[1] Here predictability of issue means that a distance function on the values of this issue can be easily defined using common sense knowledge. I.e., price issue has good predictability ($10 is most likely to be closer to $20 than to $30) and color has poor predictability.

strategy always concedes on every issue that prevents it from making unfortunate steps in case of strong mismatch in issues weights between opponents. From the other side, such concession algorithm does not allow trade-offs between issues and thus brings the bids further from the Pareto-frontier (see fig. 1c). This effect has an impact on the Trade-off strategy causing multiple unfortunate steps.

ABMP is not affected by the opponent's strategy because of its concession algorithm. However, it is essential for a negotiation strategy efficiency to be able to make trade-offs between issues that are unequally important for the opponents. Thus, such concession strategy pushes bids away from the Pareto-frontier.

4 Conclusion

It is important to gain a better understanding of the negotiation dance, the exchange of offers between parties, in a more formal way. In order to do so a toolbox for analyzing this exchange needs to be developed. Such a toolbox, elements of which were outlined in this extended abstract, may provide a basis for relating and explaining the moves of negotiating parties to five key factors that shape the outcome of a bilateral negotiation with incomplete information: (i) knowledge about the negotiation domain (e.g. the market value of a product or service), (ii) one's own and one's opponent's preferences, (iii) process attributes (e.g. deadlines), (iv) the negotiation strategies, and (v) the negotiation protocol.

Many challenges remain for developing the toolbox envisaged. One important extension of the toolbox is to introduce *benchmark problems* for bilateral negotiation that can be used to evaluate automated negotiation systems. Additional experimental data is required to refine the concepts and to develop new concepts that need to be included in the toolbox. Also such data may provide insights into relating experimental results to the key factors identified above.

References

[1] Bosse, T., Jonker, C.M.: Human vs. Computer Behaviour in Multi-Issue Negotiation. In: Proceedings of the 1st International Workshop on Rational, Robust, and Secure Negotiations in Multi-Agent Systems, pp. 11–24. IEEE Computer Society Press, Los Alamitos (2005)

[2] Faratin, P., Sierra, C., Jennings, N.R.: Negotiation Decision Functions for Autonomous Agents, Int. Journal of Robotics and Autonomous Systems 24(3-4), 159–182 (1998)

[3] Faratin, P., Sierra, C., Jennings, N.: Using Similarity Criteria to Make Negotiation Trade-Offs. Journal of Artificial Intelligence 142(2), 205–237 (2003)

[4] Fisher, R.R., Ury, W.L.: (and for the latest edition B. Patton), Getting to Yes: Negotiating Agreement Without Giving In. Penguin Books, 1981, 1992 (2003)

[5] Hindriks, K.V., Jonker, C.M., Tykhonov, D.: Negotiation Dynamics: Analysis, Concession Tactics, and Outcomes, Submitted to: 2007 IEEE/WIC/ACM International Conference on Intelligent Agent Technology IAT (2007)

[6] Jonker, C.M., Treur, J.: An Agent Architecture for Multi-Attribute Negotiation. In: Proc. of the 17th Int. Joint Conference on AI (IJCAI'01), pp. 1195–1201 (2001)

[7] Lin, R., Kraus, S., Wilkenfeld, J., Barry, J.: An Automated Agent for Bilateral Negotiation with Bounded Rational Agents with Incomplete Information. In: Proceedings of the 17th European Conference on Artificial Intelligence (ECAI'06), pp. 270–274 (2006)

[8] Mnookin, R.H., Peppet, S.R., Tulumello, A.S.: Beyond Winning: Negotiating to Create Value in Deals and Disputes. Harvard University Press, Cambridge (2000)

[9] Raiffa, H., Richardson, J., Metcalfe, D.: Negotiation Analysis: The Science and Art of Collaborative Decision Making. Harvard University Press, Cambridge (2002)

[10] Thompson, L.L.: The Heart and Mind of the Negotiator. Pearson Prentice Hall (2005)

Multi-agent Learning Dynamics: A Survey

H. Jaap van den Herik, D. Hennes, M. Kaisers, K. Tuyls*, and K. Verbeeck

Adaptive Agents Group, MICC, Maastricht University, The Netherlands
k.tuyls@micc.unimaas.nl

Abstract. In this paper we compare state-of-the-art multi-agent reinforcement learning algorithms in a wide variety of games. We consider two types of algorithms: value iteration and policy iteration. Four characteristics are studied: initial conditions, parameter settings, convergence speed, and local versus global convergence. Global convergence is still difficult to achieve in practice, despite existing theoretical guarantees. Multiple visualizations are included to provide a comprehensive insight into the learning dynamics.

1 Introduction

This article surveys the dynamics and performance of state-of-the-art value iteration and policy iteration reinforcement learning algorithms in multi-agent games. In particular this work studies initial conditions, parameter settings, convergence speed, and local versus global convergence in a wide variety of cooperative and competitive games.

Single-agent reinforcement learning (RL) has been studied extensively in the past [3,13]. It guarantees convergence to the optimal policy assuming sufficient learning cycles and a stationary environment.

Learning and adaptation in a multi-agent context recently has gained a great deal of interest in the Artificial Intelligence research community [1,6,10,11,12,16,4,17,20]. Accomplishing a certain task in highly uncertain environments, in which multiple agents operate, calls for multi-agent learning techniques. These agents involved are not only situated in a non-stationary environment but also need to deal with incomplete information and communication limits. In such non-stationary environments the Markov property does not hold which makes all proofs of convergence inapplicable when considering algorithms from single-agent learning based on the Markov assumption.

Reinforcement learning techniques are subdivided in value iteration and policy iteration. Q-learning and learning automata are examples of each class respectively. Value-based learners estimate a state-action value function that determines the utility of performing a given action in a given state [13]. Once the outcome is established the value function is used to derive a policy that describes the behavior of the agent. Contrary to the value based approach, policy iterators as learning automata learn directly in the policy space.

---

* Corresponding author.

M. Klusch et al. (Eds.): CIA 2007, LNAI 4676, pp. 36–56, 2007.

The remainder of this paper is organized as follows. Section 2 introduces the state-of-the-art learning algorithms. Section 3 explains the wide variety of games on which the algortihms are tested and concisely explains concepts such as Nash equilibrium and Pareto optimality. We continue in Section 4 with an elaboration on the performance criteria and the method of visualizing the learning dynamics. Section 5 covers the obtained results. A discussion and future research opportunities follow in Section 6. Section 7 concludes the article.

2   State-of-the-Art Learning Algorithms

In this section we shortly describe two different multi-agent reinforcement learning algorithms, viz. value based learners and policy based learners.

The value iteration reinforcement learning algorithms considered are Q-learning and two recent adaptations, i.e., Lenient Q-learning and FMQ-learning [4,7,8]. The policy iteration algorithms considered are Learning Automata and a number of its variants [6,15]. In particular finite action-set learning automata (FALA) and parameterized learning automata (PLA) are studied.

2.1   Value-Based Learners

We start by explaining independent Q-learning, because this is the basis of state-of-the-art value-based algorithms.

Q-Learning. Q-learning was initially introduced for single-agent environments. Each learning step refines a utility-estimation function (the value function) for state-action pairs and generates a new policy from the estimated values to draw the next action to execute. The algorithm bootstraps its estimate for the state-action value $Q_{t+1}(s, a)$ at time $t + 1$ upon its estimate for $Q_t(s', a')$ with s' the state where the learner arrives after taking action a in state s:

$$Q_{t+1}(s, a) \leftarrow (1 - \alpha)Q_t(s, a) + \alpha(r + \gamma \, max_{a'}Q_t(s', a')) \tag{1}$$

with α the usual step size parameter, γ a discount factor and r the immediate reinforcement.

In the single agent case this algorithm is able to learn optimal behavior in stationary environments [19]. The choice of the action selection mechanism is of utmost importance: it generates actions given the estimated state-action values. An ϵ-greedy action selection assigns the best action with probability $(1 - \epsilon)$ and some random action with probability ϵ.

With the Boltzmann distributed exploration mechanism, an action is selected with a probability given by:

$$p_j = \frac{e^{Q_i(s_j) \cdot \tau^{-1}}}{\sum_k e^{Q_i(s_k) \cdot \tau^{-1}}}$$

where an initially high temperature τ promotes exploration and decreasing temperature over time leads to strong exploitation in the final phase.

```
        iteration t            iteration t + 1
      T |10 -10 - - -         T |10 -10   - - -
      M | 0  7  0 6 0         M | -   -    -  - -
      B | 5  0  0 - -         B | 5   0    0  - -
```

Fig. 1. Lenient Q-learning reward register for $L = 5$, example from CG. Q-value of M is updated with $r_i = 7$, maximum of five rewards (left). The next step clears the register (right).

Convergence guarantees are usually lost in multi-agent environments, since other agents' actions can make the environment appear as non-stationary from the viewpoint of a single agent. Still it is possible and sometimes even useful, as we will show later on, to use Q-learning in a multi-agent environment. Agents are called independent when it assumed that they can neither observe the other agents actions nor the rewards they received for them; the agents only act upon the experience collected by experimenting with the environment.

Lenient Q-Learning. In a cooperative multi-agent learning environment it is a good idea to forgive mistakes, especially in the initial learning period. Consider the example of learning in soccer as in [7]. In the initial phase of learning both agents lack the skill for good actions, so even a perfect forward pass may frequently be not rewarded. This leads to the agents converging to actions that work well with a variety of opponents' strategies but it often results in suboptimal behavior.

In order to handle this problem lenient Q-learning collects L rewards for an action before it updates the estimation based on the maximum. Lower rewards are discarded and only the highest reward is used for the update which implies that only $\frac{1}{L} \cdot iterations$ learning steps are executed. Figure 1 depicts the update schematically.

For a detailed description of this algorithm we refer to [7,8].

FMQ-Learning. The FMQ - learner keeps track of the highest reward for each action and its frequency so far [4]. It is used to alter the policy generation which is not based on the Q-values anymore, but on the function $ev(Q_i(s_j))$. Let F be the parameter that describes the persistence to seek the maximal encountered reward $r_i^*(s_j)$ that was observed with frequency $f_i(s_j)$ so far.

$$ev(Q_i(s_j)) = Q_i(s_j) + F \cdot f(s_j) \cdot r_i^*(s_j)$$

The higher F the more the algorithm will alter the policy. This variation works best in combination with another FMQ learner; the ideas is that policies quickly agree on an optimum even if it is surrounded by penalties as it is in the climbing game. If F is large it enforces a quick decision for one action.

2.2 Policy-Based Learners

Rather than building estimated values for states or state-action pairs, policy-based reinforcement learners directly search the policy space for the optimal policy. The two policy-based learners that are considerd in this study are both learning automata algorithms, i.e. *finite action-set learning automata* (FALA) and *parameterized learning automata* (PLA). Both are model free, stateless and independent learners. While these restrictions are not negligible, they allow for simple algorithms that can be discussed analytically. Convergence for learning automata in single and specific multi-agent cases, such as games, has been proved in [5].

Finite Action-Set Learning Automata. The class of finite action-set learning automata (FALA) considers only automata that optimize their policies over a finite action-set $A = \{1, \ldots, k\}$ with k some finite integer. One optimization step, called *epoch* from here on, is divided into two steps: action selection and policy update. At the beginning of an epoch t the automaton draws a random action $a(t)$ according to the probability distribution $\pi(t)$, called policy. Based on the action $a(t)$ the environment responds with a reinforcement signal $r(t)$, called reward. Hereafter the automaton uses the reward $r(t)$ to update $\pi(t)$ to the new policy $\pi(t + 1)$.

The update rule for FALA is given below.

If $i = a(t)$ then

$$\pi_i(t + 1) = \pi_i(t) + \alpha r(t)(1 - \pi_i(t))$$
$$- \beta(1 - r(t))\pi_i(t)$$

$$\text{(2)}$$

otherwise

$$\pi_i(t + 1) = \pi_i(t) - \alpha r(t)\pi_i(t)$$
$$+ \beta(1 - r(t))[(k - 1)^{-1} - p_i(t)]$$

Here α and β are in $[0, 1]$ are the reward and penalty parameters respectively. Depending on α and β, the update scheme is referred to as *linear reward-penalty* (L_{R-P}) if $\alpha = \beta$, for $\beta = 0$ it is called *linear reward-inaction* (L_{R-I}), and if β is chosen to be small compared to α it is called *linear reward-ϵ-penalty* $(L_{R-\epsilon P})$.

Assuming that r is continuous (called *S-model*[5]) and in the range $[0, 1]$, (2) does indeed give a probability distribution satisfying the following two constraints: $\sum_{i=1}^{k} \pi_i(t + 1) = 1$ and $\forall i \ \pi_i(t + 1) \in [0, 1]$.

Parameterized Learning Automata. A learning automaton following the update rule given in (2) is only guaranteed to converge locally [15]. In order to find the global optima the learning algorithm has to be refined.

One solution to this problem is adding a randomization term to the learning rule. Superimposing noise directly on the probability distribution would violate the constraints. Therefore the algorithm presented in [14] uses a probability generating function g mapping an internal state vector u to a valid probability distribution π. This class of algorithms is called parameterized learning automata (PLA). The update rule from [14] for PLA simplifies to

$$u_i\left(t+1\right) = u_i\left(t\right) + \alpha r\left(t\right) \frac{\delta \ln g}{\delta u_i}\left(u\left(t\right), a\left(t\right)\right)$$
$$+ \alpha h'\left(u_i\left(t\right)\right) + \sqrt{\alpha} s_i\left(t\right) \tag{3}$$

where

$$h\left(x\right) = \begin{cases} -K\left(x - L\right)^{2n} & x \geq L \\ 0 & |x| \leq L \\ -K\left(x + L\right)^{2n} & x \leq -L \end{cases} \tag{4}$$

and $h'(x) = \frac{\delta h(x)}{\delta x}$. Furthermore α is a positive learning parameter and $s_i\left(k\right)$ is a set of IID random variables drawn from a normal distribution with zero mean and variance σ^2.

The difference $u\left(k-1\right) - u\left(k\right)$ is composed of three terms, a gradient, a bound, and a random term. The gradient term includes the probability generating function given by

$$g\left(u, i\right) = \frac{\exp u_i}{\sum_{j=1}^{k} \exp u_j} \tag{5}$$

According to π an action $a(t)$ is selected at every epoch t. The second term uses function h to ensure that the state vector remains within the bound $|u| \leq L$. Constants L, K, and n are all positive; L and K are real values whereas n is an integer. The last term superimposes noise to prevent the algorithm from getting stuck in local optima.

Next, Definition (5) is used to work out the gradient in (3):

$$\frac{\delta \ln g}{\delta u_i}\left(u\left(t\right), a\left(t\right)\right) = \frac{\delta}{\delta u_i} \ln \left(\frac{\exp u_{a(t)}\left(t\right)}{\sum_{j=1}^{k} \exp u_j\left(t\right)}\right)$$
$$= \frac{\delta}{\delta u_i}\left(u_{a(t)}\left(t\right) - \ln\left(\sum_{j=1}^{k} \exp u_j\left(t\right)\right)\right)$$
$$= \frac{\delta u_{a(t)}}{\delta u_i} - \frac{\exp u_i\left(t\right)}{\sum_{j=1}^{k} \exp u_j\left(t\right)}$$
$$= \frac{\delta u_{a(t)}}{\delta u_i} - \pi_i\left(t\right)$$
$$= \begin{cases} 1 - \pi_i\left(t\right) \; if \; i = a\left(t\right) \\ -\pi_i\left(t\right) \; otherwise \end{cases}$$

This results in the following update rule, similar to the form seen in (2) for the (L_{R-I}) scheme:

If $i = a(t)$ then

$$u_i(t + 1) = u_i(t) + \alpha r(t)(1 - \pi_i(t))$$
$$+ \alpha h' + \sqrt{\alpha} s_i(t)$$

otherwise

$$\text{(6)}$$

$$u_i(t + 1) = u_i(t) - \alpha r(t) \pi_i(t)$$
$$+ \alpha h' + \sqrt{\alpha} s_i(t)$$

3   Testbed of Games

This section provides background information on multi-agent games used as a benchmark for multi-agent reinforcement learning. Starting with a brief introduction to games, Section 3.1 concisely explains solution concepts, namely Nash equilibrium, Pareto efficiency, and maximum social welfare. Examples of various games for two and more players are provided through Sections 3.2 to 3.4.

3.1   Games

Normal form games are stateless games that make the assumption that players act simultaneously. Each player i participating in the game has a set of actions A^i available. When all agents have played an action they receive a numerical reward r^i.

Since normal form games are stateless, the behavior of player i can be described by a single probability distribution π^i over its action-set A^i. This distribution is called a strategy or policy. If $\pi_j^i = 1$ for any $j \in A^i$ then player i follows a *pure strategy* otherwise a *mixed strategy*. Furthermore let $R^i(\pi^1, \ldots, \pi^n)$ be the expectation of payoff r^i for agent i given the strategies $\pi^1, \ldots, \pi^n$.

Based on the notion of the expected payoff R and the strategy profile $\pi = (\pi^1, \ldots, \pi^n)$ this section gives a formal definition of Nash equilibrium, Pareto efficiency and maximum social welfare profiles.

Definition 1: *Nash equilibrium*
A strategy profile $\pi = (\pi^1, \ldots, \pi^n)$ is a Nash equilibrium if for all players i the following condition holds.

$$R^i(\pi) \geq R^i(\pi^1, \ldots, \pi^{i-1}, \tilde{\pi}^i, \pi^{i+1}, \ldots, \pi^n) \ \forall \tilde{\pi}^i$$

Hence no player can improve its payoff by exclusively changing its strategy to some $\tilde{\pi}$ given fixed strategies for all other agents.

Definition 2: *Pareto efficiency*
A strategy profile π is Pareto efficient (or Pareto optimal) if there is no $\tilde{\pi} \neq \pi$ such that

$$R^i(\tilde{\pi}) \geq R^i(\pi) \ \forall i$$

and for some i

$$R^i(\tilde{\pi}) > R^i(\pi).$$

Thus a Pareto efficient solution implies that no player can improve its expected payoff without making at least one other player worse off.

Definition 3: *Maximum social welfare profile*
The social welfare of an interactive situation is defined by the sum of individual rewards. Hence, the maximum social welfare can be denoted by:

$$\max_{\pi} \omega(\pi) = \max_{\pi} \sum_{i=1}^{n} R^i(\pi)$$

Furthermore, the strategy profile π^* with

$$\pi^* = \arg\max_{\pi} \omega(\pi)$$

is called a *maximum social welfare profile*. In cooperative games Pareto efficient solutions and maximum social welfare profiles are of major interest, whereas in competitive situations Nash equilibria are studied.

The next subsections introduce seven normal form games.

3.2   2 x 2 Matrix Games

Normal form games with two players each choosing from two actions are called 2 x 2 matrix games. The family of 2 x 2 games can be subdivided into three categories according to there payoff matrices [9]: (a) games with one pure equilibrium, (b) games with one mixed equilibrium and (c) games with two pure equilibria and one mixed equilibrium. This subsection presents one example for each class.

The *Prisoners' Dilemma* (PD) is a well studied category (a) game in which the players may *confess* (C) or *deny* (D) [2]. The payoff matrix of the PD game is given below. The single pure Nash equilibrium is located in the bottom-right corner, corresponding to both players playing action C.

	D	C
D	3,3	0,5
C	5,0	1,1

Matching Pennies (MP) is a 2 x 2 game belonging to category (b) and defined by the following payoff matrix:

	H	T
H	1,-1	-1, 1
T	-1, 1	1,-1

Both players chose simultaneously for one side of a penny, either they play *Head* (*H*) or *Tail* (*T*). If both pennies show the same face player 1 keeps the coins; for a mismatch player 2 gets rewarded. The mixed equilibrium is reached if both players play strategies $(0.5, 0.5)$ which means that a player selects action H and T each with probability 0.5.

Category (c) is covered by the next example, the *Bach or Stravinsky* (BoS) game, also referred to as the Battle of the Sexes. In this strategic situation the players want to visit a concert together. They can chose between *Bach* (*B*) or *Stravinsky* (*S*) but no communication is allowed. Player 1 prefers B whereas player 2 has a preference for S. Strategies corresponding to the joint action pairs (B, B) and (S, S) form pure equilibria; the mixed equilibrium is defined by the strategies $(\frac{2}{3}, \frac{1}{3})$ and $(\frac{1}{3}, \frac{2}{3})$ for player 1 and 2 respectively. Miscoordination results in a zero payoff for both. The payoff matrix of the BoS game is given below.

	B	S
B	2,1	0,0
S	0,0	1,2

3.3 Penalty and Climbing Game

Games with more than one equilibrium are studied to investigate convergence to local or global optima. The BoS game contains multiple pure equilibria, however it is not possible to point out the best due to a conflicting interest. The mixed equilibrium is a fair solution but not in the least optimal with respect to social welfare. Therefore games with equal payoff for both players are studied; these games are called symmetric games. The *penalty game* and the *climbing game* are examples of symmetric games. Payoff matrices for both games are given in Figure 2.

In the penalty game players have to coordinate their actions in order to yield high payoffs (joint actions $(1, 1)$ and $(3, 3)$). Miscoordination leads to punishment by negative rewards. Furthermore the joint action $(2, 2)$ is also an equilibrium but not Pareto efficient.

Two equilibria can be found in the climbing game. Joint actions corresponding to positive non-zero payoffs are points of attraction, where the joint actions $(1, 1)$ and $(2, 2)$ form equilibria. Once again the Pareto optimal Nash equilibrium is surrounded by negative rewards to punish miscoordination. Learners have to virtually climb up to reach the maximum reward.

$$
\begin{bmatrix} 10 & 0 & -10 \\ 0 & 2 & 0 \\ -10 & 0 & 10 \end{bmatrix}
\begin{bmatrix} 11 & -10 & 0 \\ -10 & 7 & 6 \\ 0 & 0 & 5 \end{bmatrix}
$$

Fig. 2. Payoff matrices for the penalty (left) and the climbing game (right)

3.4   Guessing and Dispersion Game

So far the introduced games cover interactions between two players. Since MAS with only two agents are barely seen in practice this subsection defines symmetric games with n players and n actions where n can be any finite integer.

The *guessing game* is well suited for generalization up to n players. Each player i selects an action a^i from the same action set $A = \{1, \ldots, n\}$ synchronously. If all players 'guess' the same action the reward will be maximal; if a player exclusively chooses for an action his reward will be minimal. This game is called a coordination game; all players have to coordinate, to 'guess', the same action in order to achieve the maximum payoff.

Closely related is the *dispersion game* also called anti-coordination game. The goal in this particular game is to disperse over the entire action set as much as possible. Just like in the previous game an action set of size n is assumed; which means the maximal dispersion outcome (MDO) is reached if all players choose for sole actions.

Since these two games can be easily scaled up to more than just two players it is more convenient to use a payoff function instead of matrices. Based on the sum of players selecting the same actions these functions are given in (7) and (8) for the guessing game and the dispersion game respectively.

The number of players selecting action j is defined as

$$S(j) = \sum_{i=1}^{n} id(a^i, j)$$

where

$$id(i, j) = \begin{cases} 1 \ if \ i = j \\ 0 \ otherwise \end{cases}.$$

Thus the payoff function of the guessing game can be denoted by

$$r^i = \frac{S(a^i)}{n}. \tag{7}$$

For the dispersion game the following payoff function applies:

$$r^i = \begin{cases} 1 \ if \ S(a^i) = 1 \\ 0 \ otherwise \end{cases} \tag{8}$$

The payoff differs between the case where one player occupies an action slot alonw (payoff equals 1) and the situation where at least two players have selected the same action (payoff equals 0). Note that the reward $r^i \in [0, 1]$ for all agents $i = 1, \ldots, n$.

4   Methodology

In order to explore the performance of a team of learners to its full extent it must be studied under various initial conditions and parameter profiles. The following four subsections present the different means that we will use to approach the given task in Section 5.

4.1   Policy Trajectory Plot

A rather simple way of displaying evolving policies are trajectory plots. In a 2 x 2 game $\pi_2^1 = 1 - \pi_1^1$ and $\pi_2^2 = 1 - \pi_1^2$. Therefore the strategy profile $\pi = \left(\pi^1, \pi^2\right)$ can be reduced to the pair $\left(\pi_1^1, \pi_1^2\right)$ without losing information. The trajectory of this pair is recorded during one single or multiple runs and plotted in a 2D space. To indicate the direction of convergence grayscales are used for trajectory plots in Section 5. With increasing number of epochs t the brightness of the trajectory changes from light to dark.

For the penalty and the climbing game the cardinality of the action-sets equals 3 for both agents. Therefore using the same transformation as above the strategy profile can only be reduced to a 4-tuple and is not displayable in the 2D space. However, the policy trajectories $\pi^1(t)$ and $\pi^2(t)$ can be plotted separately using two simplex plots. The three vertices of a simplex correspond to the pure policies $(1, 0, 0)$, $(0, 1, 0)$ and $(0, 0, 1)$.

4.2   Directional Field Plot

A second visual method to analyze learning dynamics are directional field plots. Again a reduced strategy profile (see Subsection 4.1) is used for analyzing 2 x 2 games. A team of learners start at regular grid points over $[0, 1]^2$:

$$\left(\pi_1^1(t_0), \pi_1^2(t_0)\right) \in [0, 1] \times [0, 1]$$

The velocity field of this team can then be denoted by

$$\frac{d(v, u)}{dt} = \frac{\left(\pi_1^1(t_0 + \Delta t) - \pi_1^1(t_0), \pi_1^2(t_0 + \Delta t) - \pi_1^2(t_0)\right)}{\Delta t}$$

where Δt is the number of epochs conducted at every grid point and v and u denote the respective strategies of both players. The velocities are displayed by arrows pointing in the direction of $\frac{d(v, u)}{dt}$. The length of the arrows indicates the absolute value $\left\| \frac{d(v, u)}{dt} \right\|$.

4.3   Convergence

While 2 x 2 games can easily be studied by graphical analysis, higher dimensional games like the penalty and the climbing game require analytical means. The third method studies convergence in respect to spatial and temporal measures. To measure spatial convergence a metric over the space of strategy profile needs to be defined. Let

$$d(\pi, \gamma) = \max_i \max_j |\pi_j^i - \gamma_j^i| \tag{9}$$

be the distance of two strategy profiles π and γ. Then a strategy profile π has converged to an optimal π^* if the distance $d(\pi, \pi^*)$ is less than the threshold ϵ at any point in time from epoch T on.

Definition 4: A strategy profile π is called ϵ-*converged* to π^* in T epochs if the following condition holds:

$$d(\pi(t), \pi^*) < \epsilon \ \forall t, \ t \geq T \tag{10}$$

Note that (9) is used in favor of other metrics since it applies to multi-agent situations without losing the intuitive explanation of threshold ϵ in (10): If a strategy profile is ϵ-*converged* each single action probability diverges at most ϵ from its desired value for all agents.

By means of ϵ-*convergence*, Section 5 studies the convergence to Nash equilibria and Pareto optimal solutions. Multiple runs are conducted in order to estimate the percentage-wise convergence to different points of attraction and the convergence time T. The experiments are repeated to determine confidence intervals.

4.4   Cumulative Reward Plot

The coordination and anti-coordination games (see Subsection 3.4) are cooperative and therefore the maximum social welfare can be used as a performance threshold. Both payoff functions (7) and (8) for the coordination games yield the same maximum social welfare of value n. If the cumulative reward $\sum_{i=1}^{n} r^i(t) = n$ a maximum social welfare profile is being played in t. Section 5 shows cumulative reward plots to indicate if agents successfully converge to these desired profiles.

5   Results

In this section we present the results obtained with the value-iteration learners and the policy-iteration learners in all three type of games. We summarize all learning performances and for some of the most interesting cases we provide visualizations of the learning dynamics. Concerning Learning Automata, the emphasis is placed on FALA, including various update schemes; the explanations only relate to PLA if the results differ significantly.

Not only successful settings are shown, but also conditions and parameters under which the algorithms may fail to converge optimally. Furthermore, a variety of visualizations allow the reader to receive an intuitive impression of the learning dynamics.

5.1   Simple Games: 2 × 2

Policy-Iteration Learners. Table 1 summarizes the learning performance of FALA in 2 x 2 games. The confidence intervals for mean estimates are obtained by gathering 101 samples each averaging 20 runs of a particular game with $T_{max} = 5\,000$ epochs. Thus T-values in the table approaching T_{max} indicate that the learners have not converged (see (10)).

The L_{R-I} update scheme converges in the PD and BoS game to the pure equilibria but fails to find the mixed one in the MP game. Results for the L_{R-P} confirm the finding that the basin of attraction coincides with the equilibrium in the MP game and is located near the mixed one in the BoS game. However, high values for T indicate that both situations are unstable. Thus it cannot be guaranteed that a team of learners stays in a equilibrium once it is reached. Due to the penalty term the action selection persists stochastic, and the team may jump out of the ϵ-convergence region once in a while.

Table 1. Convergence performance of FALA in 2 x 2 games. 95% confidence intervals for mean estimates of of ϵ-convergence percentage with $\epsilon = 0.1$ and mean convergence time T.

FALA L_{R-I} $\alpha = 0.01$, $\beta = 0$

	Nash eq.	Convergence %	T
PD	$(0,0)$	$99.4\% \pm 0.3\%$	1468.8 ± 31.3
MP	$(\frac{1}{2},\frac{1}{2})$	$2.4\% \pm 0.7\%$	4408.7 ± 44.2
	$(0,0)$	$49.2\% \pm 2.3\%$	
BoS	$(1,1)$	$50.8\% \pm 2.3\%$	559.5 ± 17.3
	$(\frac{2}{3},\frac{1}{3})$	$0.0\% \pm 0.0\%$	

FALA L_{R-P} $\alpha = \beta = 0.01$

	Nash eq.	Convergence %	T
PD	$(0,0)$	$0.0\% \pm 0.0\%$	3895.0 ± 35.9
MP	$(\frac{1}{2},\frac{1}{2})$	$99.1\% \pm 0.4\%$	4821.1 ± 4.1
	$(0,0)$	$0.0\% \pm 0.0\%$	
BoS	$(1,1)$	$0.0\% \pm 0.0\%$	4550.4 ± 11.2
	$(\frac{2}{3},\frac{1}{3})$	$19.8\% \pm 1.7\%$	

FALA $L_{R-\epsilon P}$ $\alpha = 0.01$, $\beta = 0.001$

	Nash eq.	Convergence %	T
PD	$(0,0)$	$0.0\% \pm 0.0\%$	942.2 ± 19.7
MP	$(\frac{1}{2},\frac{1}{2})$	$62.0\% \pm 2.3\%$	4813.0 ± 3.9
	$(0,0)$	$47.5\% \pm 1.9\%$	
BoS	$(1,1)$	$48.4\% \pm 1.9\%$	652.6 ± 22.8
	$(\frac{2}{3},\frac{1}{3})$	$0.0\% \pm 0.0\%$	

FALA $L_{R-\epsilon P}$ $\alpha = 0.01$, $\beta = 0.0001$

	Nash eq.	Convergence %	T
PD	$(0,0)$	$100.0\% \pm 0.0\%$	1301.9 ± 24.9
MP	$(\frac{1}{2},\frac{1}{2})$	$7.1\% \pm 1.2\%$	4909.2 ± 2.5
	$(0,0)$	$50.3\% \pm 2.2\%$	
BoS	$(1,1)$	$49.7\% \pm 2.2\%$	559.7 ± 15.4
	$(\frac{2}{3},\frac{1}{3})$	$0.0\% \pm 0.0\%$	

The $L_{R-\epsilon P}$ is a trade-off between the previous schemes. Depending on the ratio of reward factor α and penalty factor β, convergence to pure equilibria or near mixed ones is reached. For the parameter setting $\alpha = 0.01$, $\beta = 0.001$ the learners do not ϵ-converge to the equilibrium in the PD, though the T is comparable small to T_{max}. Considering the corresponding field plot (see Figure 3) it becomes clear that the team does converge but to a point near $(0.2, 0.2)$ which is outside the ϵ-convergence region.

It is worth noting that all results obtained for L_{R-I} and $L_{R-\epsilon P}$ in Figure 3 and Table 1 can be reproduced using PLA with appropriate learning rates and zero temperature for L_{R-I} and small values of σ for $L_{R-\epsilon P}$.

These findings are convincingly illustrated by the dynamics of the learning algorithms in Figure 3. All field plots are rendered using small learning rates and parameters $\Delta t = 10$, $r = 100$ as explained in Subsection 4.2. The plots help to localize basins of attraction and show when these coincide with Nash equilibria. We do not show the dynamics of the PD game.

Value-Iteration Learners. Table 2 summarizes the convergence behavior of the studied Q-learning algorithms in simple games. Confidence intervals are computed from 101 samples that average over 20 runs each. Q-values are initialized to corresponding policies that follow a uniform distribution over the policy space.

Table 2 shows convergence of FMQ and lenient Q to the Nash equilibrium in the PD game. Both learners converge to the Pareto optimal strategy (D, D) for all runs that do not converge to the Nash equilibrium (FMQ 25.4% and lenient Q-learner 14.6%). (D, D) is also the maximum social welfare profile.

The MP game yields one mixed NE where both players mix both actions equally. Figure 4 visualizes the learning behavior of the three learners in the MP.

The Battle of Sexes game yields two pure and one mixed equilibrium. Figure 5 shows the learning dynamics of the three learners in this game. All learners converge to the pure Nash equilibrium under $\tau = 0.1$, but not if the temperature is increased to $\tau = 0.5$.

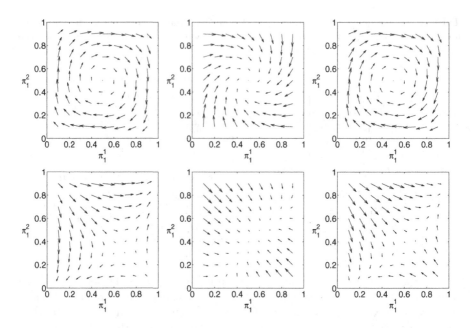

Fig. 3. Overview of directional field plots for FALA in MP (top) and BoS (bottom). Columns correspond to different update schemes; left column: $\alpha = 0.01$, $\beta = 0$ (L_{R-I}), center column: $\alpha = \beta = 0.01$ (L_{R-P}) and right column: $\alpha = 0.01$, $\beta = 0.001$ ($L_{R-\epsilon P}$).

Table 2. ϵ-near convergence with $\epsilon = 0.1$ to equilibria in 2x2 games analyzed after $I = 2000$ iterations. All learners use $\alpha = 0.01$, $\tau = 0.1$ for PD and BoS, $\tau = 0.5$ for PM. Equilibria are given as $(\pi_1(a_{11}), \pi_2(a_{21}))$. Indicated are 95% confidence intervals for convergence percent and mean convergence time.

<table>
<tr><th colspan="4">Q-learner</th><th colspan="4">FMQ-learner $F = 3$</th></tr>
<tr><td></td><td>NE</td><td>Convergence</td><td>T</td><td></td><td>NE</td><td>Convergence</td><td>T</td></tr>
<tr><td>PD</td><td>$(0,0)$</td><td>$99.4\% \pm 0.4\%$</td><td>1080.1 ± 8.0</td><td>PD</td><td>$(0,0)$</td><td>$74.6\% \pm 1.9\%$</td><td>5.2 ± 0.6</td></tr>
<tr><td>MP</td><td>$(\frac{1}{2},\frac{1}{2})$</td><td>$92.0\% \pm 1.3\%$</td><td>$1862.9 \pm 3.2$</td><td>MP</td><td>$(\frac{1}{2},\frac{1}{2})$</td><td>$78.7\% \pm 1.8\%$</td><td>$1893.3 \pm 2.4$</td></tr>
<tr><td></td><td>$(0,0)$</td><td>$50.0\% \pm 2.4\%$</td><td></td><td></td><td>$(0,0)$</td><td>$50.6\% \pm 2.2\%$</td><td></td></tr>
<tr><td>BoS</td><td>$(1,1)$</td><td>$50.0\% \pm 2.4\%$</td><td>129.2 ± 2.2</td><td>BoS</td><td>$(1,1)$</td><td>$49.4\% \pm 2.2\%$</td><td>2.5 ± 0.2</td></tr>
<tr><td></td><td>$(\frac{2}{3},\frac{1}{3})$</td><td>$0.0\% \pm 0.0\%$</td><td></td><td></td><td>$(\frac{2}{3},\frac{1}{3})$</td><td>$0.0\% \pm 0.0\%$</td><td></td></tr>
</table>

<table>
<tr><th colspan="4">Lenient Q-learner $L = 3$</th></tr>
<tr><td></td><td>NE</td><td>Convergence</td><td>T</td></tr>
<tr><td>PD</td><td>$(0,0)$</td><td>$85.4\% \pm 1.4\%$</td><td>186.3 ± 6.9</td></tr>
<tr><td>MP</td><td>$(\frac{1}{2},\frac{1}{2})$</td><td>$81.9\% \pm 1.6\%$</td><td>$1547.9 \pm 8.8$</td></tr>
<tr><td></td><td>$(0,0)$</td><td>$48.3\% \pm 2.1\%$</td><td></td></tr>
<tr><td>BoS</td><td>$(1,1)$</td><td>$51.7\% \pm 2.1\%$</td><td>128.0 ± 4.7</td></tr>
<tr><td></td><td>$(\frac{2}{3},\frac{1}{3})$</td><td>$0.0\% \pm 0.0\%$</td><td></td></tr>
</table>

5.2   Penalty and Climbing Games

Policy-Iteration Learners. The results for the penalty and the climbing game are summarized by Table 3. The L_{R-P} has not converged to pure policies in these games and is therefore omitted in the overview. Suboptimal convergence for all

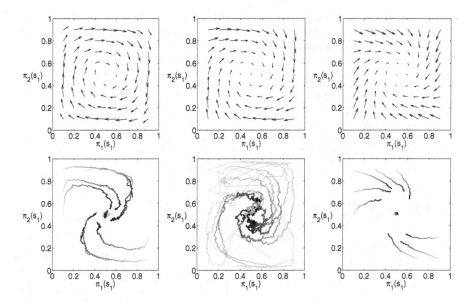

Fig. 4. Directional field plots (top row, $I = 10$) and trajectories (bottom row, $I = 600$) in the MP for the Q-learner (left), FMQ ($F = 3$, center) and lenient Q-learner ($L = 3$, right) under $\alpha = 0.01$ and $\tau = 1$

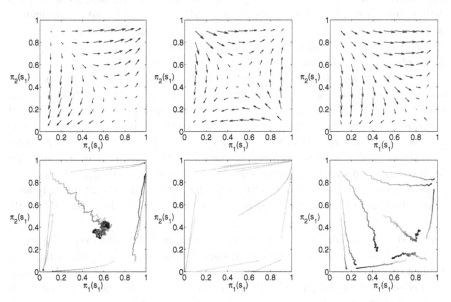

Fig. 5. Directional field plots ($I = 20$, top row) and example trajectories (bottom row) for Q-learner (left), FMQ ($F = 3$, center) and lenient Q-learning ($L = 3$, right) in BoS under $\tau = 0.5$ and $\tau = 0.01$

Table 3. Learning performance in the penalty game (PG) and the climbing game (CG). 95% confidence intervals for mean estimates of ϵ-convergence percentage with $\epsilon = 0.1$ to different joint actions and convergence time T.

FALA L_{R-I} $\alpha = 0.01$, $\beta = 0$

	Joint action	R	Convergence %	T
PG	(1,1)	10	46.1% ± 2.3%	
	(2,2)	2	5.3% ± 1.1%	1176.7±
	(3,3)	10	47.6% ± 2.3%	36.7
CG	(1,1)	11	36.9% ± 1.9%	
	(2,2)	7	15.7% ± 1.6%	2623.0±
	(2,3)	6	10.7% ± 1.3%	
	(3,3)	5	4.8% ± 1.0%	58.3

FALA L_{R-I} $\alpha = 0.01$, $\beta = 0.001$

	Joint action	R	Convergence %	T
PG	(1,1)	10	49.3% ± 2.2%	
	(2,2)	2	0.0% ± 0.0%	1247.5±
	(3,3)	10	49.9% ± 2.3%	37.1
CG	(1,1)	11	34.5% ± 2.0%	
	(2,2)	7	0.3% ± 0.2%	3347.1±
	(2,3)	6	0.0% ± 0.0%	
	(3,3)	5	0.0% ± 0.0%	73.8

FALA L_{R-I} $\alpha = 0.01$, $\beta = 0.0001$

	Joint action	R	Convergence %	T
PG	(1,1)	10	49.0% ± 2.2%	
	(2,2)	2	3.2% ± 0.8%	1160.6±
	(3,3)	10	46.4% ± 2.1%	37.9
CG	(1,1)	11	39.7% ± 2.3%	
	(2,2)	7	15.1% ± 1.5%	2735.8±
	(2,3)	6	6.2% ± 1.0%	
	(3,3)	5	2.0% ± 0.7%	79.1

PLA $\alpha = 1$, $\sigma = 0.05$, $L = 1.8$, $K = 0.5$, $n = 1$

	Joint action	R	Convergence %	T
PG	(1,1)	10	41.5% ± 2.0%	
	(2,2)	2	0.5% ± 0.3%	4927.5±
	(3,3)	10	41.1% ± 2.1%	3.4
CG	(1,1)	11	52.0% ± 1.9%	
	(2,2)	7	10.7% ± 1.2%	4943.8±
	(2,3)	6	6.5% ± 1.0%	
	(3,3)	5	3.8% ± 0.8%	3.5

other learners are rare in the penalty game, whereas the climbing game is quite challenging. For the first time parameterized learning automata significantly outperform standard FALA. The bound L is chosen small to keep exploration on a constant level whereas the learning rate is set to a high value to approach decisively high payoff situations.

The simplex plots in Figure 6 show example runs for FALA and PLA under two initial conditions. The first condition studies the learning dynamics starting from a strategy profile near $((0,0,1),(0,0,1))$ corresponding to common payoff 5. From this point the learners virtually have to climb up in order to reach the optimal solution, which is in this case a Pareto optimal Nash equilibrium and a maximum social welfare profile. The second initial condition is near another Nash equilibrium yielding a payoff equal to 7. This point challenges the learners as well; in order to improve the common payoff both agents simultaneously have to switch to action 1. If only one agent switches the reward reduces to 6, 0 or -10.

Value-Iteration Learners. The results for regular Q-learning and FMQ learning from [4] as well as the results for lenient Q-learning from [8] are confirmed. Table 4 compares the algorithms' performances in both games. All results of this subsection refer to the games with penalties $c = p = 10$. Confidence intervals are calculated from 101 samples that average over 20 runs.

Experiments in penalty games make use of an iteration dependent temperature and a learning rate of $\alpha = 0.9$. The experiments use a decay factor $s = 0.006$ and an initial temperature $\tau^0 = 500$.

$$\tau^t \leftarrow (\tau^0 - 1) \cdot e^{-s \cdot t} + 1 \tag{11}$$

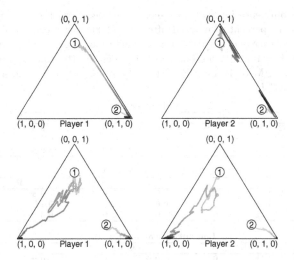

Fig. 6. Simplex plots for FALA (top) and PLA (bottom) in the climbing game shown with two initial conditions each. Convergence after $t = 200$ epochs with $\alpha = 0.1$, $\beta = 0.01$ ($L_{R-\epsilon P}$) for FALA and $\alpha = 0.3$, $\sigma = 0.2$, $L = 4$, $K = n = 1$ for PLA.

Table 4. Average ϵ-near convergence over 2020 runs ($\epsilon = 0.1$) to the maximum social welfare policy for all learners in the CG and PG with penalties $c = p = 10$. All learners under $\alpha = 0.9$ and decreasing τ, $F = 10$ and $L = 10$ for CG and $L = 5$ for PG.

Learner	CG	PG
Q	21.8%	79.6%
FMQ	98.9%	100.0%
Lenient Q	99.9%	99.3%

Table 5 lists the confidence intervals of ϵ-near convergence with $\epsilon = 0.1$ to pure strategy profiles in percentages.

The strategy profile π^* corresponding to (T, L) is a Pareto efficient Nash equilibrium. Furthermore, it is the maximal social welfare profile and as such the desired point of convergence for cooperative players. It also yields the highest individual payoff, so it is as well the best strategy profile for independent learners.

The climbing game cannot be solved satisfactorily by the regular Q-learner. It can be observed, that Q-learning converges to π^* in about 21.8%. Both adaptations outperform this by far, FMQ with $F = 10$ achieves 98.9% while lenient Q-learning with $L = 10$ achieves 99.9%.

Example trajectories of the FMQ-learner are visualized in Figure 7.

5.3 Scaling: Dispersion and Guessing Games

Policy-Iteration Learners. The following scaling experiments are conducted to give an intuition about how well learning automata scale with respect to the number of agents in coordination and anti-coordination games. Therefore the

Table 5. 95% Confidence intervals for ϵ-near convergence with $\epsilon = 0.1$ in CG to pure strategy profiles in percent. Analyzed after $I = 2000$ iterations with $\alpha = 0.9$ and decreasing τ. Q-learner (top, 43.1% not converged to any pure strategy profile), FMQ-learner (middle, $F = 10$, 0.1% n.c.) and lenient Q-learner (bottom, $L = 10$, 0.1% n.c.). Player 1 chooses T, M or B and player two chooses L, C or R. The maximal social welfare profile is (T, L).

Q-learner

	L		C		R	
T	21.8 ±	1.9	0 ±	0	0 ±	0
M	0 ±	0	0.2 ±	0.2	6.0 ±	1.1
B	0 ±	0	0 ±	0	28.9 ±	1.9

FMQ-learner $F = 10$

	L		C		R	
T	98.9 ±	0.4	0 ±	0	0 ±	0
M	0 ±	0	0.6 ±	0.3	0.1 ±	0.1
B	0 ±	0	0 ±	0	0.3 ±	0.3

Lenient Q-learner $L = 10$

	L		C		R	
T	99.9 ±	0.2	0 ±	0	0 ±	0
M	0 ±	0	0 ±	0	0 ±	0
B	0 ±	0	0 ±	0	0 ±	0

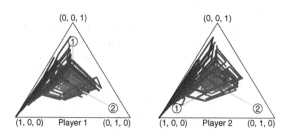

Fig. 7. Two example trajectories for both FMQ-learners with $F = 3$ in the CG show convergence to the global optimum (T, L) starting close to (B, L) in (1) and (M, C) in (2). Initial high exploration causes large policy shifts while eventual exploitation allows convergence.

emphasis lies not on statistical sampling but rather on an intuitive understanding of example runs.

Figure 8 shows the cumulative reward plot for FALA using the $L_{R-\epsilon P}$ update scheme in the dispersion game as well as PLA in the guessing game. In both cases the learners converge to a maximum social welfare profile. Note that the dispersion game has $n!$ distinct maximum social welfare profiles whereas the guessing game has only n. Convergence time T (see (10)) for the two example runs are $T \approx 600$ and $T \approx 6000$ respectively.

For the dispersion game the same facts apply. The payoff function (8) sharply rewards only the case in which an agent has exclusively selected an action. Again the agent has to be decisive in order to learn this action quickly. Furthermore randomness is required to escape from a zero reward situation where no exclusive action slot has been found yet.

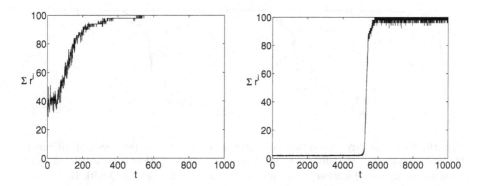

Fig. 8. Cumulative reward plot for FALA in the dispersion game (left) and for PLA in the guessing game (right) with both 100 agents. Learning parameters are $\alpha = 0.1$, $\beta = 0.01$ $(L_{R-\epsilon P})$ and $\alpha = 1$, $\sigma = 0.005$, $L = 10$, $K = n = 1$ for FALA and PLA respectively.

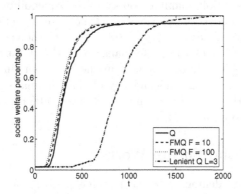

Fig. 9. Social welfare percentage over iterations in the GG for different learners ($n = 100$, $\alpha = 1$ and $\tau = 0.01$; averaged over 10 runs). All learners except the lenient Q-learner converged to the suboptimal solution with two equally sized groups once.

Value-Iteration Learners. The GG with n players has n maximal social welfare profiles while the DG has $n!$ maximal social welfare profiles. As $n!$ is much larger than n the DG can be solved much faster than the GG.

In the GG all agents try to group as quickly as possible. Convergence to suboptimal solutions, e.g. two groups with equally many agents, are not uncommon. Figure 9 shows the speed of convergence for different learners in the guessing game. An increase of the FMQ persistence F shifts the grouping process to an earlier iteration. However, there is a point of diminishing returns. Furthermore, increasing F does not seem to increase the qualitative convergence while lenient Q-learning slows down the learning process but converges to a maximal social welfare equilibrium.

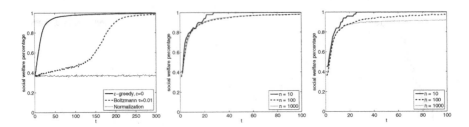

Fig. 10. Social welfare percentage over iterations in the DG (averages of 10 runs). Q-learner under $\alpha = 0.1$, different action selection methods (left, $n = 1000$), different numbers of agents: Q-learner with ϵ-greedy (center, $\epsilon = 0$) and FMQ with Boltzmann distribution (right, $F = 10$, $\tau = 0.01$).

Figure 10 (left) visualizes the impact of the exploration method. Furthermore, scalability of the Q-learner with an ϵ-greedy action selection is compared to the FMQ heuristic with a Boltzmann action selection. An equilibrium can be found within reasonable numbers of iterations using the ϵ-greedy action selection. This action selection method actually allows to scale up to thousand agents without significant deterioration of the performance over the iterations, the lines for $n = 100$ and $n = 1000$ almost coincide in the corresponding plot. FMQ also scales well but has a stronger dependency of the maximal convergence on the number of agents. However, a performance of above 90% is achieved within the first 50 iterations even for $n = 1000$.

6   Discussion and Future Work

The obtained results demonstrate the learning performance of all algorithms considered. Graphical tools help to understand the learning dynamics of the algorithms. Adequate parameter settings for convergence to equilibria have been shown. Overall, it can be observed that all results are quite sensitive to the parameter settings.

In general, low temperatures and accompanying high exploitation lead to convergence to pure strategy profiles while higher temperatures that impose more exploration allow convergence to mixed equilibria. Furthermore, higher convergence to mixed equilibria is achieved by smaller learning rates. In contrast, high learning rates can be applied to overcome penalties in cooperative coordination games. FMQ-learning with high persistence F drives the learning process to pure strategy profiles within few iterations if the temperature is low. Lenient Q-learning finds mixed solutions but requires many iterations to converge. Contrary, FALA are more robust with respect to parameter changes. Scaling down the learning rate generally results in better convergence performance although increasing the required number of iterations. The $L_{R-\epsilon P}$ scheme gives a good trade-off between the two extremes L_{R-I} and L_{R-P}. It unites the ability to find mixed policies with high percentage of global convergence. This result is also confirmed by the various visualizations used in this work.

An second observation from our research is that formal criteria may fail under practical conditions. Thus empirical results as conducted in this work are essential. The PLA superimpose noise on the learning update in order to overcome convergence to suboptimal solutions and for this theoretical guarantees can be found [5]. However, for a challenging task, such as the climbing game, we could not reach it in an experimental setting. Although this may be caused by the limited number of epochs, the central issue is clearly the high dimensional parameter space. The PLA update rule comprises five parameters that all have to be tuned to fit the environment.

Scaling experiments in the Dispersion Game reveal high performance of the Q-learner with an ϵ-greedy action selection and FMQ with a Boltzmann action selection under low temperatures. High exploitation imposed by these action selection methods is required to facilitate a quick dispersion over the actions. For the same reason, a high learning rate is needed for the LA algorithms that also indicate good scaling potential. The Guessing Game requires quick grouping but also more exploration to avoid suboptimal solutions with several, approximately equally sized groups. This implies that a trade-off needs to be chosen between fast convergence and optimal convergence. However, it should be noted that this work presents only example runs for the two coordination games and therefore cannot give any general conclusion on scaling. In future investigations we would like to test the scalability of FALA and PLA more extensively.

7   Conclusion

This research has experimentally studied the learning performance of state-of-the-art value-based and policy-based iterators in multi-agent games. In particular Q-learning, Lenient Q-learning, FMQ, FALA, and PLA are surveyed. A variety of competitive and cooperative games have served as a testbed to analyze their learning performance. Furthermore, various visualization methods are used and interconnected to reveal the complex learning dynamics of LA in games.

From the performances of independent reinforcement learners we may conclude that the learners are highly dependent on the correct parameter tuning. For the value-based methods, high temperatures enhance exploration and enable the convergence to mixed equilibria, while small temperatures enforce exploitation and increase the probability of convergence to pure strategy profiles. Stability of the learning process can be supported by small learning rates and a temperature that decreases over time. In the context of penalty games, the adaptations FMQ and lenient Q-learning outperform the regular Q-learner significantly and converge to the global optimum. For the policy-based methods, results show that the $L_{R-\epsilon P}$ scheme maintains a good trade-off between convergence performance and robustsness.

References

1. Claus, C., Boutilier, C.: The dynamics of reinforcement learning in cooperative multiagent systems. In: Proceedings of the National Conference on Artificial Intelligence, vol. 15, pp. 746–752 (1998)

2. Gibbons, R.: A Primer in Game Theory. Harvester Wheatsheaf (1992)
3. Kaelbling, L.P., Littman, M.L.: Reinforcement learning: A survey. Journal of Artificial Intelligence Research 4, 237–285 (1996)
4. Kapetanakis, S., Kudenko, D.: Reinforcement learning of coordination in cooperative multi-agent systems (2002)
5. Narendra, K., Thathachar, M.A.L.: Learning Automata An Introduction. Prentice-Hall, Englewood Cliffs, NJ (1989)
6. Nowé, A., Verbeeck, K., Peeters, M.: Learning automata as a basis for multi agent reinforcement learning. In: Tuyls, K., 't Hoen, P.J., Verbeeck, K., Sen, S. (eds.) LAMAS 2005. LNCS (LNAI), vol. 3898, pp. 71–85. Springer, Heidelberg (2006)
7. Panait, L., Tuyls, K.: Theoretical advantages of lenient q-learners: An evolutionary game theoretic perspective. In: AAMAS 2007 (2007)
8. Panait, L., Sullivan, K., Luke, S.: Lenience towards teammates helps in cooperative multiagent learning. In: Proceedings of the Fifth International Joint Conference on Autonomous Agents and Multi Agent Systems – AAMAS-2006, ACM Press, New York (2006)
9. Redondo, F.V.: Game Theory and Economics. Cambridge University Press, Cambridge, MA (2001)
10. Sen, S., Sekaran, M., Hale, J.: Learning to coordinate without sharing information. In: Twelfth National Conference on Artificial Intelligence, pp. 426–431 (1994)
11. Shoham, Y., Powers, R., Grenager, T.: If multi-agent learning is the answer, what is the question (2006)
12. Stone, P.: Multiagent learning is not the answer. It is the question. To appear in Artificial Intelligence (2007)
13. Sutton, R.S., Barto, A.G.: Reinforcement Learning: An Introduction. MIT Press, Cambridge, MA (1998)
14. Thathachar, M.A.L., Phansalkar, V.V.: Learning the global maximum with parameterized learning automata. IEEE Transactions on Neural Networks 6(2), 398–406 (1995)
15. Thathachar, M.A.L., Sastry, P.S.: Varieties of learning automata: An overview. IEEE Transactions on Systems, Man, and Cybernetics-Part B: Cybernetics 32(6), 711–722 (2002)
16. Tuyls, K., Parsons, S.: What evolutionary game theory tells us about multiagent learning. Preprint submitted to Elsevier (2007)
17. Tuyls, K., t Hoen, P.J., Vanschoenwinkel, B.: An evolutionary dynamical analysis of multi-agent learning in iterated games. Autonomous Agents and Multi-Agent Systems 12, 115–153 (2006)
18. Verbeeck, K., Nowé, A., Parent, J., Tuyls, K.: Exploring selfish reinforcement learning in repeated games with stochastic rewards, pp. 239–269 (2006)
19. Watkins. Learning from delayed rewards. PhD thesis, King's College, Oxford (1989)
20. t Hoen, P.J., Tuyls, K., Panait, L., Luke, S., la Poutré, H.: An overview of cooperative and competitive multiagent learning. In: LAMAS 2005. LNCS (LNAI), vol. 3898, pp. 1–46. Springer, Heidelberg (2006)

An Architecture for Hybrid P2P Free-Text Search

Avi Rosenfeld[1,2], Claudia V. Goldman[3], Gal A. Kaminka[2], and Sarit Kraus[2]

[1] Department of Industrial Engineering
Jerusalem College of Technology, Jerusalem, Israel
[2] Department of Computer Science Bar Ilan University, Ramat Gan, Israel
[3] Samsung Telecom Research Israel, Herzliya, Israel
{rosenfa,galk,sarit}@cs.biu.ac.il, c.goldman@samsung.com

Abstract. Recent advances in peer to peer (P2P) search algorithms have presented viable structured and unstructured approaches for full-text search. We posit that these existing approaches are each best suited for different types of queries. We present PHIRST, the first system to facilitate effective full-text search within P2P networks. PHIRST works by effectively leveraging between the relative strengths of these approaches. Similar to structured approaches, agents first publish terms within their stored documents. However, frequent terms are quickly identified and not exhaustively stored, resulting in a significantly reduction in the system's storage requirements. During query lookup, agents use unstructured searches to compensate for the lack of fully published terms. Additionally, they explicitly weigh between the costs involved with structured and unstructured approaches, allowing for a significant reduction in query costs. We evaluated the effectiveness of our approach using both real-world and artificial queries. We found that in most situations our approach yields near perfect recall. We discuss the limitations of our system, as well as possible compensatory strategies.

1 Introduction

Full-text searching, or the ability to locate documents based on terms found within documents, is arguably one of the most essential tasks in any distributed network [5]. Search engines such as Google [16] have demonstrated the effectiveness of centralized search. However, classic solutions also demonstrate the challenge of large-scale search. For example, a search on Google for the word, "a", currently returns over 10 billion pages [16].

In this paper, we address the challenge of implementing full-text searches within peer-to-peer (P2P) networks. Our motivation is to demonstrate the feasibility of implementing a P2P network comprised of resource limited machines, such as handheld devices. Thus, any solution must be keenly aware of the following constraints: **Cost** - Many networks, such as cellular networks, have cost associated with each message. One key goal of the system is to keep communication costs low. **Hardware limitations** - we assume each device is limited in the amount of storage it has. Any proposed solution must take this limitation into consideration. **Distributed** - any proposed solution must be distributed equitably. As we assume a network of agents with similar hardware composition, no one agent can be required to have storage or communication requirements grossly beyond that of other machines.

M. Klusch et al. (Eds.): CIA 2007, LNAI 4676, pp. 57–71, 2007.
© Springer-Verlag Berlin Heidelberg 2007

To date, three basic approaches have been proposed for full-text searches within P2P networks [15]. Structured approaches are based on classic Information Retrieval theory [2], and use inverted lists to quickly find query terms. However, they rely on expensive publishing and query lookup stages. A second approach creates super-peers, or nodes that are able to locally interact with a large subset of agents. While this approach does significantly reduce publishing costs, it violates the distributed requirement in our system. Finally, unstructured approaches involve no publishing, but are not successful in locating hard to find items [15].

In this paper we present PHIRST, a system for **Peer-to-Peer Hybrid Restricted Search for Text**. PHIRST is a hybrid approach that leverages the advantages of structured and unstructured search algorithms. Similar to structured approaches, agents publish terms within their documents as they join or add documents to the P2P network. This information is necessary to successfully locate hard-to-find items. Unstructured search is used to effectively find common terms without expensive lookups of inverted lists. Another key feature in PHIRST is its ability to restrict the number of peer addresses stored within inverted lists. Not only does this insure that the hardware limitations of agent nodes are not exceeded, it also better distributes the system's storage. We also present a full-text query algorithm where nodes explicitly reason based on estimated search costs about which search approach to use, reducing query costs as well.

To validate the effectiveness of PHIRST, we used a real web corpus [11]. We found that the hybrid approach we present used significantly less storage to store all inverted lists than previous approaches where all terms were published [5,15]. Next, we used artificial and real queries to evaluate the system. The artificial queries demonstrated the strengths and limitations of our system. The unstructured component of PHIRST was extremely successful in finding frequent terms, and the structured component was equally successful in finding any term pairs where at least one term was not frequent. In both of these cases, the recall of our system was always 100%. The system's performance did have less than 100% recall when terms of 2 or more words of medium frequency were constructed. We present several compensatory strategies for addressing this limitation in the system. Finally, to evaluate the practical impact of this potential drawback, we studied real queries taken from IMDB's movie database (www.imdb.com) and found PHIRST was in fact effective in answering these queries.

2   Related Work

Classical Information Retrieval (IR) systems use a centralized server to store inverted lists of every document within the system [2]. These lists are "inverted" in that the server stores lists of the location for each term, and not the term itself. Inverted lists can store other information, such as the term's location in the document, the number of occurrences for that term, etc. Search results are then returned by intersecting the inverted lists for all terms in the query. These results are then typically ranked using heuristics such as TF/IDF [3]. For example, if searching for the terms, "family movie", one would first lookup the inverted list of "family", intersect that file with that of "movie", and then order the results before sending them back to the user.

The goal of a P2P system is to provide results of equal quality without needing a centralized server with the inverted lists. Potentially, the distributed solution may have advantages such as no single point of failure, lower maintenance costs, and more up-to-date data. Toward this goal a variety of distributed mechanisms have been proposed.

Structures such as Distributed Hash Tables (DHTs) are one way to distribute the process of storing inverted lists. Many DHT frameworks have been presented, such as Bamboo [13], Chord [9], and Tapesty [14]. A DHT could then be used for IR in two stages: publishing and query lookups. As agents join the network, they need to update the system's inverted lists with their terms. This is done through every agent sending a "publish" message to the DHT with the unique terms it contains. In DHT systems, these messages are routed to the peer with the inverted list in LogN hops, with N being the total number of agents in the network [9,13]. During query lookups, an agent must first identify which peer(s) store the inverted lists for the desired term(s). Again, this lookup can be done in LogN hops [9,13]. Then, the agent must retrieve these lists and intersect them to find which peer(s) contain all of the terms.

Li et al. [5] present formidable challenges in implementing both the publishing and lookup phases of this approach in large distributed networks. Assuming a word exists in all documents, its inverted list will contain N entries. Thus, the storage requirements for these inverted lists are likely to exceed the hardware abilities of agents in these systems. Furthermore, sending large lists will incur a large communication cost, even potentially exceeding the bandwidth limitation of the network. Because of these difficulties, they concluded that naive implementations of P2P full-text search are simply not feasible.

Several recent developments have been suggested to make a full text distributed system viable. One suggestion is to process the structured search starting with the node storing the term with the fewest peer entries in its inverted list. That node then forwards its list to the node with the next longest list, where the terms are locally intersected before being forwarded. This approach can offer significant cost savings by insuring that no agent can send an inverted list longer than the one stored by the **least** common term [15]. Reynolds and Vahdat also suggest encoding inverted lists as Bloom filters to reduce their size [12]. These filters can also be cached to reduce the frequency these files must be sent. Finally, they suggest using incremental results, where only a partial set of results are returned allowing search operations to halt after finding a fixed number of results, making search costs proportional to the number of documents returned.

Unstructured search protocols provide an alternative that is used within Gnutella and other P2P networks [1]. These protocols have no publishing requirements. To find a document, the searching query sends its query around the network, until a predefined number of results have been found, or a predefined TTL (Time To Live) has been reached. Assuming the search terms are in fact popular, this approach will be successful after searching a fraction of the network. Various optimizations have again been suggested within this approach. It has been found that random walks are more effective than simply flooding the network with the query [8]. Furthermore, one can initiate multiple simultaneous "walks" to find items more quickly, or use state-keeping to prevent "walkers" from revisiting the same nodes [8]. Despite these optimizations, unstructured searches have been found to be unsuccessful in finding rare terms [1].

In super-peer networks, certain agents store an inverted list for all peer documents for which it assumes responsibility. Instead of publishing copies over a distributed DHT network, agents send copies of their lists to their assigned super-peers. As agents are assumed to have direct communication with its super-peers, only one hop is needed to publish a message, instead of the LogN paths within DHT systems. During query processing, an agent forwards its request to its super-peer, who then takes the intersection between the inverted lists of all super-peers. However, this approach requires that certain nodes have higher bandwidth and storage capabilities [15] – something we could not assume within our system.

Hybrid architectures involve using elements from multiple approaches. Loo et al. [6,7] propose a hybrid approach where a DHT is used within super-peers to locate infrequent files, and unstructured query flooding is used to find common files. This approach is most similar to ours in that we also use a DHT to find infrequent terms and unstructured search for frequent terms. However, several key differences exist. First, their approach was a hybrid approach between Gnutella ultrapeers (super-peers) and unstructured flooding. We present a hybrid approach that can generically use any form of structured or unstructured approaches, such as random walks instead of unstructured flooding or global DHT's instead of a super-peer system. Second, in determining if a file was common or not, they needed to rely on locally available information from super-peers, and used a variety of heuristics to attempt to extrapolate this information for the global network [6]. As we build PHIRST based on a global DHT, we are able to identify rare-items based on complete information. Possibly most significantly, Loo et al. [7] only published the files' names, and not their content. As they considered full text search to be infeasible for the reasons previously presented [5], their system was limited to performing searches based on the data's file name, and not the text within that data. As our next section details, we present a publishing algorithm that actually becomes cheaper to use as subsequent nodes are added. Thus, PHIRST is the first system to facilitate effective full-text search even within large P2P networks.

3   PHIRST Overview

First, we present an overview of the PHIRST system and how its publishing and query algorithms interconnect. While this section describes how information is published within the Chord DHT [9], PHIRST's publishing algorithm is generally presented in section 4 so it may be used within other DHT's as well. Similarly, section 5 presents a query algorithm (algorithm 2) which generally selects the best search algorithm based on the estimated cost of performing the search algorithms at the user's disposal. The selection algorithm is generically written such that new search algorithms can be introduced without affecting the algorithm's structure. Only later, in algorithm 3 do we present how these costs are calculated specific to the DHT and unstructured search algorithms we used.

In order to facilitate structured full-text search for even infrequent words, search keys must be stored within structured network overlays such as Chord. Briefly, Chord uses consistent hash functions to create an m-bit identifier. These identifiers form a circle modulo 2^m. The node responsible for storing any given key is found by using a preselected hash function, such as SHA-1, to compute the hash value of that key. Chord then routes the key to the agent whose Chord identifier is equal to or is the successor (the next existent node) of that value [9]. For example, Figure 1 is a simple example with an identifier space of 8, and 3 nodes. Assuming the key hashes to a value of 6, that key needs to be stored on the next node within the circular space, or node 0. Assuming the key hashes to 1, it is stored on node 1.

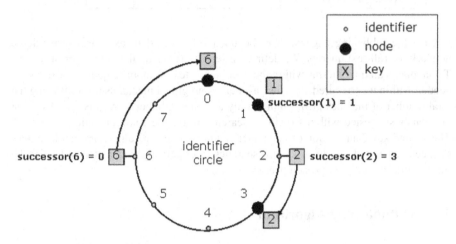

Fig. 1. An example of a Chord ring with m=3. Figure based on Chord paper [9].

The hashing quality within the Chord algorithm has several important qualities. First, it creates important performance guarantees, such as LogN average search length. Furthermore, nodes can be easily added (joins) or removed (disjoins) by inserting them into the circular space, and re-indexing only a fraction of the pointers within the system. Finally, the persistent hashing function used by Chord has the quality that no agent will get more than O(LogN) keys than the average [9]. We refer the reader to the Chord paper for further details [9].

However, the DHT's performance guarantees only balancing the number of keys stored per node, but not the number of addresses stored in the inverted lists for each key. For example, Table 1, gives an example of the inverted lists for five words. Common words, such as "a" and "the" within the table, will produce much long inverted lists, than uncommon words such as "aardvark" and "zygote". Due to space restrictions we will only present up to the first 7 inverted entries for each word, out of a potential length of N rows. Balancing guarantees only apply to the number of words (out of N), but not the size of each inverted list (the length of that row). Because word distribution within documents typically follow Zipf's law, some of the words within documents occur very frequently while many others occur rarely [4]. In an extreme example, one node may be responsible for storing extremely common words such as "the" and "a", while other nodes are assigned only rare terms. Thus, one key contribution of this paper is a publishing algorithm that can equitable distribute these entries by allowing agents to cap the number of inverted list entries they will store.

Table 1. Example of several words (keys within the DHT), and their inverted lists

Word (key)	Address1	Address2	Address3	Address4	Address5	Address6	Address7
a	111-1111	111-1112	111-1113	111-1114	111-1115	111-1116	111-1117
aardvark	111-4323						
the	111-1111	111-1112	111-1113	111-1114	111-1115	111-1116	111-1117
zoo	123-4214	123-9714	333-9714				
zygote	548-4342						

Once the publishing stage has been begun, a distributed database exists to search the network for full-text queries. We define the search task as finding a number of results, T, that match all query terms within the documents' text. Capping a query at T results is needed within unstructured searches, as there is no global mechanism for knowing the total number of matches [15]. Finding only a limited number of results has also been previously suggested within structured searches to reduce communication costs [12]. The second key contribution of this paper is a novel querying algorithm that leverages between structured and unstructured searches to effectively find matches despite the limit in the amount of data each peer stores.

4   The Publishing Algorithm

Every time an agent joins the network, or an existing agent wishes to add a new document, it must publish the words in its document(s) as described in Algorithm 4. First, the agent generates a set of max terms it wishes to add (line 1). Similar to other studies [15] we assume that the agent preprocesses its document to remove extraneous information such as HTML tags and duplicate instances of terms. Stemming, or reducing each word to its root form, is also done as it has been observed to improve the accuracy of the search [15]. Furthermore, as we detail in the Experimental Results section (section 6), stemming also further reduces the amount of information needed to be published and stored. The publishing agent, ID_{Source}, then sends every unique term, $Term_i$, to be stored in an inverted list on peer ID_{DEST} (lines 3-4). The keys being stored are these words that are sent, with each word either creating a new inverted list, or being added to an existing file. In addition to these terms, the agent also updates a counter of the total number of documents contained between all agents within the system (line 4). For simplicity, let us assume this global counter is stored on the first agent, ID_1. We will see that this value is needed by the query algorithm described below.

PHIRST's publishing algorithm enforces an equitable term distribution by only storing inverted lists until a length of d. For every term node, $Term_i$ out of a total of $received$ terms, ID_{DEST} is requested to store it must decide if it should fulfill that request. As lines 6 and 7 of the algorithm detail, assuming agent ID_{DEST} currently has fewer than d entries for $Term_i$, it adds the value ID_{Source} to its list (or creates an inverted list if this is the first occurrence). Either way, nodes log that a certain number of $COUNTER$ instances of that term exist (lines 8). This information is used by the query algorithm to determine the global frequency of this term. Because we limit each node to only storing d out of a possible N terms, the storage requirements of the system

---

Algorithm 1. Publishing Algorithm(Document Doc)

1: Terms $\Leftarrow$ Preprocessed words in Doc
2: **for** $i = Term_1$ to $Term_{max}$ **do**
3:    PUBLISH($Term_i, ID_{Source}, ID_{DEST}$)
4: PUBLISH(DOC-COUNTER+1, ID_1)
5: **for** $i = Term_1$ to $Term_{received}$ **do**
6:    **if** SIZE($ID_{DEST}, Term_i$) < d **then**
7:       ADD-Term($Term_i, ID_{Source}$)
8:    UPDATE-Counter($Term_i, COUNTER$)

---

are reduced to $d*N$ from $N*N$. As we set $d << $ N, we found this savings to be quite significant.

Theoretically, additional information about each term may be published, such as the position that term occurred or how many instances of that term existed within the document and aggregate this and similar information into a rating for the term it is about to publish. This information may be especially important when more than d instances of that term exist. The receiving agent, ID_{DEST}, could then decide which d term instances to store by continuously sorting scores of the terms it has, and maintaining only those with the top d highest rating. In a similar vein, if more than d instances of $Term_i$ exist, it may be advantageous to store the d most recent documents, especially if turnover exists within nodes.

The performance guarantees of DHT's such as Chord insure the publishing algorithm runs with fairly low cost. Because each node, ID_{Source}, needs LogN hops to find the agent, ID_{DEST}, responsible for storing that term's inverted list, the total number of messages needed to publish a document is of order $O(max * logN)$ where max is the number of terms in that document. Note that the publishing algorithm described here sends all terms, even those which in fact do not need publishing because they already contained d terms.

5   The Search Algorithm

The search algorithm is called once any agent wishes to conduct a distributed full-text search. As Algorithm 2 describes, this process operates in two stages. First, we retrieve the global frequencies of all search terms (line 1) and sort all terms from least to most frequent (line 2). This value can be calculated through looking up the frequency of that term ($COUNTER$), and dividing this number by the total number of documents ($DOC - COUNTER$). Finding these values requires one lookup of the value of $DOC - COUNTER$ (assumed to be stored on agent ID_1 in the publishing algorithm), as well as a lookup for the frequencies of each term from the agent storing term $Term_i$. Referring back to algorithm 4 note that the peer storing $Term_i$ has a counter with this value even if more than d instances of this term occurred.

Once the frequency of all terms are known, the algorithm then reasons about which algorithm to select. This process iteratively calls the tradeoff function which we define below (algorithm 3). If unstructured search is deemed less costly, all terms are immediately searched for simultaneously (lines 7–10). This type of search can either

---

Algorithm 2. Hybrid Search Algorithm(String $Query_1 \ldots Query_{max}$)

1: space $\Leftarrow \infty$ {Used for initialization to all P2P nodes}
2: Retrieve Frequencies of $Query_1 \ldots Query_{max}$
3: **Term** $\Leftarrow$ Sorted Query Terms Least to most Frequent {Term is an array}
4: **for** $i = Term_1$ to $Term_{max}$ **do**
5:      Frequency $\Leftarrow$ Product of Frequencies($Term_i \ldots Term_{max}$)
6:      Tradeoff $\Leftarrow$ Calculate-Tradeoff(space, $Term_i \ldots Term_{max}$, Frequency)
7:      **if** Tradeoff > 0 **then**
8:          **while** Found $<$ T AND NOT Exhausted(space) **do**
9:              Search-Unstruct(space, $Term_i \ldots Term_{max}$)
10:         Break
11:     **else**
12:         space $\Leftarrow$ List($Term_i$) $\cap$ space
13:         **if** $i = Term_{max}$ **then**
14:             **if** space $>$ T **then**
15:                 return first T list entries
16:             **else**
17:                 return all list entries

---

terminate because T matches have been found or the search space has been exhaustively searched. If structured search is deemed less costly, that term's inverted list is requested, and the search space is intersected with that of the new term (line 12). Assuming we have reached the last term (lines 12-17) we return the first T matches found after all terms were successfully intersected. Once the structured search identifies that fewer matches than T matches were found (line 15) it returns all list entries (line 17).

This algorithm has several key features. First, the search process is begun starting with the least frequent term. This is done following previous approaches [15] to save on communication costs. We denote the inverted list length of the **least** common search term as length($Term_1$) where length is a function that returns the size of an inverted list and $Term_1$ is the first term after the terms are sorted based on frequency. Each successive peer receives the previously intersected list, and locally intersects this information with that of its term (line 13). The result of this process is that intersected lists become progressively smaller (or at worse case stay the same size) with the maximum information any peer can send being bounded by length($Term_1$). Second, one might question why agents do not immediately return the entire inverted list of the terms they store, instead of first returning the term's frequency. This is done because the information gained from this frequency information, such as bounding search costs to the size of the least frequent term, far outweighs the search costs involved with processing the query in two stages. Finally, as the search goal is to return T results, the last node within a structured search does not need to return its entire inverted list. Instead, it only needs to send the first T results (or failure or NULL as in line 17 if under T results exist). Because of this, the maximal search cost will be of order $(max\text{-}1) * $ length($Term_1$) $+$ T where max is the number of terms in the search query.

Arguably the most important feature of this algorithm is its ability to switch between using structured and unstructured searches midway through processing the query terms. Even if structured search is used for the first term(s), the algorithm iteratively calls the tradeoff algorithm (algorithm 3) after each term. Once the algorithm notes that unstructured search is cheaper, it immediately uses this approach to find all remaining terms. For example, assume a multi-word query contains several common and uncommon words. The algorithm may first take the intersection of the inverted lists for all infrequent words to create a list f. The algorithm may then switch to use unstructured search within f to find the remaining common words.

Similarly, note that this approach lacks a TTL (Time To Live) for its unstructured search. We assume unstructured searches are to be used only when the expected cost of using an unstructured search is low (see algorithm 3 below). We expect this to occur when the unstructured search will terminate quickly, such as when: (i) the search terms are very common from the onset or (ii) unstructured search is used to find the remaining common terms after structured search generated an inverted list of f terms.

We now turn to the search specific mechanism needed to identify which search types will have the higher expected cost. This tradeoff depends on T, or the number of search terms wanted, the costs specific to using the different types of searches, and d or the maximal number of inverted list entries published for each term. Algorithm 3 details this process as follows:

Algorithm 3. Calculate-Tradeoff(Space, $Term_i \ldots Term_{num}$, Frequency)

1: Expect-Visit $\Leftarrow$ T / Frequency {Number of nodes Unstructured search will likely visit}
2: COSTS $\Leftarrow C_U$*(Expect-Visit) - C_S *(Sending(query-terms))
3: **if** COSTS > 0 **then**
4:     RETURN 1 {pure unstructured search}
5: **else if** COSTS < 0 AND Size($Term_i$) < d **then**
6:     RETURN -1 {pure structured search for this term}
7: **else**
8:     space $\Leftarrow$ List($Term_i$) $\cap$ space
9:     RETURN 1 {Use unstructured afterwards because of lack of more values}

First, the algorithm calculates the expected cost of conducting an unstructured search. The expected number of documents that will be visited in an unstructured search before finding T results is: T / term-frequency (line 1). For example, if we wish to find 20 results, and the frequency of the term(s) is 0.5, this search is expected to visit 40 documents before terminating. We can compare this value to that of using a structured search, whose cost is also known, and is proportional to the length of the inverted lists that need to be sent. We assume there is some cost, C_U associated with conducting an unstructured search on one peer. We also assume that some cost C_S is associated with sending one entry from the inverted list (line 2). Because the cost of unstructured search is C_U * T / Frequency, and the cost of structured search is bounded by C_S * ((max-1) * length($Term_i$) + T), the algorithm can compare the expected cost of both searches before deciding how to proceed (lines 3-6).

For many cases, a clear choice exists for which search algorithm to use. Let us assume that $C_U = C_S = 1$, and assume that all documents have been indexed, or $d=DOC - COUNTER$. When searching for common words, the cost of using the unstructured search is likely to be approximately T. Processing the same query with structured search will be approximately the number of documents ($DOC - COUNTER$)

or a number much larger than T. Conversely, for infrequent terms, say with one term occurring only T times, the cost of an unstructured search will be $DOC - COUNTER$ or a number much larger than T, while the structured search will only cost a maximum of T * max-1 + T. Finally, structured search is also the clear choice for queries involving one term. Note that in these cases, no inverted lists need to be sent (max-1=0), and only the first T terms are returned. The cost of using unstructured search will be greater than this amount (except for the trivial case where the frequency of the term is 1.0).

There are two reasons why the most challenging cases involve queries with terms of medium frequency. In these cases, the cost of using both the structured and unstructured searches are likely to be similar. However, the expected frequency of terms is not necessarily equal to their actual frequency. For example, while the words "new" and "york" may be relatively rare, the frequency of "new york" is likely to be higher than the product of both individual terms. As a result, the PHIRST approach is most likely to deviate from the optimal choice in these types of cases.

A second challenge results from the fact that we only published up to d instances of a given term. In cases where inverted lists were published without limitation, e.g. d equals N ($DOC - COUNTER$), the second algorithm contains only two possible outcomes – either the expected cost is larger for using structured search, or it is not. However, our assumption is that hardware limitations prevent storing this number of terms, and d must be set much lower than N. As a result, situations will arise where we would **like** to use inverted lists, but as these files have incomplete indices, this approach will fail in finding results in position d+ε. While other options may be possible, in these cases our algorithm (in lines 7-9) takes the d terms from the inverted lists, and conducts an unstructured search for all remaining terms. In general, we found this approach will be effective so long as the T < d, or the relationship, T < d << N exists. We further explore the impact of this limitation in the next section.

6   Experimental Results

In this section we present experimental results used to validate the effectiveness of the algorithms in this paper. As our research goal was to check if PHIRST is appropriate for medium sized newsgroups, we chose a corpus of 2000 real movie websites to conduct our experiments [11]. The results from the publishing experiments demonstrate that PHIRST actually becomes more feasible as more documents and agents are added to the network. We also created two types of query experiments. In one group we created artificial queries based on the frequency of words. This experiment demonstrated the theoretical strengths and weaknesses of PHIRST. We also studied real movie queries based on the Internet Movie Database (www.imdb.com). These experiments demonstrated that any weakness in PHIRST is likely to be insignificant in handling real queries.

6.1   Publishing Experiments

Recall that the publishing algorithm is based on storing a maximum of d entries in a given term's inverted list. We simulated the publishing process to study how this parameter affected the average number of stored inverted entries with and without term stemming. Figure 2 displays the average number of inverted terms (Y-axis) in groups of 50, 250, 500, 1000 and 2000 agents (X-axis). We assumed that every agent published 1 document taken from the movie corpus [11]. In the left graph, we used the Paice stemming algorithm [10] on each term before storing it. The right graph published each term without stemming. In both graphs we also ran the publishing algorithm with d=25 and 75.

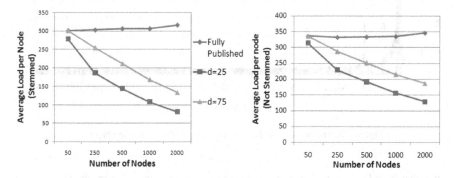

Fig. 2. Comparing publishing requirements of full publishing versus publishing limited to d=75

Several interesting results can be seen from this graph. First, on average stemming saved approximately 50 words per document. This is because stemming lumps similar words, reducing the number of unique words occurring per document. Second, note the publishing algorithm has progressively larger storage savings as the number of nodes grows. Assuming d=N, all terms will be stored, and no publishing gain will be realized by using the PHIRST approach. However, assuming d is kept fixed, the more documents that are added, the gap between d and N grows. This results in progressively more words exceeding the d threshold, and no longer needing to be stored. As a result, the publishing algorithm becomes **more** scalable the more nodes that are added, making full text search feasible even in very large P2P networks.

Table 2. Average number of inverted list entries if 1 document published for every 2 agents.

Number of Nodes	50	250	500	1000	2000
Fully Published	150.43	151.51	153.13	153.1265	157.8343
d=25	138.84	93.106	72.17	53.97	40.605
d=75	150.43	127.14	105.72	84.38	67.035

Finally, in this experiment we assumed each node had 1 document to publish. We also ran this approach with more dense (e.g. 2 documents per node) or more sparse (e.g. 1 document every 2 nodes) network assumptions. As one would expect, the number of terms each node stores is proportional to the total number of nodes. For example, Table 2 shows the sparse assumption of 1 document published for every two nodes. These values are identical to those in Figure 2 * 0.5.

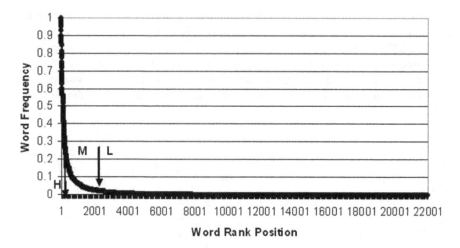

Fig. 3. Distribution of words by rank order within a movie corpus

We also found a Zipfian distribution of terms with a long tail of infrequent terms (see Figure 3). Similar distributions have been found in P2P systems for items such as file frequency [6,7] and term frequency [4]. The storage saving results we found were from words with frequencies greater than d, or those terms towards the head of this distribution.

6.2   Query Experiments

We first conducted query experiments based on artificial queries chosen based on term frequency. Figure 3 displays the rank order of all words within the 2000 word corpus. We considered words of high frequency if they appeared in 30% or more of the documents. There were 200 words in this category. Note that high frequency words are not just "stop" words like "the", "and", or "a", but can be specific to the corpus. For example, these words included movie specific terms such as "character", "play", and "plot". At the other extreme, we define low frequency words as those appearing 50 times or less (frequency 2.5% or less). The large majority of terms were within this category due to the long tail of the term distribution. Finally, we assume medium frequency words are those between these extremes.

We created paired terms (2 terms) of all permutations of these categories. This involves words both with high frequency (HH), both of low frequency (LL), both of medium frequency (MM), low high combinations (LH), low medium combinations (LM), and medium high combinations (MH). Note that the order of the words does not impact the query algorithm as terms are first sorted by the query algorithm based on their frequency. For example, the low medium category (LM) is consequently equivalent to the medium low one (ML).

Next, we generated 1000 artificial queries from each category. We studied how many results were returned from each of 4 search algorithms. The Structured Search (**SS**) method published all terms and sent these indices between agents as necessary during queries. The Unstructured Search (**US**) used no publishing and used a random walk

Table 3. Comparing cost levels of SS, US, TTL, and PHIRST methods in LL, LM, LH, MM, MH, and HH artificial queries

	SS	US	TTL=100	PHIRST
LL	1466	2000000	100000	1466
LM	2206	2000000	100000	2142
LH	3177	1987754	100000	2010
MM	20732	1865474	99953	13256
MH	60188	234211	95624	18075
HH	871986	19746	20077	19995

approach to find query results. The TTL=100 method used an unstructured search, but terminated after visiting 100 agents. Finally, the hybrid PHIRST approach implemented the publishing and query algorithms described in this paper. In these experiments we used a value of $d=75$ in the PHIRST method.

Table 3 displays the average number of nodes visited (in the case of unstructured search) and / or the inverted list entries sent (for structured search) in finding 20 matches from each query (T=20). For simplicity, we assume that the costs of visiting nodes through unstructured search, and sending inverted list entries are equal, or $C_U = C_S$. As expected, we find that Structured Search (SS) is most expensive in finding common terms; where Unstructured Search (US) is most effective. Conversely, SS is most effective in finding rare terms. The hybrid PHIRST approach operates similarly to SS in finding rare terms (LL) and US in finding common items (HH). Note that in middle categories (for example MH) this approach sent the least amount of information. PHIRST saves costs by only sending a maximum of d entries even when structured search is deemed necessary. Furthermore, this approach switches between the SS and US methods as needed, saving additional costs.

The results in Table 4 display the number of query results returned from each search algorithm. This result underlies the potential strengths and weakness within the PHIRST method. Despite the lower costs of PHIRST, this approach was overall equally effective in returning the query results. When word combinations were frequent, the unstructured search component of the PHIRST method still found these results (thus MH was still successful). At the other extreme, assuming the word frequency of any term was less than d, at least one term was fully indexed. In these cases, complete recall was also guaranteed if structured search is used on the indexed term(s) followed by unstructured search to find all remaining terms. In these experiments, all terms taken from the L category were in less than d documents (e.g. L values had 50 or fewer instances while $d=75$), resulting in full recall for all of these categories (LL, LM, and LH) as well. As predicted in section 5, the query algorithm did have slight trouble in finding series of terms of medium frequency. Note that the PHIRST method did return slightly fewer results in the MM case (870 versus 874).

We found that this limitation was negligible in answering real world queries once d was significantly higher than T. To verify this claim we used the 1000 most popular real movie keywords taken from the Internet Movie Database Internet Movie Database[1]

---

[1] (http://www.imdb.com/Search/keywords).

Table 4. Comparing recall levels of SS, US, TTL, and PHIRST methods in LL, LM, LH, MM, MH, and HH artificial queries

	SS	US	TTL=100	PHIRST
LL	3	3	0	3
LM	68	68	2	68
LH	1167	1167	47	1167
MM	874	874	93	870
MH	4626	4626	1180	4626
HH	5000	5000	4997	5000

Table 5. Comparing recall levels of SS, US, TTL, and PHIRST methods with regard to different numbers of results (T)

	SS	US	TTL=100	PHIRST
T=5	4592	4592	2138	4587
T=20	15598	15598	3712	15252
T=50	30347	30347	4534	28154
T=2000	105649	105649	5254	35087

Table 6. Comparing cost levels of SS, US, TTL, and PHIRST methods with regard to different numbers of results (T)

	SS	US	TTL=100	PHIRST
T=5	57680	591841	86578	12006
T=20	68696	1181515	97735	24976
T=50	83435	1567039	99269	38744
T=2000	158737	2000000	100000	68610

taken from October 25, 2006. These queries were typically between 1 and 4 words (mean 1.94).

Table 5 compares the number of results found from these queries with SS, US, and TTL=100 methods, and the PHIRST method with d=75 with variable values for T. Note that the PHIRST algorithm found nearly all results (99.89%) when only 5 results were requested (T=5). PHIRST held up fairly well even when 20 matches (T=20) were required with 97.78% of all matches found. The recall of the PHIRST approach dropped with T (92.77% at T=50, and only 33.23% at T=N). This confirms the claim that in real queries the recall of the PHIRST approach will be nearly 100% for $T \ll d$ (e.g., T=5), but performed poorly once $T \gg d$ (e.g., T=N).

Table 6 displays the search costs for finding these real queries for the 4 algorithms described in this paper assuming $C_S = C_U = 1$, and each agent stored only one document. We again found the PHIRST approach had significantly lower search costs that all three of the other approaches. Again, observe that the advantage to the PHIRST approach is most effective when d>>T. If T=5, the PHIRST approach has nearly 1/5 the cost of the next best method (SS) (with a high recall of 99.89%). If T=20, its cost is still nearly 1/3 that of the next best method (SS) (recall still high at 97.78%). If T=N, the cost advantage of the PHIRST approach is under 1/2 from the next best method (TTL=100) (recall only 33.23%).

7 Conclusion

In this work we present PHIRST, a hybrid P2P search approach that leverages the strengths of structured and unstructured search. We present a P2P publishing algorithm that insures that no agent can hold more than d entries in its inverted list of a given term. This ensures that no one agent is required to hold disproportional amounts of data. PHIRST is highly scalable in that every agent typically stores fewer entries as the number of agents grows. This allows us to partially index all words in the corpus while keeping storage costs low. We also present a querying algorithm that selects the best search approach based on global frequencies of all words in the corpus. This allows us to choose the best method based on estimated cost. PHIRST uses unstructured search to compensate for the lack of published inverted list terms and structured search to location rare terms.

References

1. Chawathe, Y., Ratnasamy, S., Breslau, L., Lanham, N., Shenker, S.: Making gnutella-like p2p systems scalable. In: SIGCOMM '03, pp. 407–418 (2003)
2. Gravano, L., García-Molina, H., Tomasic, A.: Gloss: text-source discovery over the internet. ACM Trans. Database Syst. 24(2), 229–264 (1999)
3. Joachims, T.: A probabilistic analysis of the Rocchio algorithm with TFIDF for text categorization. In: Proceedings of ICML-97, pp. 143–151 (1997)
4. Joung, Y.-J., Fang, C.-T., Yang, L.-W.: Keyword search in dht-based peer-to-peer networks. In: ICDCS '05, pp. 339–348. IEEE Computer Society, Los Alamitos (2005)
5. Li, J., Loo, B., Hellerstein, J., Kaashoek, F., Karger, D., Morris, R.: On the feasibility of peer-to-peer web indexing and search. In: IPTPS. 2nd International Workshop on Peer-to-Peer Systems (2003)
6. Loo, B.T., Hellerstein, J.M., Huebsch, R., Shenker, S., Stoica, I.: Enhancing p2p file-sharing with an internet-scale query processor. In: Proceedings of VLDB, pp. 432–443 (2004)
7. Loo, B.T., Huebsch, R., Stoica, I., Hellerstein, J.M.: The case for a hybrid p2p search infrastructure. In: Voelker, G.M., Shenker, S. (eds.) IPTPS 2004. LNCS, vol. 3279, p. 2. Springer, Heidelberg (2005)
8. Lv, Q., Cao, P., Cohen, E., Li, K., Shenker, S.: Search and replication in unstructured peer-to-peer networks. In: ICS '02, pp. 84–95 (2002)
9. Morris, R., Karger, D., Kaashoek, F., Balakrishnan, H.: Chord: A Scalable Peer-to-Peer Lookup Service for Internet Applications. In: ACM SIGCOMM 2001, pp. 149–160 (2001)
10. Paice, C.D.: Another stemmer. SIGIR Forum 24(3), 56–61 (1990)
11. Pang, B., Lee, L., Vaithyanathan, S.: Thumbs up? sentiment classification using machine learning techniques. In: EMNLP '02, pp. 79–86 (2002)
12. Reynolds, P., Vahdat, A.: Efficient peer-to-peer keyword searching. In: Middleware, pp. 21–40 (2003)
13. Kubiatowicz, J.: Handling churn in a DHT. In: USENIX '04, pp. 127–140 (2004)
14. Zhao, B.Y., Huang, L., Stribling, J., Rhea, S.C., Joseph, A.D., Kubiatowicz, J.D.: Tapestry: a resilient global-scale overlay for service deployment. IEEE Journal on Selected Areas in Communications 22(1), 41–53 (2004)
15. Yang, Y., Dunlap, R., Rexroad, M., Cooper, B.F.: Performance of full text search in structured and unstructured peer-to-peer systems. In: IEEE INFOCOM (2006)
16. http://www.google.com

Multi-agent Cooperative Planning and Information Gathering

Fariba Sadri

Imperial College London
`fs@doc.ic.ac.uk`

Abstract. In this paper we propose a multi-agent architecture, made of co-operative information agents, where agents can share with one another their knowledge of the environment and expertise in planning for achieving goals. In particular we consider how through communication such agents can incrementally learn partial and full plans. Such information exchange is particularly useful in the case of situated agents which have diverse abilities and expertise and which have partial views of their environments. It is also useful in the case of agent systems where agents collaborate towards achieving joint or individual goals. We describe an agent model based on abductive logic programming and give detailed protocols and policies of communication. We then define formally what it means for such information exchanges to be *effective*, and prove results regarding termination and effectiveness of dialogues based on the formalized policies.

1 Introduction

Intelligent information agents are finding their way in many diverse applications, from the more traditional technologies such as industrial processes and monitoring to the more recent, such as web technologies and ambient intelligence. There are several features that are common to many such complex applications. The environment is dynamic and complex. It is unlikely that any one entity (artificial agent/human) has complete knowledge about the environment. Moreover, agents are expected to monitor the environment and events, and react and adapt to changes as well as pursue goals. Because of the complexity and the distributed nature of the environment, no one entity may have complete pre-specified knowledge or expertise about what to do in all circumstances. But, collectively, with information co-operation, the agents may be able to cope. Agents need to learn on the fly.

In this paper we consider how agents can learn new plans and information about the environment through communication and cooperation with each other. The same techniques can also be used for agent-human communication and co-operative information gathering. Our definition of a plan for a goal is a partially ordered set of actions that, if executed in such a way that respects the partial order it would achieve the goal. As a concrete scenario one can consider an architecture for cooperative information systems to deal with the information space in the internet and large scale intranet computing environments [20]. In such an architecture agents can model

M. Klusch et al. (Eds.): CIA 2007, LNAI 4676, pp. 72–88, 2007.

information mediators and brokers, each of which has its own goals and its own expertise, and they cooperate by sharing their expertise. Plan in this context will be "query plans" [7, 10] constructed by the mediator agents with help from each other and from the brokers.

Our work differs substantially from the mainstream of multi-agent planning [see for example 22], which focuses on coordination, goal refinement, filtering and task allocation. We focus, instead, on cooperative planning through information sharing by communication. We consider a system composed of situated agents that are capable of communication with each other. The agents jointly or individually pursue their goals and recognise and adapt to changes in their environments. They can plan how to attempt to achieve their goals and execute actions. They can also exhibit reactive behaviour. Such reactive behaviour includes responding to messages they receive from other agents, observing and recording changes in their environments, and reacting to such changes by executing actions or modifying their plans. An agent model that incorporates these activities and capabilities is called the KGP model and has been described in [4, 11]. We describe an extended model called *KGP+* which takes inspiration from and augments both the KGP model, and the abductive logic-based agent model of [13]. However, the policies and the formal results we give in this paper can be seen in the abstract, parametric on the concrete agent model used, and, in particular, the planning facility within the agent model. To give a concrete description and facilitate a future implementation we present KGP+ as an extension of the KGP model.

In addition to the functionalities of KGP agents, KGP+ agents can share information about plans and knowledge about their environments. They also have self-updating functionalities to update their knowledge bases, for example to update their plan libraries when they learn new information through communication.

We describe a language for communication, public protocols and private interaction policies aimed at gathering information. Other work, for example [18, 19], uses an abductive agent approach similar to ours. But these assume that agents already have plans for achieving their goals to start with, and communicate with other agents only to try to obtain resources that are needed in their plans. We do not assume that each agent already has a plan for its goals. Instead we explore how communication can be directed towards acquiring information, to help agents plan for their goals. We concentrate on agents asking for and passing on information to each other about environments, about how to achieve goals, and about how to break down a (macro) action into manageable (atomic) actions. These allow KGP+ agents to incrementally build plan libraries through multiple dialogues with different agents. This is particularly useful in a system of heterogeneous agents where agents have different expertise and abilities, and where they are situated in dynamic diverse environments where each agent may have only a partial view of the environment. Since the KGP+ agent model and the interaction policies are specified formally they lend themselves well to formal verification and analysis. We prove results for termination and effectiveness of dialogues.

The paper is structured as follows. Section 2 gives the background on abductive logic programming and KGP agents, and preliminary definitions. Section 3 describes the KGP+ agents and Section 4 details the formal results. Section 5 extends the framework described in section 3 for further functionality and Section 6 summarises the related work and concludes.

2 Background and Preliminaries

2.1 Abductive Logic Programming

An abductive logic program (ALP) [12] is a tuple $<P, Ab, IC>$ where P is a logic program, consisting of clauses of the form *Head if Body*, where *Head* is an atom, and *Body* is a conjunction of literals. All variables in *Head* and *Body* are assumed universally quantified over the whole clause. *Ab* is a set of abducible predicates. *IC* is a set of integrity constraints of the form $L_1, ..., L_n \rightarrow H_1, ..., H_m$, where the L_i are literals, the H_i are atoms, possibly *false*, and the "," on both sides of $\rightarrow$ denotes "*and*"[1]. All variables occurring only in the H_i are assumed existentially quantified on the right side of $\rightarrow$, and all others are assumed universally quantified over the whole integrity constraint.

Given an ALP and a query Q (conjunction of literals signifying goals or observations), the purpose of abduction is to find a set of atoms in the abducible predicates that, in conjunction with P, *entails* Q, while *satisfying* the *IC*, wrt. some notion of *entailment* and *satisfaction*. Such a set of abducibles is an *abductive answer* to Q wrt the ALP. The notions of ALP and abductive answer have been extended to incorporate constraints [9, 16], for example temporal constraints.

2.2 The KGP Agent Model

KGP agents [11] have knowledge, goals and plans. They can decide and plan for goals, execute actions and make observations. The KGP model has a knowledge base composed of various modules, including KB_{plan}, for planning, KB_{re}, for reactivity, and KB0, for storing dynamic information about the agent's action executions and observations in the environment. It is assumed that KB0 is included in all the other knowledge modules. KB_{plan} and KB_{re} are abductive logic programs. The knowledge base modules are accessed by a collection of Capabilities, including Planning and Reactivity. These two, respectively, allow agents to plan for their goals and react to observations, which include messages received from other agents, and properties observed in the environment and in respect to other agents.

Capabilities are utilised in a set of Transitions that describe how the state of the agent changes. The transitions include Passive and Active Observation Introduction and Action Execution. Actions are timed action operators, such as *open-door(t)*.[2] KGP agents have three types of actions, physical, sensing and communicative. Passive and Active Observation Introduction Transitions record observations. These can be passive, just opening the senses to see what arrives, or they can be active, actively seeking some information from the environment. The Planning and Reactivity Capabilities have been specified in abductive logic programming and have been

---

[1] We differ slightly here from the standard syntax of *IC* in ALP, in order to simplify the syntax without losing generality.

[2] For simplicity, sometimes we call action operators, without the times, actions as well.

implemented by an abductive proof procedure called CIFF [9], which combines abductive reasoning with constraint satisfaction.

As an illustration, and because we will use it later in the paper and it also helps with understanding abduction, we give a specification of the Planning Capability, denoted as $\models^T_{plan}$ as follows:

$$KB_{plan}, G \models^T_{plan} (As, TCs).$$

This specification is a modified version of the one in the KGP model: here, for our purposes, we have assumed that agents plan *fully* for one goal at a time. The specification means that given an agent's KB_{plan}, and a goal G, at time T, (As, TCs) is a plan for G wrt KB_{plan}, where As is a set of timed action operators and TCs is a set of temporal constraints. G is a positive or negative property. Abductive planning is described in [16]. (As, TCs) is such that:

1. TCs specifies a partial order on the actions in As,
2. TCs allows all actions in As to be executed in the future of T, and
3. according to KB_{plan}, the execution of the actions in As, respecting the temporal constraints in TCs, is expected to ensure that G holds after the actions.

(Of course in an unpredictable environment other events can prevent the success of plans. This is one reason why the model contains the Passive and Active Observation Introduction Transitions to check, if necessary, whether or not the agent's actions have succeeded.)

For example KB_{plan}, web-connection(T1) $\models^T_{plan}$ ([turn-machine-on(T2), enable-broadband(T3), launch-web-server(T4)], [T2<T4, T3<T4, T4<T1]) denotes that one way to establish a web connection is to turn the machine on and enable broadband, in whatever order, and then to launch a web browser.

We assume the notation $KB_{plan}, G \models^T_{plan} \perp$ denotes that the agent's KB_{plan} provides no plan for G, and the notation $KB_{plan}, G \models^T_{plan} \varnothing$ denotes that G already holds according to KB_{plan} (which includes information about the environment in KB0).

The exact contents of KB_{plan} is not crucial to the rest of the paper. But to give a concrete context briefly, KB_{plan} is an abductive theory <P_{plan}, Ab_{plan}, IC_{plan}>, based on the event calculus (EC) [14]. It allows to write meta-logic programs which "talk" about object-level concepts of fluents, events (that we interpret as action operations), and time points. EC allows to represent a wide variety of phenomena, including operations with indirect effects, non-deterministic operations, and concurrent operations [21]. P_{plan} consists of two parts: *domain-independent* and *domain-dependent* rules. The basic domain-independent rules are:

holds_at(F, T2) if happens(O,T1), initiates(O, T1, F), T1<T2, not clipped(T1, F, T2)
holds_at(not(F),T2) if happens(O,T1), terminates(O,T1,F), T1<T2, not declipped(T1,F,T2)
holds_at(F, T) if initially(F), 0≤T, not clipped(0, F, T)
holds_at(not(F), T) if initially(not(F)), 0≤T, not declipped(0, F, T)
clipped(T1, F, T2) if happens(O,T), terminates(O, T, F), T1≤T<T2
declipped(T1, F, T2) if happens(O,T), initiates(O, T, F), T1≤T<T2

These rules, in effect, state that a property (fluent) F holds at a time if an earlier actions has initiated it or it holds initially, and it has not been terminated since; analogously for negated properties not(F). The domain-dependent rules define *initiates, terminates, initially,* and *precondition,* e.g.

initiates(go(X,L1,L2),T, at(X,L2)) if holds_at(mobile(X),T)
terminates(go(X,L1,L2),T, at(X,L1)) if holds_at(mobile(X),T), L1≠L2
initially(at(bob,(1,1)))   initially(mobile(bob))

Roughly speaking, the predicates *initiates, terminates* represent the cause-effects links between actions and fluents in the modeled world. The above indicate that the action of going from one location L1 to some other location L2 initiates the agent (robot) X being at location L2, and terminates X being at location L1, provided that X is mobile. Moreover, some agent bob is initially at location (1,1) and mobile. The conditions of the rules defining *initiates* and *terminates* can be seen as preconditions for the effects of the operator *go* to take place. Preconditions for the executability of operators are specified by means of a set of rules defining the predicate *precondition*, e.g.

precondition(go(X,L1,L2), at(X,L1))   precondition(go(X,L1,L2), adjacent(L1,L2))
precondition(go(X,L1,L2), free(L2))

namely the preconditions of the operator go(X,L1,L2) are that X is at the initial location L1, L1 is adjacent to L2 and location L2, X is moving to, is free.

There are additional domain-independent (bridge) rules allowing agents to draw conclusions from the contents of KB0, which represents the "narrative" part of the agent's knowledge, i.e. the direct information the agent has about the changing environment. These rules contain

holds_at(F, T2) if observed(F, T1), T1≤T2, not clipped(T1, F, T2)
happens(O, T) if executed(O, T)

The first deals with the case where an agent makes observations in the environment (via the Passive or Active Observation Transitions). A clause *observed(F, T)*, recorded in KB0, indicates that the agent has observed property *F* holding at time *T*. The second deals with the case of observing that an action has been executed. We do not give the rest of the bridge rules here. In addition to these, P_{plan} also contains rules such as

happens(O, T) if assume_happens(O,T)

that facilitate planning through abduction. The abducibles, Ab_{plan}, in P_{plan}, include the predicate *assume_happens*. IC_{plan}, the set of core integrity constraints of P_{plan}, is the following set

holds_at(F,T), holds_at(not(F),T) → false
assume_happens(O,T), precondition(O,P) → holds_at(P,T)
assume_happens(O, T), not executed(O, T), time_now(T') → T > T'

The integrity constraints in IC_{plan} prevent plans which are unfeasible. The first integrity constraint makes sure that no plan is generated with inconsistent consequences. The second integrity constraint makes sure that, if a plan requires an action at a certain time point, its preconditions are also planned for to hold at that time point. The last integrity constraint makes sure that each action introduced in a plan which has not been executed yet, is indeed executable in the future. IC_{plan} can also contain domain-dependent integrity constraints.

Example 1: If we add *initially(free(L))*, namely that all locations are initially free, and appropriate clauses for the property *adjacent* to the above, the following can be generated as a plan for bob to be at location (3,2): ((go(bob(1,1), (1,2), T1), go(bob(1,2), (2,2), T2), go(bob(2,2), (3,2), T3)), (T1<T2, T2<T3)). However, if while planning it is observed that location (2, 2) is not free then a different plan will be generated, for example : ((go(bob(1,1), (2,1), T1), go(bob(2,1), (3,1), T2), go(bob(3,1), (3,2), T3)), (T1<T2, T2<T3)).

The Planning Capability extracts the plans from the abductive answers that are returned for the goal wrt P_{plan} which consist of instances of the predicate "assume_happens" together with temporal constraints. Notice that the Planning Capability ensures that the plan that is generated is feasible and consistent, in the sense that it satisfies all the integrity constraints in P_{plan}. This property remains unchanged in KGP+ agents who accumulate planning knowledge "on the fly".

   The KGP model also provides a *goal selection operator*, which, for the purposes of this paper, we represent as a predicate *sg(G, T)* to indicate that at time T goal G is selected (to be planned for). If G is selected at time T, then according to the agent's KB, G does not hold already. The KGP+ agent model, described in Section 3, extends the KGP model to provide the features that are required for cooperative planning and information gathering. We end this section by giving a collection of preliminary definitions required in the rest of the paper.

Definition. An *agent system* is a finite set AS of ground terms, each representing the name of a KGP+ agent. AS contains at least 2 terms. Agents in an agent system can communicate with each other, and we assume that they share an ontology for communication. This ontology is given below.

Definition. The *Communication Language*, L, for KGP+ agents consists of utterances of the form *tell(X, Y, Content, T, Id)*, where X and Y are, respectively, the sender and the receiver agents, T is the time of the utterance, Id is a unique dialogue identifier, and *Content* is one of the following:

1. enquire-plan(Goal, Constraints)
2. inform-plan(Goal, Constraints, Plan)
3. inform-plan(Goal, Constraints, no-more)
4. inform(Goal)
5. end

In the above *Goal* is a property, for example *web-connection* or *at(bob(3,2))*, *Constraints* is a set of atoms, and *Plan* is a tuple *(As, TCs)*, where *As* is a set of timed action operators and *TCs* is a set of temporal constraints.
   The utterance with content (1) enquires if the receiver knows of a plan for *Goal* that avoids all elements in the set *Constraints*. For example the content *enquire-plan(at(bob(3,2)), (at(bob(2,2)), at(bob(2,3)))* is an enquiry for a plan to get bob to location (3,2) whilst avoiding it visiting locations (2,2) and (2,3). The utterance with content (2) is used to inform that *Plan* is a plan, satisfying the *Constraints*, for achieving *Goal*. Content (3) is used to inform that the sender has no (more) such plans, and content (4) is used to inform that *Goal* already holds. Content (5) is used to end the (current) dialogue.

A dialogue will be defined formally later in the paper. Below is an informal example of one:

Example 2:
- Agent a asks agent b if it knows how to achieve a goal G.
- Agent b answers with a plan P1 for G. The plan requires a resource R, which a does not have.
- So a asks b if it knows of another plan for G not requiring resource R.
- Agent b answers with another plan P2 for G which does not require resource R. But P2 requires execution of an action A, and a does not have enough information about how to execute A.
- So a asks another agent c for information about how to execute action A.
- Agent c provides a with information about action A, in effect breaking down action A, via a plan, to "smaller" actions that a knows about.
- Agent a, who has now learned how to achieve goal G, by combining the information it has learned from agents b and c, ends the dialogues.

Definition. Given language L we define $F(L)$, the set of all *final utterances*. These are all utterances with content of the form 3, 4 or 5.

3   KGP+ Agents

KGP+ agents share the Capabilities of KGP agents, have knowledge bases, goals, plans, and can perform actions and react to communications they receive from other agents. But there are differences between KGP and KGP+ agents, detailed below. To motivate the design we give an informal overview.

Overview: Agent a has to achieve a goal G, and does not have enough information about doing so. Agent a seeks information from other agents. It first chooses an agent to contact. For this it uses information it maintains in a knowledge base (KB_{sys} below) about other agents. This knowledge base is based on the history of past communications and allows a to avoid agents it already knows have no useful knowledge about G. Having chosen an agent, a sends a request for information, and through a dialogue, it refines its request incrementally, taking into account its abilities in executing actions, until it either gets useful information or determines that that agent cannot help any further. In both cases it will record whatever new useful information it has learned, and in the latter case, it continues to consult other agents, if any remain, in an attempt to fill in any remaining gaps in the required information. Below we detail the architecture that makes this type of scenario possible.

3.1 Actions

In addition to the 3 types of actions of KGP agents, KGP+ agents have a 4th type of updating their own knowledge bases. This action is represented as *update(Name-of-KB, Info)*. The execution of this action not only leads to the insertion of a record of its execution in KB0, but also updates the knowledge base named in the action by the information represented in *Info*. This action is used by KGP+ agents to update their knowledge bases when they learn new information from other agents.

3.2 Selection Operators

In addition to the goal selection operator sg, KGP+ agents have a plan selection operator, $sp(G, C, T, P)$, that selects one answer to the planning problem available via the Planning Capability. More formally sp has the following property for each agent a:

$$sp(G, C, T, P), neg(C, C') \rightarrow KB^a_{plan}+C', G \models^T_{plan} P,$$

where G is a goal, C is a set of atoms, T is a time point, P is a plan, and $neg(C, C')$ generates a set C' of denials[3], one denial for each element in C, namely a denial $A \rightarrow false$ for each A in C. Thus P is a plan for G that takes into account the information in KB^a_{plan} and the additional integrity constraints in C'.

3.3 Knowledge Base

The knowledge base of a KGP+ agent is similar to that of KGP agents, but with three main differences: the contents of the KB_{plan} and KB_{re} of a KGP+ agent are different and a KGP+ agent has an additional knowledge base module, KB_{sys}. The KB_{sys} of an agent contains information about the other agents in its agent system, with whom the agent can communicate, and its accumulated knowledge of what they know (or do not know) as judged by its communications with them.

3.3.1 KB$_{plan}$

In addition to the EC theory given in Section 2, the KB_{plan} of KGP+ agents includes a *plan library* which is incrementally augmented through information cooperation. Moreover their KB_{plan} also contains information about their own abilities in executing actions.

The KB_{plan} of an agent a, represented as KB^a_{plan}, is an ALP $<P,Ab,IC>$, where P contains clauses of the form $able(a, A)$ to mean that a has the ability of performing action represented by action operator A. All KGP+ agents have the ability of updating their own knowledge bases. So P in KB^a_{plan} includes a clause

> able(a, update(Name-of-KB, Information)).

Plan libraries in P contain domain-dependent clauses of the form

C1   holds_at(G, T) if perform(A_1, T_1),....,perform(A_n, T_n), Constraints(T, T_1,...,T_n)[4].

The A_i are action operators, and *Constraints* is a (possibly empty) conjunction of temporal constraints, involving T, T_1, ..., T_n. P also contains the following domain-independent clause

C2   perform(A, T) if able(a, A), happens(A, T).

Clause C1 indicates that performing actions A_i, i=1,..n, respecting the temporal constraints, leads to G holding. C2 indicates that performing action A amounts to doing A, provided agent a is able to do it. *Ab*, the set of abducible predicates in

---

[3] A denial is an integrity constraint of the form $L_1, ...,L_n \rightarrow false$.

[4] To allow conditional plans, this syntax can be augmented with an additional conjunction of conditions to be checked in the environment or via the agent's knowledge base before selecting the plan from the library.

KB^a_{plan}, contains the predicate *assume_happens* (as before). All KGP+ agents are equipped with the domain-independent parts of P in their KB_{plan}, but they may differ in the contents of their domain-dependent parts. Thus they may have different abilities in action execution and different expertise in their planning knowledge.

3.3.2 KB0
Similarly to the KGP model, the KB0 of KGP+ agents records information about the agent's action executions and observations. For ease of notation, we assume that when an agent a executes a communication action *tell(a, X, Content, T, ID)*, it is recorded in its KB0 as *tell(a, X, Content, T, ID)*. Similarly when it receives a communication *tell(X, a, Content, T, ID)* it is recorded in its KB0 as *tell(X, a, Content, T, ID)*.[5]

3.3.3 KB_{sys}
The KB_{sys} of a KGP+ agent a, denoted KB^a_{sys}, contains information about the agents with whom a can communicate. It contains ground clauses of the form *agent(X)*, for all such agents X, apart from a, itself, which are in its agent system(s). It also contains the two clauses

asked(X, G) if tell(a, X, enquire-plan(G, C), T, Id)

has-no-useful-information(X, G) if tell(X, a, enquire-plan(G,C),T, Id)

The first represents that a has already asked X for information about achieving G. The second represents that X does not have information about achieving G that is useful to a, because X, itself, has asked a how it could achieve G. Agents use the information in their KB_{sys} to decide which agent to communicate with when they need information. This knowledge base will also record any further available information that allows selection of agents to seek cooperation from, such as information about their expertise, reliability and response times.

3.3.4 KB_{re}
The KB_{re} of KGP+ agents is an ALP $<P, Ab, IC>$. KB_{re} has reactive rules similar to those of KGP agents, which we have not given as they do not have an impact on the paper. But in addition KB_{re} of KGP+ agents contains their individual communication policies for information gathering. Here we focus on these policies. The policies are designed to be compliant with the following protocol of interaction between any 2 agents a and b:

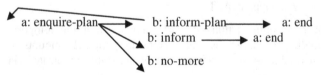

The *IC* component of KB_{re} contains policies of the form $U_i, C \rightarrow U_{i+1}$, KBupdate

---

[5] Strictly speaking, the first should be represented as *executed(tell(a, X, Content, ID), T)*, and the latter as *observed(tell(X, a, Content, ID), T)*.

where U_i and U_{i+1} are utterances, C is a (possibly empty) conjunction of conditions other than utterances and *KBupdate*, which may be missing from the policy, refers to an action of updating a knowledge base module.

These policies are used in the following way. If the agent receives an utterance U_i and conditions C are satisfied wrt the agent's KB, Capabilities and selection operators, then the agent will generate the utterance U_{i+1}, and execute the action specified in *KBupdate*. An operational model for similar policies is given in [18]. In the KGP model the conditions C only refer to the knowledge base of the agent. We allow a richer language for the policies, whereby C can also refer to the selection operators of the agents. Also in the KGP the *KBupdate* component does not exist.

The set *Ab* of abducibles in KB_{re} consists of *assume_happens* and *tell*[6]. The communication policies of a KGP+ agent *a*, in *IC*, are as follows. We first give the policies and then give an explanation for each one. The policies impose that a response has to be made within 1 time unit of receiving an utterance. This is arbitrary; any finite time lapse will do as far as our formal results in Section 4 are concerned.

Policies for answering requests for information from other agents:

I1: tell(X, a, enquire-plan(G, C), T, Id), sp(G, C, T, P), P≠⊥, P≠∅ → tell(a, X, inform-plan(G, C, P), T+1, Id)

I2: tell(X, a, enquire-plan(G, C), T, Id), sp(G, C, T, P), P=⊥ → tell(a, X, inform-plan(G, C, no-more), T+1, Id)

I3: tell(X, a, enquire-plan(G, C), T, Id), sp(G, C, T, P), P=∅ → tell(a, X, inform(G), T+1, Id)

Policies for responding to offers of information from others:

I4: tell(X, a, inform-plan(G, C, no-more), T, Id), agent(Y), not asked(Y, G), not has-no-useful-information(Y, G), newId(NId) → tell(a,Y,enquire-plan(G, C), T+1, NId)

I5: tell(X, a, inform-plan(G, C, P), T, Id), P ≠ no-more, unable(a, P, P1), add(C, P1, C') → tell(a, X, enquire-plan(G, C'), T+1, Id)

I6: tell(X, a, inform-plan(G, C, P), T, Id), P ≠ no-more, all-able(a, P) → tell(a, X, end, T+1, Id), do(update(KB^a_{plan}, (G,P)), T+1)

I7: tell(X, a, inform(G), T, Id), → tell(a, X, end, T+1, Id), do(update(KB0, observed(G, T)), T+1)

Policies for requesting information

I8: sg(G, T), KB^a_{plan}, G $\models^T_{plan}\bot$, agent(X), not asked(X,G), not has-no-useful-information(X, G), newId(NId) → tell(a, X, enquire-plan(G, []), T, NId)

(I1), (I2) and (I3) deal with a request from X for a plan for G that avoids the elements in C. (I1) states that if a has a plan that it can choose (so its Planning Capability entails at least one suitable plan) then a informs X of this chosen plan. (I2) states that if a has no knowledge of such a plan, then a informs X that it has no (more) suitable plans for G. (I3) states that if a knows that G already holds then it can inform X about this.

---

[6] We can use the notation happens(tell(X,Y,Content, Id), T) so that we have one uniform abducible in all KB modules, namely *assume_happens*. But we prefer not to do this for ease of reading the policies.

(I4) states that if an enquiry to agent X has failed to inform a of a (suitable) plan for G then a asks another agent, if any is left amongst the ones it knows about, whom it has not already asked about G and who may have useful information about achieving G. The predicate *newId* is for housekeeping purposes; it generates a new identifier for the utterance (in effect to start a new *dialogue* – defined later).

(I5) states that if X informs a of a plan for G but there are some actions in the plan that a is not able to perform, then a asks X for another plan that excludes these actions. *add(C, P1, C')* is defined so that C'=C∪{assume_happens(A): A in P1}.

(I6) states that if X informs a of a plan for G and a is able to perform all its actions (all-able(a, P)), a ends the dialogue and updates its plan library in KB^a_{plan} with the new plan. (I7) states that if X informs a that G already holds then a ends the dialogue and records this new information, as an observation in its KB0. (I8) states that if a has a goal G for which it cannot generate a plan, then a enquires if another agent has a plan for G. The choice of the agent is done as in (I4). The P component of KB_{re} has definitions for the auxiliary predicates used above in (I1) – (I8) including *all-able* and *unable*. These definitions are straight-forward and we do not give them here.

Let **POL** represent the policies (I1)-(I8) above, and all the related auxiliary definitions in P in KB_{re}. Below we give a simple example to illustrate the use of the policies.

3.4 An Example

Let a, b, c be KGP+ agents in an agent system AS. Suppose a has a goal of obtaining *web-connection*, but a and b have no knowledge about achieving this goal. But let KB^c_{plan} contain the following clauses:

holds_at(web-connection,T) if perform(enable-broadband, T1), perform(open-browser, T2), T1<T2, T2=T

holds_at(web-connection,T) if perform(enable-ISP, T1), perform(dial-up-ISP,T2), perform(open-browser, T3), T1<T2, T2<T3, T3=T,

i.e. c knows of two ways of achieving *web-connection*, one through broadband and one through ISP dialing. Then the following "sequence of dialogues" (to be defined formally later) can occur between a, b, c:

tell(a, b, enquire-plan(web-connection, [])), 10, id1) : a asks b for information about achieving *web-connection*

tell(b,a,inform-plan(web-connection,[], no-more), 11, id1): b has no information about this

tell(a, c, enquire-plan(web-connection, [])), 12, id2): a asks c for the same information (with a different dialogue identifier)

tell(c,a,inform-plan(web-connection,[],([enable-broadband(T1),   open-browser(T2)], [T1 <T2, T2=T])), 13, id2): c answers with a plan consisting of first enabling broadband and then opening browser

tell(a, c, enquire-plan(web-connection, [enable-broadband]), 14, id2): a does not have the ability to enable broadband, so a asks c for a different plan without this requirement

tell(c, a, inform-plan(web-connection, [enable-broadband], ([enable-ISP(T1), dial-up-ISP(T2), open-browser, T3)], [T1<T2, T2<T3, T3=T])), 14, id2): c provides another plan that avoids having to enable broadband.

tell(a, c, end, 15, id2): This new plan is good for a, so a ends the dialogue with c.

After this sequence KB^a_{plan} will have the new information:

holds_at(web-connection,T) if perform(enable-ISP, T1), perform(dial-up-ISP,T2), perform(open-browser, T3), T1<T2, T2<T3, T3=T,

and KB^a_{sys} will imply the new information

asked(b, web-connection) and asked(c, web-connection).

KB^b_{sys} and KB^c_{sys} will both imply the new information

has-no-useful-information(a, web-connection).

Note that integrity constraints (I4) and (I8) in POL allow parallel "dialogues" (conversations) amongst agents. So an agent can hold dialogues with several agents at the same time as it attempts to accumulate information about how to achieve its goal(s). If this parallelism is not desired then the conditions in (I4) and (I8) can be modified to allow the selection of *one* (suitable) agent to communicate with at a time. The choice can be arbitrary or it can be based on the past history of communications providing information about the "helpfulness/usefulness" of other agents. We do not address this any further in this paper.

4 Formal Results

In this Section we give a set of formal definitions and use them to prove formal properties of communications amongst KGP+ agents. We show that "dialogues" and "sequences of dialogues" amongst KGP+ agents are finite (Properties 4.2, 4.3) and effective in allowing agents to share their planning expertise and knowledge about the environment (Theorems 4.1, 4.2, 4.3, 4.4, 4.5).

Let AS be an agent system. In the results below we assume that the Planning Capability of all agents in AS can produce only a finite number of alternative plans for any goal, and each in finite time.

Definition. A *dialogue* D generated by POL between 2 KGP+ agents X and Y in a system of agents AS is an ordered list of utterances $[U_1, U_2, U_3, ...]$ such that:

1. for each U_i either X is the sender and Y, the receiver, or Y is the sender and X, the receiver,
2. if U_i is uttered by X then U_{i+1}, if it exists, is uttered by Y and vice versa,

3. for every pair U_i, U_{i+1}, $i \geq 1$, U_{i+1} can be sent by Z, Z=X or Z=Y, only if there is a policy U_i, $C \rightarrow U_{i+1}$, *KBupdate* in POL, U_i is received by Z and the condition C is satisfied by Z, and
4. all the U_i share the same dialogue identifier ID.

Definition. $D'=[U_1, U_2, ..., U_n, U_{n+1}, ..., U_{n+m}]$, $m \geq 1$, is an *extension of a dialogue* $D=[U_1, U_2, ..., U_n]$ if and only if D' is a dialogue.

Definition. A dialogue $D=[U_1, U_2, ..., U_n]$ is *terminated* if and only if U_n is in F(L), i.e. U_n is a final utterance.

Property 4.1. A dialogue $[U_1, U_2, ..., U_n]$ has no extension if it is terminated.

Proof (sketch). POL allows no continuation of a dialogue after a message has been sent that is a final utterance.

Definition. A *plan request dialogue initiated by an agent a with respect to G generated by POL* is a dialogue $[U_1, U_2, ...]$ generated by POL between a and some other agent such that U_1's sender is a, and U_1's content is of the form enquire-plan(G, C). Below we may call such a dialogue simply a *plan request dialogue*, or a *plan request dialogue initiated by a*, for brevity, if the rest of the details are irrelevant or understood.

Property 4.2. Every plan request dialogue has:

   i) a finite number of utterances, and
   ii) a finite number of extensions.

Proof (sketch). The proof follows because of Property 4.1 and because, thanks to POL, no agent will reply with the same plan for the same goal to another agent more than once.

Definition. A plan request dialogue $D=[U_1, U_2, ..., U_n]$ initiated by an agent *a* generated by POL is *successful* if and only if U_n is an utterance by *a* with content *end*. D is a *failed* dialogue if U_n is an utterance by the other agent and its content is of the form *inform-plan(Goal, Constraints, no-more)*.

The following set of results show the *effectiveness* of POL in allowing the sharing of information. After a successful plan request dialogue wrt G, the enquiring agent either knows G holds already or knows a suitable plan for achieving it, i.e.

Theorem 4.1. If D is a successful plan request dialogue wrt G initiated by X and T is the time of the last utterance in D, then there exists a plan P such that KB^X_{plan}, $G \models^{T+1} P$ and $P \neq \perp$ and all-able(X, P).

Proof (sketch). The last utterance of a successful dialogue can be generated by only two policies in POL, namely I6 and I7. The result holds via this observation and the definition of the Planning Capability.   □

Given two agents X and Y in AS, if Y has a plan for a goal G with actions all within X's abilities then every terminated plan request dialogue between X and Y, initiated by X wrt G is successful, i.e.:

Theorem 4.2. Let D be a terminated plan request dialogue wrt G between agents X and Y. Let T be the time of the first utterance in D. Then D is successful if

$$KB^Y_{plan}, G|=_{plan}{}^T P \text{ and } P \neq \perp \text{ and } P \neq \varnothing \text{ and all-able}(X, P).$$

Proof (sketch). Policies I5 and I6 together with the definition of the Planning Capability and Property 4.2 ensure this theorem.   □

The dynamic nature of the agent system allows us to prove a more interesting result. During a dialogue the knowledge of agents may change, as agents can make observations in their environments or can be occupied in multiple dialogues from which they learn new plans. Thus we can prove the following:

Theorem 4.3. Let D be a terminated plan request dialogue wrt G between agents X and Y. Let T_1 be the time of the first utterance in D and T_n the time of the last utterance. Then D is successful if either (i) for some T, $T \geq T1$, $T < T_n - 2$, KB^Y_{plan}, $G|=_{plan}{}^T P$ and $P \neq \perp$ and $P \neq \varnothing$ and all-able(X, P), or (ii) for some T, $T < T_n - 2$, observed(G, T) is in $KB0^y_T$, and for no T', $T' > T$, $T' \leq T_n - 2$, is clipped(G, T, $T_n - 2$) entailed by KB^Y_{plan}, where $KB0^y_T$ represents the KB0 of agent Y at time T.

So far we have focused on single dialogues. But POL allows the sharing of information amongst any number of agents. To show this we define "sequences of dialogues".

Definition. A *sequence of plan request dialogues initiated by agent X with respect to goal G generated by POL* between KGP+ agents in a system of agents AS is an ordered list of dialogues $[D_1, D_2, D_3,...]$ such that all the following hold:

1. Each D_i is a plan request dialogue initiated by X with respect to goal G.
2. Each D_i is between X and another agent in AS.
3. For all D_i, D_j, $i \neq j$, if D_i is a dialogue between X and Y, and D_j is a dialogue between X and Z, then $Z \neq Y$.
4. Each D_i is a terminated dialogue.
5. For each D_i, $i \geq 1$, D_{i+1} exists only if D_i is a failed dialogue.
6. For each D_i, D_{i+1}, $i \geq 1$, the time of the last utterance of D_i is before the time of the first utterance of D_{i+1}.
7. For each D_i, D_{i+1}, $i \geq 1$, there is a policy U_i, $C \rightarrow U_{i+1}$, KBupdate in POL, such that the last utterance in D_i is U_i and received by X and C is satisfied by X.

In the sequel by "a sequence of plan request dialogues initiated by X" we mean "a sequence of plan request dialogues initiated by agent X with respect to a goal G generated by POL".

Property 4.3. Every sequence of plan request dialogues generated by POL is finite.

Definition. A sequence of plan request dialogues $[D_1, D_2, ..., D_n]$ initiated by X is successful if D_n is a successful dialogue.

Theorem 4.4. If $[D_1, D_2, ..., D_n]$ is a successful sequence of plan request dialogues initiated by X wrt G, and T is the time of the last utterance in D_n, then

$$KB^X_{plan}, G|=^{T+1} P \text{ and } P \neq \perp \text{ and all-able}(X, P).$$

Proof (sketch). This follows fro m the definition of the successful sequence, the definition of the Planning Capability and policies I6 and I7.

Definition. A sequence of plan request dialogues $[D_1, D_2, ..., D_n]$ is *terminated* if $[D_1, D_2, ..., D_n, D_{n+1}]$ is not a sequence of plan request dialogues for any D_{n+1}.

If at least one agent in agent system AS has a plan for a goal G of X with actions all within X's abilities then every terminated sequence of plan request dialogues initiated by X wrt G is successful, i.e.:

Theorem 4.5. Let $SD=[D_1, D_2, ..., D_n]$ be a terminated sequence of plan request dialogues initiated by X wrt G, and let T be the time of the first utterance of D_1. SD is successful if there is an agent Y in AS and a plan P such that

$$KB^Y_{plan}, G \models^T_{plan} P, P \neq \perp \text{ and } P \neq \varnothing \text{ and all-able(X, P).}$$

5 Extension

In this Section we describe how the framework presented so far can be extended to deal with cases where no one agent in the agent system may have a complete suitable plan for a goal for some agent, but collectively, by incrementally assimilating their knowledge in an appropriate way such a plan may be found. The extension deals with further heterogeneity amongst agents, whereby, in effect, because of their different environments and abilities, some actions are considered *atomic* by some agents and *macro-actions* by others. In such situations an agent can *combine* the information it obtains from different agents into one plan with actions it is able to execute. The framework we have described so far is can accommodate this extension, as follows.

The KB_{plan} of each agent a is extended by adding to its domain-independent part the clause

perform(A, T) if not able(a, A), holds_at(macro(A), T),

namely that the agent can perform an action even though it is not able to perform it atomically, provided it can break it down to manageable actions.

We add to the domain-dependent part of KB_{plan} clauses of the form

holds_at(macro(A),T) if perform(A_1,T_1),....,perform(A_n, T_n), Constraints(T,T_1,...,T_n).

Such clauses, in effect, provide libraries for breaking down macro actions to smaller sub-actions. Constraints(T, T_1, ..., T_n) represents a (possibly empty) set of temporal constraints on the times T, T_1, ..., T_n of the actions.

KB_{re} of each agent a is modified simply by modifying (I5) as follows:

tell(X, a, inform-plan(G, C, P), T, Id), P $\neq$ no-more, unable(a, P, P1), add(C, P1, C') $\rightarrow$ tell(a, X, enquire-plan(G, C'), T+1, Id), do(update(KB^a_{plan}, (G,P)), T+1).

This allows agents to record "partial" plans that they learn, because through further communications they may learn a breakdown of the actions they are as yet unable to perform.

6 Conclusion

In this paper we propose an agent model based on abductive logic programming and equipped for communication for exchange of information about environments, actions and plans for achieving goals. To our knowledge this work is novel. It is, however, related to a number of other areas. Several papers [1, 2, 5, 6, 15] have considered self-updating logic programs, and others [3, 8, 17, 18, 19] have discussed agent interactions in the context of resource negotiation. But none of these address the plan information cooperation theme of this paper.

A large body of literature exists on multiagent planning; [22], for example provides an introduction and survey. These works, however, address other issues, such as filtering plans which are incompatible with goals of another agent, and partial global planning (PGP), where tasks are contracted based on some organizational structure, particularly useful in domains of distributed sensor networks, and resource usage optimization.

Our work provides rich scope for further extensions, for example agents asking for plans that satisfy certain temporal constraints, or agents asking each other why a particular plan has failed, in the anticipation of receiving environmental or plan repair information.

References

[1] Alferes, J.J., Brogi, A., Leite,, Pereira, L.M: An evolving agent with EVOLP. In: Proceedings of APPIA-GULP-PRODE'02 Joint Conf. on Declarative Programming (AGP'03), Reggio Calabria, Italy (2003)

[2] Alferes, J.J., Brogi, A., Leite, J.A., Pereira, L.M.: Logic Programming for Evolving Agents. In: Klusch, M., Omicini, A., Ossowski, S., Laamanen, H. (eds.) CIA 2003. LNCS (LNAI), vol. 2782, pp. 281–297. Springer, Heidelberg (2003)

[3] Amgoud, L., Parsons, S., Maudet, N.: Arguments, dialogue and negotiation. In: Proceedings of the 14th European Conference on Artificial Intelligence (ECAI-2000),, pp. 338–342. IOS Press, Berlin, Germany (2000)

[4] Bracciali, A., Demetriou, N., Endriss, U., Kakas, A., Lu, W., Mancarella, P., Sadri, F., Stathis, K., Terreni, G., Toni, F.: The KGP Model of Agency for Global Computing: Computational Model and Prototype Implementation. In: Priami, C., Quaglia, P. (eds.) GC 2004. LNCS, vol. 3267, pp. 342–369. Springer, Heidelberg (2005)

[5] Dell'Acqua, P., Pereira, L.M.: Enabling agents to update their knowledge and to prefer. In: Brazdil, P.B., Jorge, A.M. (eds.) EPIA 2001. LNCS (LNAI), vol. 2258, pp. 183–190. Springer, Heidelberg (2001)

[6] Dell'Acqua, P., Pereira, L.M.: Updating agents. In: Rochefort, S., Sadri, F., Toni, F. (eds.) Proceedings of the ICLP'99 Workshop on Multi-Agent Systems in Logic (MASL'99) (1999)

[7] Duschka, O., Genesereth, M.R.: Query Planning in Infomaster, 1997 ACM Symposium on Applied Computing. San Jose, CA (February 1997)

[8] Dutta, P.S., Jennings, N.R., Moreau, L.: Adaptive distributed resource allocation and diagnostics using cooperative information-sharing strategies. In: Proceedings of 5th Int. conf. on Autonomous Agents and Multi-Agent Systems (AAMAS'06), pp. 826–833. Hakodate, Japan

[9] Endriss, U., Mancarella, P., Sadri, F., Terreni, G., Toni, F.: The CIFF Proof Procedure for Abductive Logic Programming with Constraints. In: Proceedings of JELIA 2004, International Conference on Logics in AI, Lisbon, Portugal, September 2004. LNCS (LNAI), pp. 31–43. Springer, Heidelberg (2004)

[10] Genesereth, M.R., Keller, A.M., Duschka, O.: Infomaster: An Information Integration System. In: ACM SIGMOD Conference, ACM Press, New York (1997)

[11] Kakas, A., Mancarella, P., Sadri, F., Stathis, K., Toni, F.: The KGP Model of Agency. In: Proceedings of ECAI 2004, European Conference on Artificial Intelligence, Valencia, Spain, August 23-27, pp. 33–37 (2004)

[12] Kakas, A.C., Kowalski, R.A., Toni, F.: The role of abduction in logic programming. In: Gabbay, D.M, Higger, C.J., Robinson, J.A. (eds.) Handbook of Logic in Artificial Intelligence and Logic Programming, vol. 5, pp. 235–324. Oxford University Press, Oxford (1998)

[13] Kowalski, R.A., Sadri, F.: From Logic Programming Ttowards Multi-agent Systems. Annals of Mathematics and Artificial Intelligence 25, 391–419 (1999)

[14] Kowalski, R., Sergot, M.: A Logic-based Calculus of Events. In: New Generation Computing, vol. 4(1), pp. 67–95 (February 1986). Also in Knowledge Base Management-Systems. Thanos, C., Schmidt, J. W.(eds.) Springer-Verlag, pp. 23–51. Also in The Language of Time: A Reader. Mani, I., Pustejovsky, J., Gaizauskas, R.(eds.) Oxford University Press (2005)

[15] Leite, J., Soares, L.: Enhancing a multi-agent system with evolving logic programs. In: Inoue, K., Satoh, K., Toni, F. (eds.) Computational Logic in Multi-Agent Systems. LNCS (LNAI), vol. 4371, Springer, Heidelberg (2007)

[16] Mancarella, P., Sadri, F., Terreni, G., Toni, F.: Planning Partially for Situated Agents. In: Leite, J.A., Torroni, P. (eds.) Computational Logic in Multi-Agent Systems. LNCS (LNAI), vol. 3487, pp. 230–248. Springer, Heidelberg (2005)

[17] Rahwan, I., Ramchurn, S.D., Jennings, N.R., McBurney, P., Parsons, S., Sonenberg, E.: Argumentation-based negotiation, Knowledge Engineering Review

[18] Sadri, F., Toni, F., Torroni, P.: Resource Reallocation via Negotiation Through Abductive Logic Programming. In: Flesca, S., Greco, S., Leone, N., Ianni, G. (eds.) JELIA 2002. LNCS (LNAI), vol. 2424, pp. 419–431. Springer, Heidelberg (2002)

[19] Sadri, F., Toni, F., Torroni, P.: Minimally intrusive negotiating agents for resource sharing. In: Gottlob, G. (ed.) Proceedings of the 18th Biennal International Joint Conference on Artificial Intelligence (IJCAI 2003), Acapulco, Mexico, August 12-15, AAAI Press, Stanford, California, USA (2003)

[20] Shakshuki, E., Ghenniwa, H., Kamel, M.: An architecture for cooperative information systems. Knowledge-Based Systems 16, 17–27 (2003)

[21] Shanahan, M.P.: An Abductive Event Calculus Planner. The Journal of Logic Programming 44, 207–239 (2000)

[22] de Weerdt, M., ter Mors, A., Witteveen, C.: Multi-agent planning: An introduction to planning and coordination. In: Handouts of the European Agent Summer School, pp. 1–32

Using Distributed Data Mining and Distributed Artificial Intelligence for Knowledge Integration

Ana C.M.P. de Paula, Bráulio C. Ávila, Edson Scalabrin,
and Fabrício Enembreck

Pontifícia Universidade Católica do Paraná – PUCPR
PPGIA – Programa de Pós-Graduação em Informática Aplicada
Rua Imaculada Conceição, 1155, CEP 80215-901
Curitiba PR, Brasil
Tel./Fax: +55 41 3027-1669
{avila,scalabrin,fabricio}@ppgia.pucpr.br

Abstract. In this paper we study Distributed Data Mining from a Distributed Artificial Intelligence perspective. Very often, databases are very large to be mined. Then Distributed Data Mining can be used for discovering knowledge (rule sets) generated from parts of the entire training data set. This process requires cooperation and coordination between the processors because inconsistent, incomplete and useless knowledge can be generated, since each processor uses partial data. Cooperation and coordination are important issues in Distributed Artificial Intelligence and can be accomplished with different techniques: planning (centralized, partially distributed and distributed), negotiation, reaction, etc. In this work we discuss a coordination protocol for cooperative learning agents of a MAS developed previously, comparing it conceptually with other learning systems. This cooperative process is hierarchical and works under the coordination of a manager agent. The proposed model aims to select the best rules for integration into the global model without, however, decreasing its accuracy rate. We have also done experiments comparing accuracy and complexity of the knowledge generated by the cooperative agents.

Keywords: Distributed Data Mining, Distributed Artificial Intelligence, Cooperative Agents, Learning Agents.

1 Introduction

Data Mining [1] [3] permits efficient discovery of valid, non-obvious information in large collections of data, and it is used in information management and decision making, enabling an increase in business opportunities.

Despite the augmentation of computer processing power, much attention is given to speeding up the data mining process. Basically, speeding up approaches can be either (i) data-oriented or (ii) algorithm-oriented. In (i) the dataset is processed and the learning instances space is reduced by discretization, attribute selection or sampling. In (ii) new search strategies are studied or data are mined in a distributed or parallel fashion.

M. Klusch et al. (Eds.): CIA 2007, LNAI 4676, pp. 89–103, 2007.

According to Freitas and Lavigton [12], Distributed Data Mining (DDM) [4] [5] consists of partitioning the data being mined among multiple processors, applying the same or different data mining algorithms to each local subset and then combining the local knowledge discovered by the algorithms into a global knowledge. The authors also discuss that such global knowledge is usually different (less accurate as the number of subsets increases) from the knowledge discovered by applying the mining algorithm on the entire dataset formed from the union of the individual local datasets. Unfortunately, subsets are often too large to be merged into a single dataset and then, some distributed learning approach must be used.

DDM can deal with public datasets available on the Internet, corporate databases within an Intranet, environments for mobile computation, collections of distributed sensor data for monitoring, etc. Distributed Data Mining offers better scalability, possibility of increase in data security and better response time when compared with a centralized model.

Techniques from Distributed Artificial Intelligence [6] [7], related to Multi-Agent Systems (MAS), can be applied to Distributed Data Mining with the objective of reducing the necessary complexity of the training task while ensuring high quality results.

Multi-Agent Systems [8] [9] are ideal for the representation of problems which include several problem solving methods, multiple points of view and multiple entities. In such domains, Multi-Agent systems offer the advantages of concurrent and distributed problem solving, along with the advantages of sophisticated schemes of interaction. Examples of interaction include cooperative work towards the achievement of a shared goal.

This paper discuss about the use of Multi-Agent Systems [10] [11] in Distributed Data Mining technique taking as example a previous work developed for knowledge integration [26]. Model integration consists in the amalgamation of local models into a global, consistent one. Agents perform learning tasks on subsets of data and, afterwards, results are combined into a unique model. In order to achieve that, agents cooperate so that the process of knowledge discovery can be accelerated. The proposed model previously aims to select the best rules for integration into the global model without, however, causing a decrease in its accuracy rate.

This paper is organized as follows: section 2 discusses the relationship between Distributed Data Mining and Distributed Artificial Intelligence. The knowledge integration protocol approach is discussed in section 3 and its results are displayed and discussed in section 4. In section 5 we include some related work and conclusions are exposed in section 6.

2   Distributed Data Mining and Distributed Artificial Intelligence

Distributed Data Mining (DDM) resulted from the evolution of Data Mining as it incorporated concepts from the field of Distributed Systems. DDM deals with data which is remotely distributed in different, interconnected locations.

According to Freitas [12], distributed mining is a 3-phase process:

- Split the data to be mined into p subsets, where p is the number of processors available, and send each subset to a distinct processor;
- Each processor must apply a mining algorithm to the local dataset. Processors may run the same mining algorithm or different ones;
- Merge the local knowledge discovered by each mining algorithm into a consistent, global knowledge.

DDM systems handle different components: mining algorithms, subsystems of communication, resource management, task planning, user interfaces and others. They provide efficient access to data and distributed computational resources, while they still permit monitoring the mining procedure and properly presenting results to users. A successful DDM system must be flexible enough to fit a variety of situations. It must also dynamically identify the best mining strategy according to the resources at hand and provide a straightforward manner of updating its components.

Client-server and agent-based architectures are examples of solutions which have been proposed in the literature. All agent-based DDM systems contain one or more agents in charge of each dataset. Such agents are responsible for analyzing local data and deliberately communicate with others during the mining stage, exchanging local knowledge until a global, coherent knowledge is reached. Due to the complexities involved in maintaining complete control over remote resources, many agent-based systems utilize a supervisor agent, called facilitator, which controls the behavior of local agents. Java Agents for Meta-learning (JAM) [13] and the BODHI system [14] follow this approach.

As previously mentioned, some of the techniques proposed in Distributed Artificial Intelligence, such as Multi-Agent systems, can be applied to Distributed Data Mining as to reduce the complexity needed for training while ensuring high quality results.

Distributed Artificial Intelligence (DAI) is the union of Artificial Intelligence (AI) and techniques from Distributed Systems. A definition for DAI which is more suitable to the concept of agency is given by Jennings [15], who declared that the object of investigation for DAI lies on knowledge models and techniques for communication and reasoning, necessary for computational agents to be able to integrate societies composed of computers and people. Jennings also splits DAI into two main research areas:

Distributed Problem Solving (DPS) – divides the solution of a particular problem amongst a number a modules which cooperate by sharing knowledge about the problem and the solutions involved.

Multi-Agent Systems (MAS) – studies the behavior of a set of (possibly pre-existing) autonomous agents whose shared goal is the solution of an arbitrary problem.

A Multi-Agent System, according to works by Jennings, Sycara and Wooldridge [16] and Wooldridge [17], is a computer program with problem solvers located in interactive environments, which are capable of flexible, autonomous, socially organized actions which can, although not necessarily, be directed towards predetermined goals or targets. Multi-Agent systems are ideal for the representation of problems which include several problem solving methods, multiple points of view and multiple entities. In such domains, Multi-Agent Systems offer the advantages of concurrent and distributed problem solving, along with the advantages of sophisticated schemes of interaction. Examples of interaction include cooperation towards the achievement of a shared goal, coordination when organizing activities for problem solving, avoidance of harmful interactions and exploitation of beneficial opportunities, and negotiation of constraints in sub-problems, so that a satisfactory performance be achieved. On the flexibility of these social interactions lies the distinction between Multi-Agent systems and traditional programs, providing power and attractiveness to the paradigm of agents.

There are a number of application domains where Multi-Agent-based problem solving is appropriate, such as: manufacture, automated control, telecommunications, public transport, information management, e-commerce and games.

3 A Multi-agent System for Distributed Data Mining

This section discuss a Distributed Data Mining technique based on a Multi-Agent environment, called SMAMDD (Multi-Agent System for Distributed Data Mining) presented previously in [26]. In this work, a group of agents is responsible for applying a machine learning algorithm to subsets of data.

Basically, the process involves the following steps: (1) preparation of data, (2) generation of individual models, where each agent applies the same machine learning algorithm to different subsets of data for acquiring rules, (3) cooperation with the exchange of messages, and (4) construction of an integrated model, based on results obtained from the agents' cooperative work.

The agents cooperate by exchanging messages about the rules generated by each other, searching for the best rules for each subset of data. The assessment of rules in each agent is based on accuracy, coverage and intersection factors. The proposed model aims to select the best rules to integrate the global model, attempting to increase quality and coverage while reducing intersection of rules.

3.1 The SMAMDD Architecture

In the distributed learning system proposed, the agents use the same machine learning algorithm in all the subsets of data to be mined. Afterwards, the individual models are merged so that a global model is produced. In this approach, each agent is responsible for a subset of data whose size is reduced, focusing on improving the performance of the algorithm and considering also the physical distribution of data. Thus, each dataset is managed by an autonomous agent whose learning skills make it capable of generating a set of classification rules of type *if-then*. Each agent's competence is implemented with the environment WEKA (Waikato Environment for Knowledge Analysis), using the machine learning algorithm RIPPER [18]. The agents also make use of a validation dataset. Training and validation percentages can be parameterized and their default values are 90 and 10%, respectively. 0 presents the general architecture of the system.

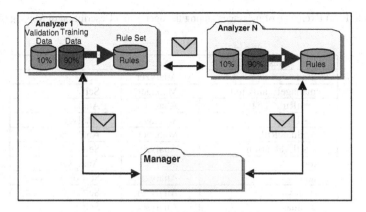

Fig. 1. System architecture: Analysers and Manager communicating. Communication can occur through the Manager or not.

In order to compare rules and rulesets, some measurements are made on the validation dataset. *Support, Error, Confidence* and *Quality* are specific measures to a rule depicted formally as the following way to a rule R predicting class *Class*:

- *Support*: number of examples covered by R;
- *Error*: number of examples covered by R but with class != *Class*;
- *Confidence = 1 – (Error / Support)*;
- *Quality = Support * Confidence*;

Coverage and Intersection are specific measures to a ruleset depicted formally as the following way to a ruleset $RS = \{R_1^{E1}, R_2^{E2}, \ldots, R_r^{Er}\}$, where E_i is the set of examples covered by rule R_i:

- Coverage = |E1| + |E2| + ... + |Er|;
- Intersection = |E' ∩ E1| + |E' ∩ E2| + ... + |E' ∩ E3|, to a Rule $R^{E'}$.

The two last measures are computed ignoring the *default rule*[1]. Otherwise the coverage of a ruleset RS' is always |DB| for any dataset DB.

Support and confidence factors are assigned to each rule, yielding the rule quality factor. In this work, the quality factor is being proposed as an evaluation metric for rules. Each *Analyzer Agent* will still maintain *intersection* and *coverage* factors, representing the intersection level among rules and the amount of examples covered by rules in the agent, respectively. At the end of the process, rules are maintained in the following format:

rule(Premises, Class,

        [Support, Error, Confidence, Coverage, Intersection, Quality,
     self_rule]).

The term self_rule indicates whether the rule has been generated in the agent or incorporated from another, in which case it is named external_rule. There is no need

---

[1] A default rule has not conditions and covers all the examples not covered previously by the other rules.

Table 1. (a) Exchange of messages among the agents. (b) Description of Messages.

(a)

Message ID	Sender	Receiver
load	Analyzer	Self
manageEvaluation	Manager	Self
waitRulesToTest	$Analyzer_i$	$Analyzer_j$
testRules	Manager	Analyzer
evaluateRules	Manager	Analyzer
startEvaluation	Analyzer	Self
test	$Analyser_i$	$Analyser_j$
add	$Analyzer_i$	Self or $Analyzer_j$
remove	$Analyzer_i$	Self or $Analyzer_j$
evaluate	Analyser	Self
finishEvaluation	Analyser	Manager
calculateRightnessTax	Manager	Self
calculateRightnessTax	Manager	Analyser
getRules	Manager	$Analyser_i$
generateReport	Manager	Self

(b)

Message ID	Action
load	Generate rules
manageEvaluation	Start the process to coordinate cooperation
waitRulesToTest	Await delivery of rules
testRules	Start the process of analysis of external rules. For each rule, calculate the following factors: support, confidence, quality, intersection and coverage. Such factors allow the generation of a *StateValue* for each rule in the agent. Each *StateValue* is stored in *NewState* and sent to the source agent in a message of type *response_test[NewState:Rule])*. The procedure *testRules* continues until all rules in the agent are tested.
evaluateRules	Request that the process of analysis of self rules not yet analyzed be started
startEvaluation	Start the analysis of local rules
test	Calculate confidence, error, support, rule quality and intersection and coverage for the rule received.
add	Add a rule to the set of selected rules
remove	Delete a rule from the set of selected rules
evaluate	Calculate confidence, coverage and intersection for the set of rules
finishEvaluation	Return to the process manageEvaluation to verify whether there exists another rule in the agent yet to be tested or if another agent can start the evaluation of its rules.
calculateRightnessTax	Request accuracy rate for remaining local rules
getRules	Request the agent's rules with best accuracy rate
generateReport	Calculate final accuracy rates, quantity and complexity of rules, and present them to the user

to test rules with the parameter value external_rule. After rules are generated and the factors mentioned are found, the process of cooperation for discovery of best rules starts. This cooperative process occurs by means of exchange of messages and is hierarchically coordinated due to the existence of a manager agent.

Agents hold restricted previous knowledge about each other, namely agent's competences and information for communication. An interface for communication with users is also embedded in the Manager agent. In addition to that, it is responsible for coordinating the process of interaction among agents.

The interaction process arises from the need to test the rules generated by an arbitrary agent against the validation set of others, where rule quality on their datasets and intersection and coverage factors obtained with the insertion of rules in their datasets are verified. Such factors are used to quantify the degree of agent satisfaction, represented by the scalar state_value, obtained with Equation 1.

$$state\_value = ((quality \times 0,33) + (coverage \times 0,17) + (intersection \times 0,033)) \tag{1}$$

Based on weights assigned to each factor in the formula above, the algorithm searches for rules with high quality and coverage factors and whose intersection factor is minimal. Weights used in this work have been determined after experiments in an attempt to obtain the highest accuracy level. Future work may approach a deeper study on values to be defined for each factor.

Thus, one must find the agent which holds the highest degree of satisfaction (state_value) for a given rule. For any given rule, the agent whose state_value is the highest must incorporate that rule into its rules set; it must also inform the agent where it came from as well as the other agents which also hold it to exclude it from their rules set. After all rules in all agents have been analyzed, they must be tested against each agent's validation set (10%). The agent's rules whose accuracy against its validation set is the highest will integrate the global model. In addition to the rules obtained at the end of that process, the accuracy of rules against the test set, the number of rules generated and their average complexity are calculated.

3.2 Cooperation Model

As seen in the previous section, there are N analyzer agents and one manager agent in SMAMDD. Their main components are: an individual knowledge base and a communication module which permits the exchange of asynchronous messages with each other. The Manager agent possesses also a module which eases the coordination tasks amongst the analyzer agents.

The human operator uses a graphical interface to start the system and visualize intermediate results generated by agents, as shown in 0. 0 partially illustrates the exchange of messages among agents in the form of a UML 2.0 diagram. Both messages and respective actions executed by each agent along the process of interaction are detailed in Tables 1a and 1b. Whenever an analyzer agent receives a *start* message it generates its ruleset based on local data running a rule-based algorithm. Then the manager chooses one analyzer A_1 to start evaluating the local rules (*evaluateRules*) and warns any other analyzer on it (*waitRulesToTest*). Then, A_1 chooses a rule r_i and asks the other analyzers for evaluation of r_i (*test*). The agent answering with the higher *stateValue* receives a confirmation to keep the rule (*add*)

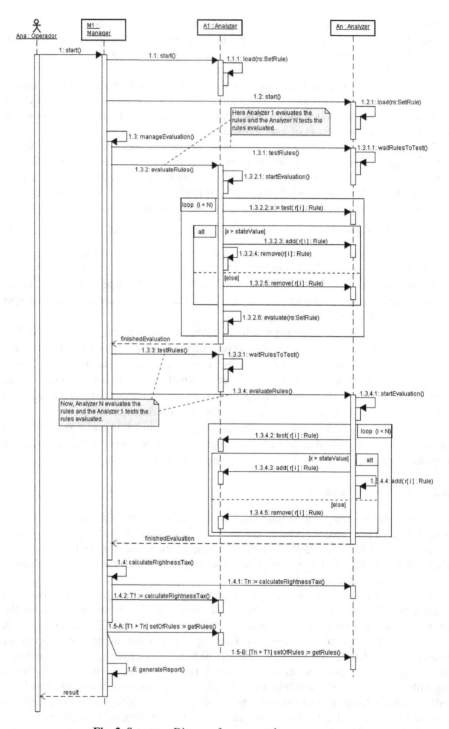

Fig. 2. Sequence Diagram for cooperation among agents

and the other agents must remove the rule from the local ruleset (*remove*). Agent A_1 continues the process sequentially with the next rules. When the evaluation of A_1 is over, it warns the manager (*finishedEvaluation*). Afterwards, manager starts a new evaluation process with A_2, A_3 and so on.

4 Results

In this section we complete the initial results presented in [26] with results regarding the complexity of the knowledge discovered with SMAMDD. With the intention of evaluating the work proposed, the following public datasets from the UCI Repository [1] have been utilized: Breast, Cancer, Vote, Zoo, Lymph, Soybean, Balance-scale, Audiology, Splice, Kr-ys-kp and Mushroom. Some combination techniques are not commonly used for learning in distributed data sets but are useful to improve the performance of the base classifiers on a centralized training data set. Such techniques generally use some heuristic for selecting instances or partitioning data. Since in distributed data mining we do not have control on the data organization and distribut-ion, the performance of these methods are not foreseen. However, studies show that the combination of classifiers generated from several samples of a centralized data set can significantly improve the accuracy of the predictions [23] [24] [25]. We briefly discuss in the next paragraphs two well-known approaches (*bagging* and *boosting*) used in the experiments.

4.1 Bagging

Quite often, training sets are not good enough to describe a given problem. In this case, the available data represent a partial view of the entire problem. Consequently, a learning algorithm will not be able to generate a good classifier because it is based on a local model. Breiman [23] discusses a procedure able to improve the performance of any algorithm by combining several classifiers: *bagging*. Bagging solves the problem of local models by selecting equal sized samples of training subsets and generating classifiers for such subsets based on a learning algorithm. Each training subset is made up of instances selected randomly but *with replacement*. This means a given instance may repeatedly appear or not at all in any training subset. Finally, the classification of a test instance is given by a vote strategy [23]. According to Breiman, bagging works very well for unstable learning algorithms like decision trees or neural networks. In such methods small changes into the training set will produce very different classifiers. However, bagging is not so efficient for stable algorithms as nearest neighbor. Spite of this, Ting and Witten contested such ideas showing that Bagging can generate also good results for stable algorithms like Naïve Bayes in large databases [25].

4.2 Boosting

We have said in the former paragraph that bagging generally works for unstable learning algorithms. This happens because learning algorithms generate very different models that are possibly complementary, even for similar inputs. Thus, the more different the data models, the larger the covered space of training instances is. However, with bagging, we do not guarantee complementary models because the instances are selected randomly. Boosting exploits this problem by providing a hill-climbing-like algorithm, guaranteeing the models to be as complementary as possible [25] [24]. There are many versions of boosting and here we discuss the general idea. To guide the generation of models, instances are initially weighed with an initial value. The weights are used for the classifier error estimation. The error is given by the sum of the weights of the misclassified instances divided by the sum of the weight of all instances. The weighing strategy forces the algorithm to pay more attention to misclassified instances (high weight). First, an equal error is assigned to all training instances. Then the learning algorithm generates the classifier and weights in all training instances are updated, increasing the weights of misclassified instances and decreasing the weights of well-classified instances. The global error of the classifier is also computed and stored. The process is repeated interactively until the generation of a small error. This procedure generates the classifiers and the weight of the training instances, where the weights represent the frequency the instances have been misclassified by the classifiers. To classify a new instance, the decision of a classifier is taken into account by assigning the weight of the classifier to the predicted class. Finally, the class with the largest weight (sum of the weights) is returned.

4.3 Empirical Results

Datasets were randomly split into n subsets each for cross-validation, yielding a total of 10 steps, iteratively, for each dataset. In SMAMDD, every partition is divided into training and validation, where the former constitutes 90 and the latter 10%. SMAMDD generates rules using only the Ripper algorithm. At the end of each step, accuracy rates, quantity of rules and complexity of rules are calculated. Table 2 presents some statistics about the datasets used in the experiments.

Table 3 presents the accuracy rates obtained by the SMAMDD and the techniques Bagging, Boosting and the RIPPER algorithm. The best results are displayed in bold.

Table 2. Statistics on datasets

Index	Dataset	Attributes	Classes	Examples
1	Breast C.	9	2	286
2	Vote	16	2	435
3	Zoo	17	7	101
4	Lymph	18	4	148
5	Soybean	35	19	683
6	Balance	4	3	625
7	Audiology	69	24	226
8	Splice	61	3	3190
9	Kr-vs-kp	36	2	3196
10	Mushroom	22	2	8124

Table 3. Average accuracy rates[2]

Datasets	SMAMDD	Ripper	Bagging	Boosting
Breast C.	**80,23 ±4,4**	71,74 ±3,8	71,39 ±3,3	72,55 ±4,8
Vote	**97,56 ±1,83**	94,88 ±2,50	95,41 ±1,12	93,99 ±1,39
Zôo	91,29 ±6,28	88,71 ±4,87	89,03 ±5,09	**94,84 ±4,36**
Lymph	**84,22 ±11,01**	74,67 ±5,36	80,44 ±7,16	83,78 ±2,78
Soybean	**95,36 ±2,14**	91,12 ±2,17	92,19 ±1,89	92,24 ±1,72
Balance	**85,37 ±4,53**	73,62 ±4,45	79,41 ±3,56	79,15 ±4,15
Audiology	**81,32 ±6,44**	73,97 ±5,53	73,23 ±7,03	75,59 ±7,54
Splice	**98,12 ±1,85**	93,31 ±0,74	95,08 ±0,66	94,67 ±0,23
Kr-vs-kp	97,04 ±2,51	98,99 ±0,32	99,21 ±0,24	**99,37 ±0,41**
Mushroom	**100,00 ±0,00**	99,93 ±0,16	100,00 ±0,00	99,93 ±0,16

Results from Table 3 are quite encouraging, as the SMAMDD system has achieved high accuracy rates in the majority of the datasets. The best results were obtained with the Balance-scale dataset, where the number of attributes is reduced.

By comparison with the boosting technique, the SMAMDD system experienced a reasonably lower accuracy performance in the Zoo dataset only. We believe that such performance loss was due to a reduced number of examples combined with a proportionally large number of attributes. Such characteristic often affects significantly the performance of non-stable algorithms such as RIPPER, producing quite different models for each subset. Consequently, concepts discovered locally rarely improve the satisfaction of neighbors and are doomed to remain limited to their original agents, having a negative impact on the agents' cooperative potential.

Even having obtained an accuracy performance higher than those of the other algorithms, the SMAMDD produced a high standard deviation in the dataset Lymph. Such distortion is due again to the large number of attributes in comparison with the number of examples.

It can be observed that some datasets presented a high standard deviation, revealing the instability of the algorithm upon the proportion of number of attributes to number of examples. With these results, it can be noticed that the SMAMDD system presents better accuracy performances when the number of attributes is small in comparison with the number of examples, i.e., the tuples space is densely populated (large volumes of data).

Table 4 compares the amount of rules generated by the SMAMDD system in comparison with the techniques bagging and boosting and the RIPPER algorithm.

From the statistics presented in Table 4, it can be noticed that the system SMAMDD has an inclination of producing a more complex model (high number of rules) when compared to the other techniques. It is also noticeable that the increase in the size of datasets, in general, causes a considerable increase in the number of rules along with higher standard deviations.

Table 5 presents the complexity of rules generated by the SMAMDD system, by the techniques bagging and boosting and by the RIPPER algorithm. The complexity

---

[2] Extracted from [**Error! Reference source not found.**].

of a rule is proportional to the number of conditions it has. Simple rules (few conditions) are preferable in relation to complex rules.

According to the statistics presented in Table 5, the complexity of the rules generated by the SMAMDD system is similar to those obtained with the other techniques. Even the largest datasets have not resulted in more complex models, demonstrating that number of attributes and volume of data have both been unable to significantly affect the complexity of rules.

Table 4. Average number of rules

Datasets	SMAMDD	Ripper	Bagging	Boosting
Breast C.	3,9 ±1,20	2,7 ±1,03	4,3 ±0,90	**2,3 ±1,40**
Vote	3,2 ±1,03	3,2 ±1,03	**2,7 ±0,48**	4,0 ±2,36
Zoo	**5,7 ±0,48**	**5,7 ±0,48**	6,4 ±0,84	6,2 ±1,40
Lymph	5,2 ±2,30	5,2 ±1,87	5,4 ±1,26	**5,0 ±2,11**
Soybean	49,3 ±18,86	25,1 ±1,29	26 ±2,54	**22,1 ±7,11**
Balance	17,2 ±7,61	11,6 ±3,24	12,4 ±3,75	**10,1 ±3,97**
Audiology	17,7 ±5,19	**13,3 ±1,64**	13,6 ±1,17	16,10 ±3,25
Splice	29,67 ±10,07	**13,67 ±5,51**	17,67 ±4,51	29,0 ±3,46
Kr-vs-kp	38,5 ±7,94	14,5 ±1,38	14,5 ±2,26	**10,0 ±4,82**
Mushroom	12,8 ±2,68	**8,6 ±0,55**	8,8 ±0,84	**8,6 ±0,55**

Table 5. Average complexity of rules

Datasets	SMAMDD	Ripper	Bagging	Boosting
Breast C.	1,84 ±0,14	**1,10 ±0,16**	1,49 ±0,12	1,41 ±0,16
Vote	1,31 ±0,60	1,31 ±0,60	**1,22 ±0,56**	1,36 ±0,65
Zoo	1,08 ±0,17	1,08 ±0,17	**1,00 ±0,16**	1,03 ±0,14
Lymph	1,24 ±0,24	**1,18 ±0,23**	1,44 ±0,23	1,27 ±0,55
Soybean	1,86 ±0,21	**1,59 ±0,06**	1,72 ±0,07	1,89 ±0,32
Balance	1,76 ±0,16	**1,71 ±0,20**	1,85 ±0,19	1,84 ±0,42
Audiology	1,61 ±0,11	**1,50 ±0,12**	1,57 ±0,14	1,66 ±0,21
Splice	3,95 ±0,16	3,52 ±0,44	**3,36 ±0,37**	3,75 ±0,31
Kr-vs-kp	3,39 ±0,24	**2,78 ±0,10**	2,89 ±0,24	3,53 ±1,86
Mushroom	1,51 ±0,11	1,37 ±0,10	**1,36 ±0,10**	1,41 ±0,11

5 Discussion and Related Work

From the collection of agent-based DDM systems which have been developed, BODHI, PADMA, JAM and Papyrus figure in the list of the most prominent and representative. BODHI [14], a Java implementation, was designed as a framework for collective DM tasks upon sites with heterogeneous data. Its mining process is distributed among local and mobile agents, the latter of which move along stations on demand. A central agent is responsible for both starting and coordinating data mining tasks. PADMA [20] deals with DDM problems where sites contain homogeneous data only. Partial cluster models are generated locally by agents in distinct sites. All local models are amalgamated into a central one which runs a second level clustering algorithm to create a global model. JAM [13] is a Java-based Multi-Agent system which was designed to be used for meta-learning in DDM. Different classification algorithms such as RIPPER, CART, ID3, C4.5, Baves and WEPBLS can be applied on heterogeneous datasets by JAM agents, which either reside in a site or are imported from others. Agents build up classification models using different techniques. Models are properly combined to classify new data. Papyrus [21] is a Java-based system for DDM over clusters from sites with heterogeneous data and meta-clusters.

In [22], an agent-based DDM system where agents possess individual models, which are built up into a global one by means of cooperative negotiation, is proposed. Only rules with confidence greater than a predetermined threshold are selected along the generation stage; the negotiation process starts just afterwards.

The SMAMDD system proposed has been designed to perform classification. An important characteristic of the system is the degree of independence, for it does not require configuration parameters and thresholds as do most of the ones aforementioned. Moreover, local models are completely analyzed, as opposed to being pruned as in [22]. Although it demands more processing time, the risk of removing a concept important for the global model is reduced.

6 Conclusion

Distributed Data Mining and Distributed Artificial Intelligence have many shared and complementary issues involving the generation of global points-of-view based on local ones. Coordination, collaboration, cooperation and specific evaluation measures are important challenges in both the areas. This paper has discussed some works that use such concepts in different ways for the accomplishment of different tasks. SMAMDD, for instance, performs model integration and yields results comparable to the ones obtained with other machine learning techniques; besides, it exhibited superior performance in some cases. The system permits discovering knowledge in subsets of data, located in a larger database. High quality accuracy rates were obtained in the tests presented, corroborating the proposal and demonstrating its efficiency.

Although promising, the use of Distributed Artificial Intelligence in Distributed Data Mining and vice-versa is a quite recent area. Particularly, some questions have yet to be better studied in future work: consolidation of the techniques developed with evaluations in databases with different characteristics and from different domains; the use of distinct classification algorithms in the generation of local models and studies concerning alternative knowledge quality evaluation metrics are yet to be done. An important issue consists in developing new strategies which permit reducing the number of interactions. Techniques of *game theory* and *coalition formation* can be useful for the detection of groups and hierarchies between learning agents. Such information can often improve the coordination, reducing the number of messages, the amount of processing and the general quality of the interaction.

References

1. Blake, C., Merz, C.J.: UCI Repository of machine learning databases - Irvine. CA: University of California. Department of Information and Computer Science (1998), http://www.ics.uci.edu/ mlearn/MLRepository.html.
2. Hand, D., Mannila, H., Smyth, P.: Principals of Data Mining. MIT Press, Cambridge, Mass (2001)
3. Han, J., Kamber, M.: Data Mining: Concepts and Techniques. Morgan Kaufman Publishers, San Francisco, CA (2001)
4. Kargupta, H., Sivakumar, K.: Existential pleasures of distributed data minig. In: Kargupta, H., Joshi, A., Sivakumar, K., e Yesha, Y.: (eds.) Data Mining: Next Generation Challenges and Future Directions, MIT/ AAAI Press (2004)
5. Park, B., Kargupta, H.: Distributed Data Mining: Algorithms, Systems, and Applications. In: Ye, N. (ed.) The Handbook of Data Mining, pp. 341–358. Lawrence Erlbaum Associates, Mahwah (2003)
6. Wittig, T.: ARCHON: On Architecture for Multi-Agent Systems. Ellis Horwood (1992)
7. Sridharam, N.S.: Workshop on Distribuited AI (Report) AI Magazine, 8 (3) (1987)
8. Jennings, N., Sycara, K., Wooldridge, M.: A Roadmap of Agent Research and Development. Autonomous Agents and Multi-Agent Systems 1, 7–38 (1998)
9. Wooldridge, M.J.: Reasoning about Rational Agents. MIT Press, Cambridge (2000)
10. Schroeder, L.F., Bazzan, A.L.C.: A multi-agent system to facilitate knowledge discovery: an application to bioinformatics. In: Proc. of the Workshop on Bioinformatics and Multi-Agent Systems, Bologna, Italy, pp. 44–50 (2002)
11. Viktor, H., Arndt, H.: Combining data mining and human expertise for making decisions, sense and policies. J. of Systems and Information Technology 4(2), 33–56 (2000)
12. Freitas, A., Lavington, S.H.: Mining very largedatabases with parallel processing. Kluwer Academic Publishers, The Netherlands (1998)
13. Stolfo, S., et al.: Jam: Java agents for meta-learning over distributed databases. In: Proceedings of the Third International Conference on Knowledge Discovery and Data Mining, pp. 74–81. AAAI Press, Menlo Park, CA (1997)
14. Kargupta, H., et al.: Collective data mining: A new perspective towards distributed data mining. In: Advances in Distributed and Parallel Knowledge Discovery, pp. 133–184. AAAI/MIT Press, Cambridge, MA (2000)
15. Jennings, N.R.: Coordination tecniques for distributed artificial intelligence. In: O'hare, G.M.P., Jennings, N.R (eds.) Foundations of distributed artificial intelligence, pp. 187–210. John Wiley & Sons, New York (1996)
16. Jennings, N., Sycara, K., Wooldridge, M.: A Roadmap of Agent Research and Development. Autonomous Agents and Multi-Agent Systems 1, 7–38 (1998)
17. Wooldridge, M.J.: Reasoning about Rational Agents. MIT Press, Cambridge (2000)
18. Cohen, W.W.: Fast effective rule induction. In: Proc. of the Twelfth Intl. Conf. on Machine Learning, pp. 115–123 (1995)
19. Schapire, R.E., Freund, Y.: The Boosting Approach to Machine Learning: An Overview. In: MSRI Workshop on Nonlinear Estimation and Classification, Berkeley, CA (2001)
20. Kargupta, H., et al.: Scalable, distributed data mining using an agent-based architecture. In: Heckerman, D., Mannila, H., Pregibon, D., Uthurusamy, R. (eds.) Proc 3rd International Conference on Knowledge Discovery and Data Mining, AAAI Press, Newport Beach, California, USA (1997)
21. Bailey, S., et al.: Papyrus: a system for data mining over local and wide area clusters and super-clusters. In: Proc. Conference on Supercomputing. ACM Press, New York (1999)

22. Santos, C., Bazzan, A.: Integrating Knowledge through cooperative negotiation - A case study in bioinformatics. In: Gorodetsky, V., Liu, J., Skormin, V.A. (eds.) AIS-ADM 2005. LNCS (LNAI), vol. 3505, Springer, Heidelberg (2005)
23. Breiman, L.: Bagging Predictors. Machine Learning 24(2), 123–140 (1996)
24. Shapire, R.E.: The Boosting Approach to Machine Learning: An Overview, MSRI Workshop on Nonlinear Estimation and Classification (2002)
25. Ting, K.M., Witten, I.H.: Stacking Bagged and Dagged Models. In: ICML, pp. 367–375 (1997)
26. de Paula, A.C.M.P., Scalabrin, E.E., Ávila, B.C., Enembreck, F.: Multiagent-based Model Integration. In: IEEE/WIC/ACM International Conference on Intelligent Agent Technology, 2006, Hong Kong. International Workshop on Interaction between Agents and Data Mining (IADM-2006) (2006)

Quantifying the Expected Utility of Information in Multi-agent Scheduling Tasks

Avi Rosenfeld[1,2], Sarit Kraus[2], and Charlie Ortiz[3]

[1] Department of Industrial Engineering
Jerusalem College of Technology, Jerusalem, Israel
[2] Department of Computer Science Bar Ilan University, Ramat Gan, Israel
[3] SRI International, 333 Ravenswood Avenue Menlo Park, CA 94025-3493, USA
{rosenfa,sarit}@cs.biu.ac.il, ortiz@ai.sri.com

Abstract. In this paper we investigate methods for analyzing the expected value of adding information in distributed task scheduling problems. As scheduling problems are NP-complete, no polynomial algorithms exist for evaluating the impact a certain constraint, or relaxing the same constraint, will have on the global problem. We present a general approach where local agents can estimate their problem *tightness*, or how constrained their local subproblem is. This allows these agents to immediately identify many problems which are not constrained, and will not benefit from sending or receiving further information. Next, agents use traditional machine learning methods based on their specific local problem attributes to attempt to identify which of the constrained problems will most benefit from human attention. We evaluated this approach within a distributed cTAEMS scheduling domain and found this approach was overall quite effective.

1 Introduction

Effectively harnessing the relative strengths of mixed human agent groups can be critical for performing a variety of complex scheduling tasks. The importance of effective coordination has been demonstrated in scheduling and planning domains such as hazardous cleanup, emergency first-response, and military conflicts [12]. Agents can quickly compute possible group compositions and assess group behavior even in dynamic and time sensitive environments [5,11,14]. This ability can be invaluable in focusing a person's attention to the most critical decisions in these environments [12].

We present the challenge of how agents can best coordinate their support decisions with people as the "Coordination Autonomy" (CA) problem. While using computer agents in these tasks can be beneficial, they are subject to several key limitations. First, we assume agents have only a partial view of the global constraints and the utility that could potentially be achieved through fulfilling these tasks. Because of this, it is impossible for agents to compute the group's utility exclusively based on their local information [5,11]. Second, we also assume there is cost associated with sending or receiving constraint information. These costs

M. Klusch et al. (Eds.): CIA 2007, LNAI 4676, pp. 104–118, 2007.

may stem from agent communication costs, or costs associated with interrupting the human operator [12]. As has been previously observed, coordinating decisions involving incomplete information and communication costs significantly increases the problem's complexity [9].

In the CA system under consideration, human operators are assumed to have access to more complete knowledge about task uncertainty because they have updated information or expert knowledge. This information can be critical in improving the group's utility by relaxing agents' rigid constraint information [12]. Specifically, we focus on how this information may affect decisions where uncertainty exists in task quality and duration. For example, a human operator in an emergency response environment may have updated information about domain weather conditions, or acquired knowledge from years of experience. Without this information, agents must plan for the worse-case scenario, e.g., that tasks with uncertainty in quality will yield the lesser utility amount, and tasks with uncertainty in duration will take the longest time. However, the user may know what task outcomes will actually be, allowing for more or higher quality tasks to be scheduled. On the one hand, this information can be invaluable in increasing the group's productivity. On the other hand, a person's time is valuable, and thus the person should only be consulted if the agent believes that the current scheduling problem is such that additional information will help.

Effectively measuring the expected value from a teammates' information is critical to the success of these types of systems [12]. Similar issues have arisen in the Information Gain measure that has been put forward by the machine learning community [6]. Information gain is typically used to learn the effectiveness of a certain attribute in classifying data, or how much additional information is gained by partitioning examples according to a given attribute. Thus, one approach may be to query the system with and without a given piece of information and build the CA application based on the resulting Information Gain.

However, there are several reasons why basic machine learning measures such as Information Gain cannot be applied to this type of problem. First, we assume there is a cost involved with agents exchanging constraint information. As a result, the cost in computing the Information Gain for a given problem may outweigh its benefit. Second, since there are a potentially large number of tasks to schedule, the size of the learning space will be very large. Sending "what if" queries for each task to a local scheduler can become resource intensive, leading to delays. Finally, sending too much constraint information can result in a lower group quality: we have previously found that in highly constrained problems, sending additional information can prevent agents from finding the optimal solution [8].

Towards addressing this problem, this paper presents an approach where agents can estimate the value of information without the resource intensive queries used in other works [12]. Our first contribution is a general, non domain-specific constraint "tightness" measure in which agents can locally measure how constrained a task is. A constraint tightness of less than one indicates that a task is underconstrained and will not benefit from any additional information.

This allows agents to locally and immediately identify that the expected utility from added information in such cases will be zero. A tightness measure of greater than one indicates that the problem **may** be affected by other constraints, and thus information **may** be of importance. However, we found the problem of determining how constrained the problem is exclusively from local information to be quite challenging. In addressing this challenge, we present a solution where agents can offline apply machine learning techniques to identify which of the remaining tasks are likely to have the highest expected utility from information.

Our next section provides the background and motivation for this work. In Section 3 we briefly describe the cTAEMS language used in quantifying the distributed scheduling tasks we studied. In Section 4 we present the domain independent constraint "tightness" measure for locally assessing what the utility of added information will be. Section 5 present how we quantified this value of information through machine learning regression and decision tree models. Section 6 provides experimental results. In Section 7, we provide a discussion about the general applicability of these results, as well as provide several directions for future research. Section 8 concludes.

2 Related Work

The goal of this paper is to quantify the expected utility to be gained from added information in distributed scheduling problems. This information can then be used to help agents decide what constraints should be forwarded to a human user for further information. Thus, this paper is linked to previous research in the field of agent-human interactions, as well as previous distributed scheduling research.

The adjustable autonomy challenge as described by Scerri et al. [13] refers to how agents can vary their level of autonomy, particularly in dealing with other types of entities such as people. They proposed a framework where agents can explicitly reason about the potential costs and gains from interacting with other agents and people to coordinate decisions. A key issue to applying their model within new domains is effectively quantifying the utility of information the entities can provide – a challenge we address in this paper.

Distributed scheduling problems belong to a more general category of distributed constraint optimization problems (DCOP) [11], and a variety of algorithms have been proposed for solving these problems [5,11,14]. In these problems, inter-agent constraints must be coordinated to find the solution that satisfies as many of these constraints as possible. However, these problem are known to be NP-complete, even if agents freely share all information at their disposal [5,9]. As such, no polynomial algorithms exist for checking how a scheduling problem's utility is affected by a given piece of information.

This paper presents a process through which agents are able to successfully quantify inter-agent interactions, such as the expected utility of sending or receiving information, exclusively with the local agent's information. Previous approaches have considered all possible group interactions, potentially creating a need to generate very large numbers of "what if" queries to obtain this information [9,12]. Our approach is significant in that agents locally estimate the utility of additional information without additional data, allowing them to reason about a limited number of interactions. This reduction allows for tractable solutions in estimating the value of information without using any queries during task execution.

In creating our approach, we draw upon previous studies that found that different categories of problem complexity exist, even within NP-complete problems [2,7]. Many instances of NP-complete problems can still be quickly solved, while other similar instances of problems from the same domain cannot. These studies have introduced the concept of a phase transition to differentiate classes of "easy" and "hard" instances of a problem [7]. Based on that body of work, we attempt to locally identify tasks which are underconstrained, or represent "easy" interactions that can be locally solved without any additional information, as well as the "hard" interactions that can potentially benefit from additional information.

However, finding such phase transitions within real-world domains is far from trivial due to the varied types of possible agent interactions [1,9]. In the theoretical graph coloring problems previously studied, phase transitions were found that were associated with the ratio of constraint clauses per variable [7]. Unfortunately, these graph coloring problems are relatively simple in that all constraints typically have equal weighting and every agent has equal numbers of constraints (edges). Thus, discovering phase transitions in experiments can be accomplished only through variation of a single parameter – the ratio of graph edges to nodes. In contrast, as we now describe, many real-world domains, such as the cTAEMS scheduling domain we have focused on, are far more complex. Novel measures are needed to quantify inter-agent actions in this and similar domains.

3   Domain Background and Description

cTAEMS is a robust, task independent language used by many researchers for studying multiagent task scheduling problems [12] based on the TAEMS language standard [4]. The cTAEMS language is composed of methods, tasks and subtasks that define a coordination problem in a Hierarchical Task Network (HTN) structure. Methods represent the most basic action an agent can perform and have associated with them a list of one or more potential outcomes. This outcome list describes what the quality (Q), duration (D) and cost (C) distributions will be for each possible result associated with the execution of that method. Tasks represent a higher level abstraction and describe the possible interrelationship between actions through what is referred to as a quality accumulation function (QAF), that indicates how the expected quality of subtasks will contribute to the overall quality of a (group) task. QAF's are of three forms: min, max or sum. In a min QAF, the total added quality is taken to be the minimum of all subtasks – it could be thought of as a logical AND relation between tasks. In a max QAF, the quality is the maximum value (or the logical OR), whereas in a sum QAF the quality is the sum of the expected quality of all subtasks. These subtasks can then be further subdivided into additional levels of subtask children, each potentially with their own QAF relationship. Finally, hard constraints between tasks or methods can be modeled in terms of what are referred to as non-local effects (NLE's) constraints between tasks, using the primitives *enable* or *disable*. Soft constraints are modeled through *facilitates* or *hinders* relationships. For example, assuming one task must occur before another, one could represent this constraint in terms of an enables relation between those two tasks. Assuming two tasks (or subtasks) cannot both be performed can be modeled in terms of a disables relation between the two tasks. System dynamics are modeled through probabilistic values for quality and duration within the HTN's tasks and subtasks. For example, a given task could have a duration of 10 with 50% probability, a duration of 4 with 25% probability, and a duration of 20 with 25% probability.

The CA system we propose takes as its input the system's distributed constraints, formulated in cTAEMS, and outputs the constraints a person should focus on. For example, Figure 1 is an example of a scheduling problem instance, described in cTAEMS. In this example, three agents, A, B, and C must coordinate their actions to find the optimal schedule for a global task T. Task T has three subtasks (A, B, and C) and these tasks are joined by a sum relationship. There are enable relationships between these tasks and thus they must be

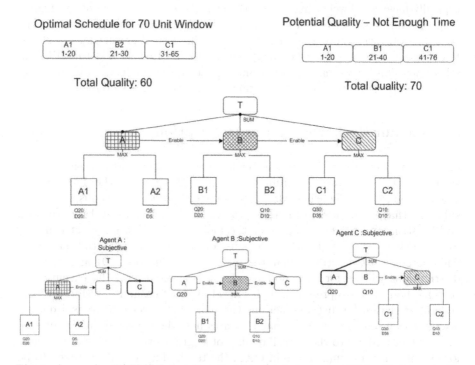

Fig. 1. A sample cTAEMS Scheduling Problem (global view middle) with 3 agents (3 subjective views on bottom)

executed sequentially. In this example, an optimal schedule would be for A to schedule method A1, B to schedule B2, and C to schedule C1. However, assuming only 70 time units exist for all three tasks, there is insufficient time for A to schedule A1, for B to schedule method B1, and for C to schedule C1. As such, one of the agents must sacrifice scheduling its method with the highest quality so the group's quality will be maximized. The group will lose 15 units of quality if A does not schedule A1, 10 units of quality if B does not schedule B1, and 20 units of quality if C does not schedule C1. Thus, B chooses B2 so the A1 and C1 can be scheduled.

Changing the cTAEMS structure, even slightly, can greatly affect the inter-agent constraints. We particularly focus on the impact uncertainty in task quality and duration will have on decisions. If a given task has a distribution of possible durations, agents supporting the automated scheduling process assume the worst-case scenario must be planned for (e.g. we must assume the task will have the smallest possible quality, or the task will take the longest possible time). Adding more precise information, such as eliminating certain duration possibilities, or identifying which outcome will definitely occur, can also greatly impact inter-agent constraints. For example, if a human operator could provide information that task B1 will take less than 15 units of time (instead of the current 20) this task could be scheduled and the group's utility will be raised. As a result, the CA system should identify task B1 as being the first task worthy of the person's attention. Novel mechanisms are required to generally find which types of constraints are most worthy of further information.

4   Locally Quantifying Scheduling Constraints

While we used the cTAEMS language to quantify scheduling constraints, the solution we are developing is meant to be as general as possible. Towards this goal, we have developed a tightness measure which we use to identify which subtasks are not constrained, and therefore cannot possibly benefit from any additional information. Our hypothesis is that general types of interactions can be quantified, similar to the phase shifts found within simpler graph coloring optimization problems previously studied [2,7]. However, novel measures of interaction difficulty are needed to help quantify interactions so phase shifts can be discovered.

Towards this goal, we present a "tightness" measure to quantify how much overlap exists between constraints. In referring to these constraints, let $G = \{A_1, A_2 \ldots, A_N\}$ be a group of N agents trying to maximize their group's collective scheduling utility. Each agent has a set of m tasks, $T = \{T_1, \ldots, T_m\}$ that can be performed by that agent. Each Task, T_i, has a time **Window** within which the task can be performed, a **Quality** as the utility that the task will add to the group upon its successful completion, and a **Duration** as the length of time the task requires to be completed. We assume that task quality and duration often have uncertainty, while task windows are typically based on the problem's structure. We model W_i as the fixed Window length for task T_i, $\{Q_{i1}, Q_{i2}, \ldots, Q_{ij}\}$ as the possible quality outcomes for T_i, and $\{R_{i1}, R_{i2}, \ldots, R_{ij}\}$ as the possible duration lengths of T_i.

Based on these definitions, we model an agent's quality *tightness* as:

$$\text{Tightness-Quality}(T_i) = \frac{Quality_{max}(T_i)}{Quality(Window(T_i))}$$

where $Quality_{max}(T_i)$ returns the maximal quality from $\{Q_{i1}, Q_{i2}, \ldots, Q_{ij}\}$, and $Quality(Window(T_i))$ returns the maximal expected quality of all **other** subtasks that share the task window of (T_i). Note that if quality uncertainty exists in these other tasks, we again assume the worse case, and the lowest quality value must be considered in computing the value of $Quality(Window(T_i))$. The measure of Tightness-Quality(T_i) can then be used to quantify what potential overlap exist between T_i and other local constraints within T_i's task window. For example, assume T_A can be fulfilled by methods A1 and A2 with A1 having a quality distribution between 10 and 20, and A2 having a quality distribution of 15 and 25. It is possible that asking the user for input about the actual quality of A1 is worthwhile as these quality distributions overlap and A1 may have a higher quality than A2 (say 20 for A1 and 15 for A2). In this example, Tightness-Quality(A1) = 1.33 indicating the quality distributions of these tasks overlap and information may be beneficial. However, a tightness under one would indicate no possible quality overlap, and thus no possible gain from information.

Similarly, we model an agent's duration *tightness* as:

$$\text{Tightness-Duration}(T_i) = \frac{Duration_{max}(T_i)}{Duration(Window(T_i))}$$

where $Duration_{max}(T_i)$ returns the maximal duration from $\{R_{i1}, R_{i2}, \ldots, R_{ij}\}$, and $Duration(Window(T_i))$ refers to the time allotted for completing task T_i. Note that within the cTAEMS problems we studied, no uncertainty existed within the time window for the tasks sharing a given window $(Window(T_i))$ so uncertainty in this value need not be considered.

As was the case within the quality tightness measure, a value of more than one indicates an overlap between T_i and other task constraints, while a value less than one indicates no possible overlap. For example, assume task T_A may last for either 10, 20, or 40 time units to be completed within a time window, $Window(T_A)$ of 25 units. According to the tightness definition, tightness Tightness(T_A) is 40/25 of 1.6. Without any additional information, we must assume that this subtask task will take the maximal time, potentially preventing that agent from performing other tasks. However, assuming information can be provided regarding the task's duration, say that T_A will only last for 10 or 20 units, the value of Tightness(T_A) drops below 1.

The tightness measures we present have two important properties: First, they are locally measurable. Agents can measure its quality or duration tightness without any additional input from other task agents or the human operators within the group. Second, they can effectively quantify the impact making a local decision will have. By definition, a tightness of 1.0 or less means that the problem is not constrained, as no task option overlaps with others options sharing that task. In such cases, agents will not obtain additional utility from more information, and the agent should not prompt the human operator for additional data. After this measure exceeds 1.0, a phase shift occurs when information **may** be helpful. However, further solutions are still needed to quantify how much the group's utility is expected to increase if this constraint is relaxed.

5   Learning the Value of Information

As the tightness measure only addresses which tasks will definitely **not** benefit from additional information, the next step in our approach is a machine learning model to suggest which tasks **will** show an increase in utility as a result of additional information. In addressing this challenge, we built two machine learning models: a regression based model where agents predict a numeric value for added information from the human operator, and a classification model where agents classify a given task as potentially benefitting or not benefiting from additional information. Alternatively, within the classification model, qualitative categories can be created, such as High, Low, and Zero impact categories instead of binary Yes and No categories.

The human-agent interface within the Coordination Autonomy (CA) application we propose will be affected by the learning model chosen. Within the regression model, the CA front-end will present the human operator a numeric field for the expected value of information that can potentially be added through the user's attention. Assuming a classification model is trained, we propose that the CA's front-end color code various tasks to represent how much information can potentially help the group. For example, referring back to Figure 1, subtasks would be colored green if information was categorized as less important while subtasks of high importance would be colored red. We believe this interface can most effectively focus the user's attention.

The procedure we adopted for training the machine learning model is outlined in algorithm 1. We used a problem generator created by Global Infotech Inc. (GITI) for the purpose of generating cTAEMs problems within the framework of the COORDINATORS DARPA program[1]. We created two sets of 50 problems (the value for X in the algorithm) where cTAEMS parameters (such as the number of tasks to be scheduled, the hierarchical structure of the tasks, the number of agents able to perform each task, the number of NLE relationships between tasks, task duration, and average task quality) were randomly generated. In the first set of problems, we generated problems with uncertainty in quality distributions, while keeping other parameters deterministic. In the second set of problems, we created problems with uncertain durations, while keeping other parameters constant. Note that these 100 total problems represent a small fraction of the total number of the thousands of problem permutations the GITI scenario generator could create. Additionally, each of these base problems had many possible permutations – these 100 total cTAEMS problems contained over 5000 subtasks where constraint information could be added.

The goal for creating these test cases was to study how user information about quality and duration uncertainty affected the group's utility. We used a previously tested cTAEMS centralized scheduler [8] to first compute the group's

---

[1] http://www.darpa.mil/ipto/programs/coordinators/

utility under the assumption that uncertainty in problems would result in the lowest probabilistic outcome. Thus, in the first set of problems, we assumed tasks would have the lowest quality, and in the second problem set the tasks would take the maximal time (line 3). We then computed what the group's utility would be if we could relax that assumption, and the user could provide information that task would have the highest possible quality, or take the shortest time (line 4). Next, we stored this information into a table along with a vector of the problem's specific parameters (e.g. problem parameters such as tightness, local NLE's, maximal duration, quality, etc.) and entered this information into the table, Table (line 5). This problem information would then be used for offline training of the machine learning model.

In computing the value of added information, we adopted a highly optimistic approach for the value of Utility($Problem_k$,j) that makes several assumptions: a) the person being contacted actually has information about the subtask j, b) the person will have information that will help the group in the maximal possible way by informing the agents that the task will have the highest quality or take the shortest duration, and c) this information can be provided without cost. Despite this oversimplification, the approach was useful for identifying the maximal upper bound for the potential utility that could be gained through added information.

Algorithm 1. Training the Information Model(Problem Set of Size X)

1: **for** $k = 1$ to X **do**
2:     Without $\Leftarrow$ Utility($Problem_k$)
3:     **for** $j = 1$ to Num(Subtasks($Problem_k$)) **do**
4:         With $\Leftarrow$ Utility($Problem_k$,j)
5:         Table[k][j] $\Leftarrow$ (With) - (Without)
6:     **end for**
7: **end for**

Interestingly, we found that only a small percentage of subtasks had any benefit from adding this type of constraint information. While each problem, k, contained at least one subtask that did benefit from added information, only 1022 subtasks, or roughly 20% of the total entries within the training data, benefited from any additional information. Thus, finding these subtasks is akin to "finding a needle in a haystack". Clearly, naive methods that query every subtasks are not appropriate, especially if there is a cost associated with generating queries or if the human is not able or willing, for whatever reason, to provide information.

6   Experimental Results

We found strong support for the usefulness of the quality and duration tightness measures in identifying the cases where information definitely did **not** help. Of the 2490 cases where a tightness value was less than 1, only 38 (1.5%) cases benefited from additional information. Next, we used the Weka machine learning package [10] to train and evaluate what the value of information would be in the remaining cases. We present the results from training and evaluating three decision tree models: a regression model based on the M5P algorithm and C45 decision trees (J48 within the Weka implementation) to create classifier models based on 2 information categories (Yes / No impact of information) and a 3 category information classification task (High, Low, and Zero impact). Note that while we present results from decision trees learning approaches, other possibilities exist. We did, in fact, train models based on Bayes Networks, Neural Networks, and SVM models and found the results to be nearly identical with those we present. In all cases, we performed 10-fold cross validation to evaluate the results.

First, we trained and evaluated the regression based model. The results from this experiment are found in Table 1. The first two rows present the results from the problem set with quality uncertainty, and the last two rows present the results from the corresponding problem set with duration uncertainty. Note that this model yielded an average correlation of 0.56 and 0.46. In comparison, we present the Naive approach which assumes all instances belong to the majority class, and information adds zero quality. While these results are certainly significant (0.56 and 0.46 being much larger than the Null hypothesis of 0.05), they do leave room for improvement as even with the tightness measure these results are far from the optimal correlation of 1.0.

Table 1. Comparing the accuracy and Mean Absolute error in regression trained model

	Learned Model	Naive (Majority)
Quality Correlation	0.56	-0.06
Quality Mean Absolute Error	1.27	2.61
Duration Correlation	0.46	-0.04
Duration Mean Absolute Error	1.67	2.00

Next, we trained a two category classification model (Yes/No categories) to find instances where information does or does not help. Table 2 presents the results. Note that the larger class (information did not help) represents over 80% of the subtasks in both problem sets, and thus even naively categorizing all subtasks within this category results in a relatively high accuracy of the model. However, as the goal is to effectively find all subtasks where information will help, this approach will find none of the desired instances. In both problem sets, the trained model had better accuracy than the naive baseline (both slightly over 84%) while still finding many of the instances where added information would help (55.20% of the instances in the quality set, 19.03% in the duration set).

Finally, we studied the three category classification task. Here, we divided the training data into a High category where information helped 10 or more units,

Table 2. Comparing the overall accuracy and number of high information instances found within a 2 category decision tree model

	Learned Model	Naive (Majority)
Accuracy (Quality)	85.04%	81.95%
Instances Found (Quality)	55.20%	0%
Accuracy (Duration)	84.13%	83.45%
Instances Found (Duration)	19.03%	0%

Table 3. Comparing the accuracy and number of high information instances found in a 3 category decision tree model with **quality** uncertainty

	Classified as High	Classified as Low	Classified As Zero	Recall	Ave. Accuracy
High	22	22	24	0.32	82.53%
Low	15	100	170	0.35	82.53%
Zero	18	86	1465	0.93	82.53%

Table 4. Comparing the accuracy and number of high information instances found in a 3 category decision tree model with **duration** uncertainty

	Classified as High	Classified as Low	Classified As Zero	Recall	Ave. Accuracy
High	13	14	19	0.28	82.54%
Low	9	83	277	0.23	82.54%
Zero	10	109	1974	0.94	82.54%

a Low category where information helped less than 10 units, and a Zero category where information did not help. The results of this experiments from the quality set are found in Table 3, and those from the duration experiments are in Table 4.

Within these tables, we present the classification confusion matrix, often presented in multi-category classification problems. We plot the number of instances found within a given category (the diagonal of table) as well as the number of instances misclassified per category. Not all misclassification errors are necessarily equally important. For example, misclassifying a High problem or Low, or a Low problem as High is likely to be less problematic that classifying a High problem as Zero. For example, within the quality experiments, only 22 of 66 High instances were classified as High, but another 22 of these instances were classified as Low. This distinction can be quite significant. The recall of the High category alone (High classified as High) is only 0.33, however, if we view classifying High either as High or Low as being acceptable, the recall jumps to 0.67.

As the machine learning models never achieved a recall near 100%, we considered creating models which were biased towards categorizing a task as benefitting from information. While this bias will result in a higher recall of this category, it will come at a cost of false positives that will lower the overall accuracy of the model. This type of approach would likely be useful if the human user is able to be prompted Q times to add information, when Q is greater than the actual number of tasks that can benefit from added information. Alternatively, this approach will also be useful if the known cost from interrupting the user is relatively low. For example, if the cost of prompting the user is 1 unit, we should be willing to ask several queries for information so additional High instances (each worth 10 units) can be found. To train this model, we followed the previous developed MetaCost approach [3] and refer the reader to their work for additional details in how the cost bias is created.

We did find that the MetaCost approach was extremely effective in increasing the recall of the desired system categories, albeit at a cost of false positives that reduced the overall accuracy. To explore this point, we used the MetaCost function to apply different weights for falsely classifying a subtask where information was useful (Yes) as belonging to the non-useful category (No). The base weights, or unbiased classification, will have equal weighting for these categories (1 to 1 weight). We found that as we increased these weights, we obtained progressively higher recall from the desired Yes category, but at an expense in overall reduction of model accuracy. We present the results of this approach from the quality experiment in a two category classification model in Table 5. For example, a 5 to 1 bias towards the Yes category found 566 of the 607 instances (or 0.93 recall). However it had a higher rate of false positives (0.32) and lower accuracy (72.61%) from the baseline (1 to 1 weights). We also applied this approach to the two category duration problem set, as well as the quality and duration 3 category classification models. As expected, in all cases the cost bias was effective in increasing the recall of the categories where information helped, albeit at a cost of more false positives.

Table 5. Exploring the tradeoff between higher recall of desired results (Yes instances found), and false positives and negatives within a 2 category decision tree model with quality uncertainty

Weight	Total Accuracy	Found	Not Found	Recall	False Reject	False Accept
1 to 1	85.04%	335	272	0.55	0.45	0.08
2 to 1	82.46%	435	172	0.72	0.28	0.15
5 to 1	72.61%	566	41	0.93	0.07	0.32
10 to 1	70.95%	584	23	0.96	0.04	0.35

7   Discussion and Future Directions

In general, we found that the tightness measure was extremely effective in finding which subtasks would **not** benefit from adding information. By locally filtering out which tasks were not constrained we were able to focus on determining whether adding information would help in the remaining subtasks. However, several key directions are possible to expand upon this work.

First, we found decision trees were overall very effective in quantifying the expected impact of adding information, and thus were helpful in recommending if the user should be contacted for additional information. In contrast, previous work on adjustable autonomy [13] found decision trees were ineffective in enabling agents to make autonomous decisions. It seems that the difference of results stems from the very different tasks considered. The previous work used the learned policy from decision trees to enable agents to act independently of people within their group. As a result, their scheduler system made several critical errors (such as canceling group meetings and volunteering people against their will for group activities) by overgeneralizing decision tree rules. In contrast, our support system never tries to make autonomous decisions, and instead took the support role of recommending what constraint(s) a person should focus on. This distinction may suggest the need to create different types of learning models for different agent-human tasks. We hope to further explore this point in the future.

Also, further work is necessary to identify general attributes where information definitively **does** add utility. Our hypothesis is that local information is sufficient for guaranteeing that a given problem is not constrained, and thus information will **not** help. However, the disadvantage to the exclusively local approach we present is that agents are less able to consider the full extent of all constraints within the problem. Because of this, we believe this approach was less affective in finding the cases where information would **definitely** help.

We have begun to study several of these directions in parallel to the work we present here. Along these lines we have studied how the tightness measure can guide agents if they should communicate all of their constraints to a centralized Constraint Optimization Problem (COP) solver [8]. The COP solver would then attempt to centrally solve the constraint problem after receiving all constraints from all agents. To address what agents should communicate, each agent viewed all of its constraints as belonging to only one task window, with all subtasks falling within this window. We found that the resulting tightness measure created three classic clusters of constraint interactions: under-constrained, constrained, and over-constrained with a clear communication policy emerging based on this measure. Under-constrained problems had low tightness and could locally be solved without sending any constraints to the COP solver. Constrained problems had a medium tightness value, and most benefited from having every agent send all of its constraints. Problems with the highest tightness value were the most constrained. In fact, these problems had so many constraints that agents sending all constraints flooded the COP solver, which was not able to find the optimal solution. In these problems, agents were again best selecting communication approaches that sent fewer constraints.

Finally, it is important to note that these research directions are complementary. We foresee applications where different tightness measures are applied to filter and predict different characteristics. Say, for example, a domain exists where agents could send constraint information freely. As we have previously found, sending too much information can prevent centralized problem solvers from finding the optimal solution [8]. One solution might be to apply the local tightness measure we present here to filter cases where information definitely will not help, and then have agents send all remaining constraints. This more limited set of constraints might be most manageable than the original set. We are hopeful that this work will lead to additional advances in this challenging field.

8   Conclusion

In this paper we presented an approach to quantifying the expected utility change from adding information to agents within distributed scheduling problems. Agents exclusively used local information about their constraints to predict whether adding information will help the group increase its utility. The significance of this work is its ability to enable agents to find which constraints will most benefit from additional human information, without using resource intensive queries required in other approaches [12]. Towards achieving this goal, we defined and used a general **tightness** measure and domain specific information from the cTAEMS distributed scheduling domain to train a regression based learning model for numerically quantifying the value of this information, as well as classifier models to identify if a given subtask should be categorized as benefiting from information or not. In general, we found that the non problem-specific tightness measures was extremely effective in finding where addition information about constraints would **not** be helpful. Domain specific cTAEMS information was moderately useful in identifying where information **would** be helpful. Finally, we presented several possible future direction of study in this challenging problem.

References

1. Brueckner, S.A., Van Dyke Parunak, H.: Resource-aware exploration of the emergent dynamics of simulated systems. In: Sven, A. (ed.) AAMAS '03: Proceedings of the second international joint conference on Autonomous agents and multiagent systems, pp. 781–788. ACM Press, New York (2003)
2. Cheeseman, P., Kanefsky, B., Taylor, W.M.: Where the Really Hard Problems Are. In: Proceedings of the Twelfth International Joint Conference on Artificial Intelligence, IJCAI-91, Sidney, Australia, pp. 331–337 (1991)
3. Domingos, P.: Metacost: a general method for making classifiers cost-sensitive. In: KDD '99: Proceedings of the fifth ACM SIGKDD international conference on Knowledge discovery and data mining, pp. 155–164. ACM Press, New York (1999)
4. Lesser, V., Decker, K., Wagner, T., Carver, N., Garvey, A., Horling, B., Neiman, D., Podorozhny, R., NagendraPrasad, M., Raja, A., Vincent, R., Xuan, P., Zhang, X.Q.: Evolution of the GPGP/TAEMS Domain-Independent Coordination Framework. Autonomous Agents and Multi-Agent Systems 9(1), 87–143 (2004)
5. Mailler, R., Lesser, V.: Solving distributed constraint optimization problems using cooperative mediation. In: Kudenko, D., Kazakov, D., Alonso, E. (eds.) Adaptive Agents and Multi-Agent Systems II. LNCS (LNAI), vol. 3394, pp. 438–445. Springer, Heidelberg (2005)
6. Mitchell, T.M.: Machine Learning. McGraw-Hill, New York (1997)

7. Monasson, R., Zecchina, R., Kirkpatrick, S., Selman, B., Troyansky, L.: Determining computational complexity from characteristic "phase transitions". Nature 400(6740), 133–137 (1999)
8. Rosenfeld, A.: A study of dynamic coordination mechanisms. Ph.D. Dissertation, Bar Ilan University (2007)
9. Shen, J., Becker, R., Lesser, V.: Agent Interaction in Distributed MDPs and its Implications on Complexity. In: Proceedings of the Fifth International Joint Conference on Autonomous Agents and Multi-Agent Systems, Japan, pp. 529–536. ACM, New York (2006)
10. Witten, I.H., Frank, E.: Data Mining: Practical Machine Learning Tools and Techniques. In (Morgan Kaufmann Series in Data Management Systems), 2nd edn., Morgan Kaufmann, San Francisco (2005)
11. Maheswaran, R.T., Tambe, M., Bowring, E., Pearce, J.P., Varakantham, P.: Taking dcop to the real world: Efficient complete solutions for distributed multi-event scheduling. In: Kudenko, D., Kazakov, D., Alonso, E. (eds.) Adaptive Agents and Multi-Agent Systems II. LNCS (LNAI), vol. 3394, pp. 310–317. Springer, Heidelberg (2005)
12. Sarne, D., Grosz, B.: Estimating Information Value in Collaborative Multi-Agent Planning Systems. In: AAMAS'07 (2007)
13. Scerri, P., Pynadath, D.V., Tambe, M.: Towards adjustable autonomy for the real world. In: JAIR, vol. 17, pp. 171–228 (2002)
14. Yokoo, M., Durfee, E.H., Ishida, T., Kuwabara, K.: The distributed constraint satisfaction problem: Formalization and algorithms. Knowledge and Data Engineering 10(5), 673–685 (1998)

Agent-Based Traffic Control Using Auctions*

Heiko Schepperle and Klemens Böhm

Institute for Program Structures and Data Organization (IPD)
Universität Karlsruhe (TH), Karlsruhe, Germany
{schepperle,boehm}@ipd.uni-karlsruhe.de

Abstract. Traffic management nowadays is one of the key challenges for cities. One drawback of traditional approaches for traffic management is that they do not consider the different valuations of waiting-time reduction of the drivers. These valuations can differ from driver to driver, e.g., drivers who are late for their job interview have a higher valuation of reduced waiting time than individuals driving home from work routinely. This also applies to trucks with urgent load, e.g., as part of a just-in-time production chain. To overcome this problem, we propose a new mechanism for traffic control at intersections called *Initial Time-Slot Auction* that is valuation-aware. It relies on agent-based driver-assistance systems to allocate the right to cross an intersection. Our evaluation shows that it does yield a significantly higher overall satisfaction.

1 Introduction

Traffic control is a key problem cities currently have to deal with [1]. Drivers are dissatisfied with high waiting times at intersections. One characteristic of traditional approaches for traffic control at intersections is that they do not consider the valuations of waiting-time reduction of the drivers. These valuations are not at all equal among drivers. For instance, drivers who do not want to miss their flight or truck drivers who must stick to their time schedule have higher valuations of such reductions than individuals driving home from work routinely. Traditional traffic-control mechanisms do not take these valuations into account. Existing mechanisms do not even let drivers inform other vehicles or the traffic-control unit about their valuations.

To increase overall satisfaction of motorists, this article investigates new mechanisms for traffic control which take the valuations of waiting time of the drivers into account. We call such mechanisms "valuation-aware" in the following. Such mechanisms require interaction among vehicles and the infrastructure to exchange valuations. In this respect, they can benefit from ongoing advances in vehicle technology.

However, designing such valuation-aware mechanisms is challenging. This is because the traffic scenario imposes several constraints that are different from

---

* This work is part of the project DAMAST (Driver Assistance using Multi-Agent Systems in Traffic) (http://www.ipd.uni-karlsruhe.de/%7Edamast/) which is partially funded by init innovation in traffic systems AG (http://www.initag.com/).

M. Klusch et al. (Eds.): CIA 2007, LNAI 4676, pp. 119–133, 2007.

the ones in conventional settings. The situation at an intersection is *highly dynamic*. Vehicles can arrive at any time. With the arrival of a new vehicle, the optimal order of vehicles crossing the intersection may change. In contrast to scheduling in communication networks, vehicles are *physical objects* with physical characteristics, e.g., they need to accelerate/decelerate, and their speed is limited. Vehicles queuing from the same direction cannot change their order, i.e., *no overtaking*.

The mechanisms envisioned must be superior to existing mechanisms, i.e., yield a higher overall satisfaction (*effectiveness*). Next, mechanisms inducing drivers to reveal their valuations truthfully may be desirable (*incentive compatibility*). Further, there should be an upper bound of the waiting time, for all drivers (*avoidance of starvation*).

This paper proposes and evaluates an approach for valuation-aware traffic control at intersections. More specifically, we make the following contributions: We first identify desiderata any valuation-aware traffic-control mechanism should fulfill. We then propose our new mechanism called *Initial Time-Slot Auction* (*ITSA*). It auctions the right to cross the intersection for a certain period of time to arriving vehicles. We call this right *time slot* in the following. *ITSA* allocates a time slot to the vehicle with the highest bid. We describe the mechanism in detail, discuss design alternatives and possible extensions of the basic variant, and say to which extent they fulfill the desiderata. We have implemented two variants in full and compare them to a *FIFO* mechanism [2,3], a state-of-the-art agent-based traffic-control mechanism from the scientific literature, using simulations. A core result of our work is that *ITSA* yields a significant reduction of valuation-weighted waiting time.

This article is part of a long-term effort that strives to design valuation-aware traffic-control mechanisms that are applicable in the real world, and this article is a first stab at the problem. For instance, we currently limit ourselves to single intersections (the fact that a vehicle typically has to cross several intersections to reach its destination is future work), and we leave aside robustness and legacy issues, i.e., "old" vehicles that are not equipped to interact with valuation-aware traffic-control mechanisms. Nevertheless, this current work has yielded valuable insights into the general problem domain and has demonstrated the potential of our approach.

Paper outline: We discuss related work in Sect. 2. Section 3 describes fundamentals of valuation-aware traffic control. In Sect. 4, we propose our new valuation-aware mechanism for traffic control and its variants. Section 5 features an evaluation, Sect. 6 concludes.

2   Related Work

This section describes related work where agents have been used for traffic management at intersections and related valuation-aware approaches. Related work regarding other aspects, e.g., auctions, is cited elsewhere in the paper.

There already exist various approaches for agent-based traffic management. We only discuss a small subset that we deem most relevant for our work. [2,3]

propose an agent-based reservation system for intersection-traffic control. It allows vehicles to reserve a time slot in advance for crossing an intersection. The system proposed is a reservation system, but not a platform where agents can trade time slots, and it is not valuation-aware. An extension of the reservation system [4] gives priority to emergency vehicles, e.g., ambulance or police cars. [5] describes an approach where trams or buses communicate their arrival to traffic lights. In both cases, only a small subset of vehicles is privileged, but the valuations of all other vehicles are ignored. [6] presents an agent-based system for urban traffic control. The authors introduce infrastructure agents that collect data and intersection agents which adapt their traffic-control policy to the data received. The model does not include agents that assist the drivers. They also do not take the different valuations of waiting-time reduction into account. *ITSA* is not the very first valuation-aware traffic control mechanism – [7,8] have recently proposed a mechanism called *Time-Slot Exchange (TSE)*. However, *TSE* only allows vehicle agents already holding a time slot to negotiate time slots with other vehicle agents. *ITSA* in turn lets the intersection agent assign time slots to vehicle agents which do not have any time slot so far. Further, *TSE* imposes strong restrictions, and only few negotiations between vehicles are successful. This clearly limits the impact of *TSE*. This is not the case with *ITSA* – it yields a significant improvement, compared to *TSE*.

3   Fundamentals

In this section we describe our traffic scenario, measures to compare valuation-aware mechanisms for traffic control, desiderata for those mechanisms, and the components of agent-based traffic management.

Traffic Scenario. In this paper we focus on individual intersections of roads. An intersection consists of several intersection lanes. Depending on the intersection lane they have chosen, vehicles can only leave the intersection in one direction. An intersection lane can intersect other intersection lanes. The intersection points are potential spots of conflict. No two vehicles are allowed to pass such a spot at the same time. The considerations in this paper are not limited to the intersection area itself, but take the neighborhood of the intersection into account as well.

Definition 1. *The* neighborhood of an intersection *consists of the lanes of the intersection area, the incoming and the outgoing lanes.*

Measures. We use the following measures to evaluate valuation-aware mechanisms.

Definition 2. *The* travel time T_t^j *of a Vehicle j is the time from its first appearance in the neighborhood until it leaves the neighborhood. The* minimal travel time $minT_t^j$ *of j is the travel time if j was the only vehicle at the intersection, observed the speed limit and any constraints from the physical world, but ignored all rules concerning the right of way (i.e., crosses red lights, does not stop at stop* signs etc.). The waiting time T_w^j of j is the difference of the travel time T_t^j and the minimal travel time $minT_t^j$.

Because the minimal travel time is a lower bound of the travel time, the waiting time is always nonnegative.

Any new mechanism should be superior to existing ones. A useful measure to compare different intersection-control mechanisms is the average waiting time.

Definition 3. *The* average waiting time *is* $\overline{T_w} = \frac{\sum_{j \in V} T_w^j}{|V|}$, *where V is the set of all vehicles.*

The average waiting time does not allow to evaluate valuation-aware mechanisms. Therefore, a more meaningful measure in our context is the waiting time weighted by the valuations of waiting-time reduction of the drivers.

Definition 4. *The* valuation $v^j(t)$ *of Driver j is the price j is willing to pay if he waits t seconds less.*

While other valuation functions might be realistic as well, we limit ourselves to linear valuation functions in this study. Thus, we denote the price per second with v^j. The valuation may be different for each vehicle, and we assume that it does not change over time.

Definition 5. *The* weighted waiting time *of Vehicle j is $vT_w^j = v^j \cdot T_w^j$.*

We use the weighted waiting times of all vehicles to compute the average weighted waiting time.

Definition 6. *The* average weighted waiting time *is* $\overline{vT_w} = \frac{\sum_{j \in V} v^j \cdot T_w^j}{|V|}$, *where V is the set of all vehicles.*

Traffic planners have goals different from the ones of individual drivers. This is because they have to consider the common welfare while drivers only consider their own benefit. This conflict also occurs in our scenario. Drivers aim at maximizing their utility. In the following, we define the utility of a vehicle agent for our particular scenario.

Definition 7. *Let b^j denote the budget of the driver of Vehicle j. His utility u^j is $u^j = b^j - v^j \cdot T_w^j$.*

In other words, the utility of drivers is the difference between their budget and their waiting time weighted by their valuations of waiting-time reduction. Although the utility u^j of the driver of Vehicle j can be negative, e.g., if the budget is zero and the driver has to wait 10 seconds, he cannot exceed his budget b^j.

Example 1. Let l and h denote two vehicles. The driver of l is not in a hurry. His valuation is low, e.g., $v^l = 0.01$. If he had to wait 10 seconds longer, his weighted waiting time would increase only by $v^l \cdot 10 = 0.01 \cdot 10 = 0.1$. The driver of Vehicle h in contrast must meet a deadline. His valuation of waiting-time reduction is high, e.g., $v^h = 1.00$. If his waiting time was reduced by 10 seconds, his weighted waiting time would be reduced by $v^h \cdot 10 = 1.00 \cdot 10 = 10$. If the driver of Vehicle l was offered to wait 10 seconds longer for one currency unit he would accept the offer. This is because his utility would increase: $u^l_{new} = b^l_{old} + 1 - v^l \cdot (T^l_{w,old} + 10) = u^l_{old} + 1 - 0.01 \cdot 10 = u^l_{old} + 0.9 > u^l_{old}$. On the other hand, the driver of Vehicle h would readily offer one currency unit to wait 10 seconds less because this would increase his utility: $u^h_{new} = b^h_{old} - 1 - v^h \cdot (T^h_{w,old} - 10) = u^h_{old} - 1 + 1.00 \cdot 10 = u^h_{old} + 9 > u^h_{old}$. Of course, this is only possible if the budget b^h exceeds one currency unit.

In our context, it is not relevant if the budget of a driver is real money, or if it is any private currency issued and distributed by the traffic authority.

Definition 8. *The* total entry budget B_e *is the sum of the budgets of all drivers when they enter the neighborhood:* $B_e := \sum_{j \in V} b^j_e$. V *denotes the set of all vehicles. The* total leaving budget B_l *is the sum of the budgets of all drivers when they leave the neighborhood:* $B_l := \sum_{j \in V} b^j_l$.

Desiderata for Valuation-Aware Traffic-Control Mechanisms. Mechanisms for traffic control at intersections give the right to cross the intersection to arriving vehicles for a certain period of time. Sometimes this right is only given if the vehicles have fulfilled some preconditions, e.g., having stopped at a stop sign. Traditional mechanisms differ regarding the scheduling algorithm, depending on time, on the direction of arrival and on the direction of departure. The valuation-aware mechanisms envisioned should encourage drivers to reveal their valuations. If the best strategy of drivers is to reveal their valuations truthfully, provided that this holds for all other drivers as well, the mechanism is *incentive compatible* [9]. Incentive compatibility is desirable. It tends to provide more efficient solutions.

Next, the mechanisms should meet the objectives effectiveness, avoidance of starvation, and zero-sum. We say that an intersection-control mechanism is *effective* if the average weighted waiting time $\overline{vT_w}$ is lower than with *FIFO*. Clearly, effectiveness means higher overall satisfaction of the drivers. *Avoidance of starvation* means that a vehicle is guaranteed an upper bound of the waiting time. Such an upper bound should be traffic-dependent: If all roads are heavily congested, this upper bound should naturally be higher than in a situation with only few vehicles.

An intersection-control mechanism is *zero-sum* if the total budget of vehicles entering the neighborhood is equal to the total budget of vehicles leaving. We believe that users would not like to see an increase of mobility costs and therefore have a preference for zero-sum.

Finally, there are two further conflicting requirements on traffic-control mechanisms at intersections, early allocation and late allocation. *Early allocation* means that the time slots should be allocated as early as possible. The reason is that a vehicle obtaining a time slot early does not have to stop and wait before entering the intersection. Instead, it can slow down and adapt its speed in

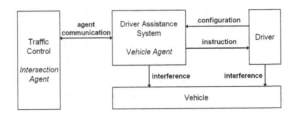

Fig. 1. Agent-based traffic control

advance so that it reaches the intersection exactly on time and can cross it without any stops. This makes driving more comfortable and is expected to reduce energy consumption. But time slots should also be allocated as late as possible (*late allocation*), to give vehicles with high valuations that arrive late a chance to acquire an early time slot. Clearly, these two requirements are in conflict. Choosing their weights depends on the preferences of the traffic planners and on the actual design of the mechanism.

Agent-Based Traffic Control. To take the different driver valuations of waiting-time reduction into account, vehicles and intersection control have to interact. But drivers should not interact with the intersection control while driving. Intelligent and autonomous driver-assistance systems make such distractions unnecessary. Agent technology is promising in this context: The autonomy property of agents [10] avoids unnecessary human interaction. We assume that every vehicle will be equipped with a platform with a standardized interface that allows for the installation of agent-based driver-assistance systems. Assuming the existence of such a generic platform within vehicles is in line with current developments in the real world. For instance, think of the so-called on-board units used by the German toll collect system for heavy trucks (http://www.toll-collect.de). To facilitate valuation-aware traffic control, intersections will have to be equipped with traffic-control units as well. These units implement the different mechanisms and use agent technology to interact with the driver-assistance systems of the vehicles.

An agent-based driver-assistance system hosts a *vehicle agent* (see Fig. 1). It can instruct the driver when to cross the intersection, and at which speed. The vehicle agent can communicate with other vehicle agents and with the *intersection agent* which represents the traffic-control unit of an intersection. The driver configures the driver-assistance system in advance to avoid distraction while driving.

Note that the traffic-control mechanisms proposed in this article are orthogonal to the extent of interference of the driver-assistance system with the driving behavior. The system could either merely instruct the driver, or, as it is the case with adaptive cruise-control systems (ACC, [11]), take control of the vehicle in certain situations. Our mechanisms would work with both alternatives.

4   Mechanisms

We first describe a generic procedure for agent-based traffic control at intersections. This will allow for a more structured presentation of our actual mechanisms. We then describe the *FIFO* mechanism. This mechanism, which has been proposed and evaluated in [2,3], will serve as a reference point for our new valuation-aware mechanism *ITSA* and its variants, which we describe subsequently. For all mechanisms we discuss to which extent they meet the desiderata incentive compatibility, avoidance of starvation, zero-sum and early and late allocation. We evaluate effectiveness in Sect. 5.

4.1   General Procedure

The following general procedure describes all agent-based mechanisms for traffic control at intersections which we currently have in mind:

(1) Vehicle and intersection agents *make contact*.
Vehicle agents have to obtain information which intersection is equipped with an intersection agent, and the intersection agent has to obtain information on vehicles that will arrive at the intersection.
(2) The vehicle agent acquires an initial time slot to cross the intersection (*initial time-slot acquisition*).
(3) If the vehicle agent is dissatisfied with the time slot acquired it can try to acquire a better one (*subsequent time-slot acquisition*) from other vehicle agents. It may try to do so repeatedly using different mechanisms.
(4) Finally, the vehicle crosses the intersection using its time slot.

The difference between Step 2 and 3 is that vehicles do not yet have a time slot before Step 2. They receive an initial time slot in this step. In Step 3 the vehicles already have received a time slot which they can now trade.

Some mechanisms do not need Step 3, e.g., those that do not allow vehicles to trade their time slots. But Step 3 is beneficial in general. Otherwise, Step 2 would have to reconcile the conflicting goals of early and late allocation. With Step 3 in turn, agents of vehicles arriving late can interact with vehicle agents which hold an earlier time slot.

While Step 3 is optional, Steps 1, 2 and 4 are part of all mechanisms described in the following.

4.2   FIFO

This subsection briefly reviews the agent-based reservation mechanism *FIFO* described in [2,3,4]. It allows vehicle agents to request time slots from an intersection agent (Step 2). It schedules the time slots in the order of request.

All vehicle agents entering the neighborhood of the intersection request a time slot from the intersection agent. If several vehicle agents ask for the same time slot, the one which requested the time slot first will obtain it.

To make a reservation, the intersection agent looks for free time slots. The intersection agent checks if the current request does not conflict with time slots reserved previously. A conflict occurs if the requested intersection lane or an intersection lane crossing the requested intersection lane is already reserved. If the desired time slot cannot be assigned, the intersection agent offers the earliest non-conflicting time slot after the desired one to the requesting vehicle agent.

Because *FIFO* is not valuation-aware, the notion of incentive compatibility does not apply. *FIFO* avoids starvation because it ensures an upper bound of waiting time. Vehicle agents do not pay any money, thus *FIFO* is zero-sum. Late allocation is not beneficial with *FIFO* since the order of arrival determines the assignment of time slots. Thus, *FIFO* should implement early allocation.

So far, we do not know any valuation-aware mechanism for intersection control, (except for [7,8] which are very recent). To evaluate the mechanisms to be described in the following, we compare them to a mechanism which is not valuation-aware. Since [2] has shown that *FIFO* outperforms traffic lights regarding average waiting time, it will be our reference point for the evaluation. We think that valuation-aware mechanisms from other domains, e.g., data routing, are not appropriate: They typically do not take physical constraints into account, like no overtaking.

Even though *FIFO* is not valuation-aware, it can be combined with other mechanisms for Step 3 of the general procedure, and such combinations could be valuation-aware. Some of these combinations are discussed in [7,8].

4.3   Initial Time-Slot Auction

In the following we propose a new mechanism which is valuation-aware. It auctions the time slots to the arriving vehicles. We call it *Initial Time-Slot Auction* (*ITSA*). This is because the auction is part of Step 2.

Basic Variant. Vehicle agents register at the intersection when entering its neighborhood. The intersection agent initiates a second-price sealed-bid auction [12] of the next available time slot. I.e., the bidder with the highest bid wins but pays only the second highest bid. Not all of the arriving vehicles can take part in the auction. Some of the vehicles already have acquired an earlier time slot. They may not take part. Further, vehicles can only take part if the vehicles driving in front already have acquired a time slot. Otherwise they cannot be sure that the vehicle in front will acquire an earlier time slot. Thus, only one vehicle per direction can bid for a time slot. We refer to them as *candidates*.

Example 2. Figure 3 shows an intersection where some vehicles already have acquired a time slot, some can acquire a time slot and some currently cannot acquire a time slot.

To implement the auction, we use the FIPA Contract Net Protocol [13] (see Fig. 2) and its terminology in the following. The intersection agent asks the candidates for proposals. It chooses the proposal with the highest offer and sends an accept-proposal to the sender of the proposal chosen and a reject-proposal to

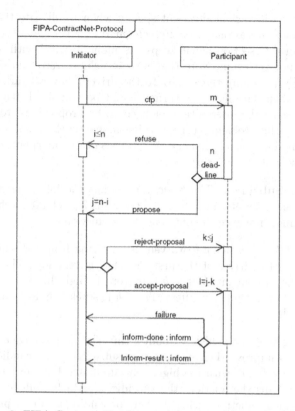

Fig. 2. FIPA Contract Net Protocol (corrected figure from [13])

all others. The vehicle agent receiving an accept-proposal (including the second highest bid, i.e., the amount of money to be paid) confirms the transaction using inform or reports a failure otherwise. For instance, if the auction of a time slot takes too long, it is possible that the winner vehicle cannot use its time slot any more because the begin is already in the past. In this case, the vehicle agent would report a failure. If the winning vehicle agent confirms the transaction, it must pay an amount corresponding to the second highest bid reported. So its budget is reduced by this amount.

The basic variant is incentive compatible. This is because we use a second-price sealed-bid auction [9,12]. It does not guarantee an upper bound of the waiting time and – at least without any extensions – does not avoid starvation.

Example 3. An example of an extension to avoid starvation is that the auction is suspended after the waiting time of a vehicle exceeds an upper bound. After the vehicle has crossed the intersection, the auction is resumed.

However, such constraints typically influence the bidding behavior, and an analysis of the interplay with incentive compatibility would be necessary. In Sect. 5, we examine whether starvation actually occurs in practice.

Because the basic variant does not specify when to allocate the time slots, both early or late allocation can be implemented.

Finally, if an intersection agent keeps the money earned with auctions, the mechanism is not zero-sum. Note that we can achieve zero-sum in a straightforward way, by returning the money to the drivers. Several variants of doing so are conceivable: The money returned could be equally distributed among all vehicles crossing the intersection, or it could be proportional to the waiting times. However, the problem with returning money is that this tends to influence the bidding behavior, too. Thus, an analysis of the interplay with incentive compatibility would be necessary as well.

Variant with Subsidies. The basic variant has the following characteristic which might yield suboptimal outcomes: The vehicles behind a vehicle without a time slot cannot influence the outcome of the auction.

Example 4. Let l and h denote two vehicles arriving from the same direction. The valuation of the driver of the first Vehicle l is very low. The valuation of the driver of the subsequent Vehicle h is extremely high. In all auctions where l is a candidate it only sends low offers. Thus, h is stuck behind l even though its valuation is high.

Vehicles waiting behind a vehicle without a time slot do not have to be inactive. They can try to improve their situation by subsidizing the candidate of their direction. The candidate with the highest accumulated bid wins the auction. If their candidate wins the auction, the subsidizing vehicles will be able to take part in a subsequent auction earlier. Instead of sending a call for proposals only to candidates the intersection agent sends it to the vehicles behind candidates as well, together with the information that their offer would only be a subsidy for the candidate of their direction. Once a candidate is chosen, the vehicles of its direction receive an accept-proposal (see also Fig. 3).

Compared to the basic variant, only one desideratum is different: Incentive compatibility is not guaranteed. A driver might hope that drivers waiting behind him subsidize him. He could be tempted to offer less than his true valuation. In our evaluation, we assume that drivers do reveal their true valuations. Even though this might be too optimistic, we do so to quantify the potential of this variant. Note that it is not clear who would offer less, after all. The drivers of vehicles waiting behind could offer less as well, because it is still unclear if they will actually benefit from subsidizing vehicles in front. As future work, we will investigate if sealed-bid combinatorial auctions are a better solution to this problem.

Both variants of *ITSA* take valuations of waiting-time reduction into account in the initial time-slot-acquisition phase already (Step 2). Nevertheless, they could also be combined with mechanisms of Step 3. Vehicles arriving late can acquire time slots which have been auctioned off before their arrival.

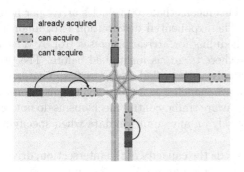

Fig. 3. Auction with subsidizing vehicles   **Fig. 4.** Intersection layout

5   Evaluation

To evaluate the benefits of the proposed mechanisms we use simulations. This is in line with other research on traffic management at intersections, e.g., [2,14,15]. We have developed a simulation framework of our own, using a space-continuous and time-discrete simulation model. This simulation framework allows simulating drivers, driver-assistance systems and vehicles at intersections. We have used the Java Agent DEvelopment Framework (JADE, http://jade.tilab.com) for the implementation of the multi-agent system. Agent-based simulation frameworks are not applicable because we do not use agents for simulation. Instead, we simulate the traffic environment for agent-based driver-assistance systems.

5.1   Settings

Our simulations have the following characteristics. We investigate a symmetric intersection consisting of four directions. Each direction has two incoming (right and left) and two outgoing (right and left) lanes. For each direction the right incoming lane allows to turn right and to go straight (both into the right outgoing lanes). The left incoming lane allows to turn left and to go straight (both into the left outgoing lanes) (see Fig. 4).

On every lane the maximum speed is $50km/h$, the speed limit within German cities. The length of incoming and outgoing lanes is $230m$. There is a virtual traffic sign $200m$ before the intersection indicating that vehicle agents can request time slots from the intersection agent. The radius of the intersection is $20m$. Thus, a vehicle crossing the intersection in straight direction drives $230 + 20 + 20 + 230 = 500m$ in total. Vehicles can accelerate with at most $3m/s^2$ and decelerate with at most $8m/s^2$.

Each simulation run simulates 40 minutes. We ignore the vehicles leaving the neighborhood during the first 10 minutes to avoid biased results (*initial phase*). We use the vehicles leaving during the next 30 minutes (*observation phase*) to compute the average waiting time $\overline{T_w}$ and the average weighted waiting time $\overline{vT_w}$. Vehicles arrive with interarrival times exponentially distributed with mean

$\frac{1}{\lambda} = 36s$ from every incoming lane. This means that 100 vehicles arrive per hour on average on each incoming lane. The exponential distribution is common in queuing theory and in stochastic scheduling for arrival processes [16]. A vehicle can choose two different directions where to go, as mentioned before. The two possible directions are chosen randomly with probability $p = 0.5$. The driver valuations of waiting-time reduction are modeled as an exponential distribution with mean $\frac{1}{\lambda} = 0.01$. Since valuation-aware traffic-control mechanisms do not yet exist in the real world, we could not rely on any empirical data when specifying this parameter.

Clearly, if the volume of traffic exceeds the capacity of the intersection, drivers will be unsatisfied with any traffic-control mechanism. With the numbers given so far, the volume of traffic is in line with the capacity of the intersection: We allow only one vehicle to cross the intersection at the same time. Thus, its capacity is $3600s/4s = 900$ vehicles per hour. This is because the crossing time is $4s$ for every vehicle independent of the direction. Our average traffic volume in turn is 800 vehicles per hour.

5.2 Mechanisms Evaluated

We have implemented two mechanisms, *ITSA* without subsidies (*ITSA⁻*) and *ITSA* with subsidies (*ITSA⁺*) in Step 2. Both mechanisms do not include any further activities in Step 3. We compare both *ITSA⁻* and *ITSA⁺* to *FIFO* without any further activity in Step 3.

In our setup, the allocation of a time slot with both *ITSA⁻* and *ITSA⁺* takes place 12 seconds in advance. I.e., a time slot is auctioned 12 seconds before it begins, three times the duration of a slot. For each auction candidates and subsidizing vehicles compute their offer by multiplying their true valuation per second v with the duration of a time slot $T = 4s$. This duration is the maximum time a vehicle needs to cross the evaluated intersection. Because budgets do not influence effectiveness, all vehicles in our experiments have budgets that are high enough to pay their valuations. In other words, vehicles can always afford to pay their bid.

The performance of a mechanism does not only depend on the average number of vehicles arriving at an intersection, but also on the distribution of arrival times. Therefore, a comparison of mechanisms is meaningful only if we compare equally initialized simulation runs. This means that every vehicle has the same type, start time and route in all simulation runs compared. We use 25 randomly chosen numbers as seeds for the runs, to compare the i-th run of *FIFO* to the ones of *ITSA⁻* and of *ITSA⁺*.

5.3 Results

We compare the differences of average waiting time $\overline{T_w}$ and of average weighted waiting time $\overline{vT_w}$ of equally initialized simulation runs between *FIFO* and *ITSA⁻* and between *FIFO* and *ITSA⁺*. For all 25 simulation runs we compute the mean, standard deviation σ, and the 99% confidence interval (CI) of the differences.

Table 1. *FIFO* vs. *ITSA⁻*

	mean	σ 99% CI
$\Delta \overline{T_w}$	-0.045 0.296 [-0.210, +0.121]
$\dfrac{\Delta \overline{T_w}}{T_w^{(FIFO)}}$	-0.005 0.024 [-0.019, 0.008]
$\Delta \overline{vT_w}$	0.048 0.024 [0.034, 0.062]
$\dfrac{\Delta \overline{vT_w}}{vT_w^{(FIFO)}}$	0.306 0.077 [0.263, 0.349]

Table 2. *FIFO* vs. *ITSA⁺*

	mean	σ 99% CI
$\Delta \overline{T_w}$	-0.232 0.245 [-0.369, -0.095]
$\dfrac{\Delta \overline{T_w}}{T_w^{(FIFO)}}$	-0.018 0.018 [-0.028, -0.008]
$\Delta \overline{vT_w}$	0.067 0.025 [0.053, 0.081]
$\dfrac{\Delta \overline{vT_w}}{vT_w^{(FIFO)}}$	0.430 0.044 [0.406, 0.455]

$T_w^{(FIFO)}$ is the waiting time with *FIFO*, $T_w^{(ITSA⁻)}$ and $T_w^{(ITSA⁺)}$ are the ones with *ITSA⁻* and *ITSA⁺*. When comparing *FIFO* and *ITSA⁻*, the absolute difference of average waiting times is $\Delta \overline{T_w} = T_w^{(FIFO)} - T_w^{(ITSA⁻)}$, and the one of average weighted waiting times is $\Delta \overline{vT_w} = vT_w^{(FIFO)} - vT_w^{(ITSA⁻)}$. The relative difference of average waiting time is $\Delta \overline{T_w}/T_w^{(FIFO)}$, and of the weighted one is $\Delta \overline{vT_w}/vT_w^{(FIFO)}$. When comparing *FIFO* and *ITSA⁺*, the notation and the formulae are analogous.

FIFO vs. ITSA⁻. Table 1 lists both the absolute and the relative differences of the average waiting time $\overline{T_w}$ and the average weighted waiting time $\overline{vT_w}$ for *FIFO* and *ITSA⁻*. *ITSA⁻* reduces the average weighted waiting time $\overline{vT_w}$ by 0.048 units, i.e., by 30.6% on average. The 99% confidence interval shows that these results are reliable. The relative difference is between 26.3% and 34.9% in 99% of all cases. *ITSA⁻* increases the average waiting time only slightly by 0.5% on average, compared to *FIFO*. Thus, *ITSA⁻* is effective regarding this waiting time. At the same time, it leaves the average waiting time nearly unchanged.

FIFO vs. ITSA⁺. Table 2 lists the absolute and the relative differences of the average waiting time and the average weighted waiting time for *FIFO* and *ITSA⁺*. *ITSA⁺* *reduces* the average weighted waiting time $\overline{vT_w}$ by 0.067 units, i.e., by 43.0% on average. The 99% confidence interval tells us that these results are very reliable. The relative difference is between 40.6% and 45.5% in 99% of all cases. *ITSA⁺* increases the average waiting time $\overline{T_w}$ only slightly, by 1.8% on average, compared to *FIFO*. In other words, *ITSA⁺* is very effective regarding average weighted waiting time, and the average waiting time remains nearly unchanged as well.

Avoidance of Starvation. To evaluate the influence of an auction on vehicles with low valuations, we compare the average waiting time $\overline{T_{w;10\%}}$ of the 10% of the vehicles with the lowest valuation. Table 3 lists both the absolute and the relative difference of the average waiting time $\overline{T_{w;10\%}}$ for *FIFO* vs. *ITSA⁻* and for *FIFO* vs. *ITSA⁺*. *ITSA⁻* increases this waiting time by factor 2.518, *ITSA⁺* by factor 1.970. These increases are not surprising because an auction affects the participants with low valuations negatively. Even though both mechanisms do

Table 3. Characteristics of the 10% of the vehicles with the lowest valuation

	FIFO vs. ITSA⁻	FIFO vs. ITSA⁺
$\Delta\overline{T}_{w;10\%}$	-23.402	-14.454
$\dfrac{\Delta\overline{T}_{w;10\%}}{\Delta T^{(FIFO)}_{w;10\%}}$	-1.518	-0.970

not guarantee avoidance of starvation, we think that their negative effects are limited. We will examine the extensions described earlier to avoid starvation as future work, to foster social acceptance of our mechanisms.

Subsumption. The results so far are promising. The variants evaluated reduce average weighted waiting time significantly, without increasing average waiting time by much. Compared to $ITSA^-$, $ITSA^+$ reduces average weighted waiting time by an additional 12.4%, while loosing incentive compatibility. To decide if this additional gain is worthwhile, we plan to carry out experiments with real users, as motivated earlier. $ITSA^+$ also improves the average waiting time $\overline{T_{w;10\%}}$ of the 10% of the vehicles with the lowest valuation. $\overline{T_{w;10\%}}$ increases much less than with $ITSA^-$. Our experiments also show that an auction mechanism which considers all vehicles at an intersection and not only the candidates can outperform variants which do not do so.

6   Conclusion

Traditional approaches for traffic management do not consider the different driver valuations of waiting-time reduction. But valuation-aware traffic control is expected to increase overall satisfaction significantly. In this article we have proposed a new mechanism for traffic control at intersections which takes the different valuations of waiting-time reduction into account, *Initial Time-Slot Auction (ITSA)*. We have proposed and discussed refinements of this mechanism which avoid starvation and allow other vehicles to subsidize vehicles in front.

As part of our evaluation, we have compared our new mechanism to the state-of-the-art *FIFO* mechanism. We have seen that all evaluated variants reduce average weighted waiting time significantly, without influencing average waiting time a lot.

An important conclusion from this study is that a valuation-aware traffic-control mechanism at intersections should include as many vehicles as possible. The refinements proposed here head for this direction.

As future work, one issue which we will investigate is vehicles crossing a sequence of intersections. In such a setting, vehicle agents will have to plan its expenses, and intersection agents may cooperate.

References

1. Schrank, D.L., Lomax, T.J.: The 2005 Urban Mobility Report. Report, Texas Transportation Institute, College Station, TX (2005)
2. Dresner, K., Stone, P.: Multiagent Traffic Management: A Reservation-Based Intersection Control Mechanism. In: AAMAS 2004, pp. 530–537. IEEE, Los Alamitos (2004)
3. Dresner, K., Stone, P.: Multiagent Traffic Management: An Improved Intersection Control Mechanism. In: AAMAS 2005, pp. 471–477. ACM, New York (2005)
4. Dresner, K., Stone, P.: Human-Usable and Emergency Vehicle-Aware Control Policies for Autonomous Intersection Management. In: Bazzan, A.L.C., Chaib-draa, B., Klügl, F., Ossowski, S. (eds.) Fourth International Workshop on Agents in Traffic and Transportation (ATT), Hakodate, Japan (2006)
5. Greschner, J.T., Gerland, H.E.: Traffic Signal Priority Tool To Increase Service Quality And Efficiency. In: Proceedings APTA Bus & Paratransit Conference, Houston, TX, pp. 138–143 (2000)
6. Roozemond, D.A., Rogier, J.L.H.: Agent controlled traffic lights. In: European Symposium on Intelligent Techniques (ESIT), 14-15 September, Aachen, Germany (2000)
7. Schepperle, H., Böhm, K., Forster, S.: Towards Valuation-Aware Agent-Based Traffic Control. In: AAMAS 2007, pp. 352–354. Honolulu, Hawai'i, USA (2007)
8. Schepperle, H., Böhm, K., Forster, S.: Traffic Management Based on Negotiations between Vehicles – a Feasibility Demonstration Using Agents. In: Ninth Workshop on Agent Mediated Electronic Commerce (AMEC IX), Honolulu, Hawai'i, USA (2007)
9. Krishna, V.: Auction Theory. Academic Press, London (2002)
10. Wooldridge, M.J.: An Introduction to Multiagent Systems. Wiley, Chichester (2002)
11. Bareket, Z., Fancher, P.S., Peng, H., Lee, K., Assaf, C.A.: Methodology for Assessing Adaptive Cruise Control Behavior. IEEE Transactions on Intelligent Transportation Systems 4(3), 123–131 (2003)
12. Vickrey, W.: Counterspeculation, Auctions, and Competitive Sealed Tenders. The Journal of Finance 16(1), 8–37 (1961)
13. FIPA – Foundation for Intelligent Physical Agents: Fipa contract net interaction protocol specification (sc00029h) (December 2002), http://www.fipa.org/specs/fipa00029/SC00029H.pdf
14. Fuerstenberg, K.C., Lages, U.: New European Approach for Intersection Safety - The EC-Project INTERSAFE. In: Intelligent Vehicles Symposium, 2005. Proceedings, IEEE, Los Alamitos (2005)
15. Hopstock, M., Ehmanns, D., Spannheimer, H.: Development of Advanced Assistance Systems for Intersection Safety. In: AMAA 2005, 9th International Conference on Advanced Microsystems for Automotive Applications, Berlin, Germany (2005)
16. Pinedo, M.: Scheduling – Theory, Algorithms and Systems, 2nd edn. Prentice Hall, Englewood Cliffs (2002)

High-Performance Agent System for Intrusion Detection in Backbone Networks

Martin Rehák[1], Michal Pěchouček[1], Pavel Čeleda[2], Vojtěch Krmíček[2],
Jiří Moninec[2], Tomáš Dymáček[2], and David Medvigy[1]

[1] Department of Cybernetics and Center for Applied Cybernetics,
Faculty of Electrical Engineering, Czech Technical University in Prague
Technická 2, 166 27 Prague, Czech Republic
{mrehak,pechouc}@labe.felk.cvut.cz
[2] Institute of Computer Science, Masaryk University
Botanická 68a, 602 00 Brno, Czech Republic
{celeda,vojtec,moninec}@ics.muni.cz

Abstract. This paper presents a design of high-performance agent-based intrusion detection system designed for deployment on high-speed network links. To match the speed requirements, wire-speed data acquisition layer is based on hardware-accelerated NetFlow like probe, which provides overview of current network traffic. The data is then processed by detection agents that use heterogenous anomaly detection methods. These methods are correlated by means of trust and reputation models, and the conclusions regarding the maliciousness of individual network flows is presented to the operator via one or more analysis agents, that automatically gather supplementary information about the potentially malicious traffic from remote data sources such as DNS, whois or router configurations. Presented system is designed to help the network operators efficiently identify malicious flows by automating most of the surveillance process.

1 Introduction

With the increasing number of Internet users, provided services and current generation of online multimedia services, the speed of most Internet links has increased significantly. This increase rendered many past methods for intrusion detection obsolete, and current backbone networks lack efficient technology for real-time detection of malicious traffic. While it may be technically possible to perform traffic analysis by means of existing signature matching techniques running on dedicated high-performance hardware, the high rate of false positive, and high cost of associated human supervision makes systematic surveillance uneconomical.

To address the above situation, and to enable the operators of backbone and large enterprise networks to analyze current threats in near real-time, we present a design of an autonomous system able to detect malicious traffic on high-speed networks and to alert the operators efficiently. While the system reasoning is

M. Klusch et al. (Eds.): CIA 2007, LNAI 4676, pp. 134–148, 2007.

based on intelligent agents and multi-agent methods, the network traffic data acquisition and preprocessing in both dedicated adaptive hardware and specialized software is essential for project success. This is due to the fact that the traditional agent techniques are not well suited for efficient low-level traffic processing.

In the work presented in this paper, we aggregate the network data to capture the information about network flows, unidirectional components of TCP connections (or UDP, ICMP equivalent) identified by shared source and destination addresses and ports, together with the protocol, and delimited by the time frame used for data acquisition (see Section 3.1). This information provides no hint about the content of the transmitted data, but by detecting the anomalies in the list of flows acquired over the monitoring period, we can detect the anomalies and possible attacks, albeit with limited effectiveness.

2   Related Work

In order to detect an attack from the flow information on the backbone level, especially without any feedback from the affected hosts, we have to analyze the patterns in the traffic data, compare them with normal behavior and conclude whether the irregularity corresponds to a known attack profile or not. This approach to Network Intrusion Detection, typically based on the flow information captured by network flow monitor is currently an important field of research into *anomaly based intrusion detection*. Numerous existing systems, based on traffic volume analysis modeled by Principal Component Analysis (PCA) methods [1], models of entropy of IP header fields for relevant subsets of traffic [2,3], or just count of the flows corresponding to the selected criteria [4] offer each a particular valid perspective on the network traffic.

The MINDS system [4] represents the flow by basic NetFlow aggregation features (srcIP, srcPrt, dstIP, dstPrt, protocol) and complements them by the number of the flows from the same srcIP, to the same dstIP and their combinations with dstPrt and srcPrt respectively. These properties are assessed both in time and number of connections defined windows, to account for slow scanning. The system proposed by Xu et al. [2] for traffic analysis on backbone links also uses the NetFlow based identity 5-tuple. The context of the single connection is defined by the normalized entropy of srcPrt, dstPrt and dstIP dimensions of the set of all connections from the srcIP of the flow in the current time frame. Another perspective anomaly detection mechanism can be based on the observation of traffic volumes in high-speed network. Lakhina et al. [5] uses statistical modeling to identify the anomalous origin-destination-aggregated flows. The method is based on Principal Component Analysis. In another work of the same authors [3], the PCA method is used to model the normal and residual entropy, to remove the systematic elements of the data before clustering. The clustering then emphasizes the anomalous characteristics of the traffic.

The above listed systems are appropriate for direct deployment on backbone network, with a specific limitations of MINDS that was designed to protect a large, open network common in university environment. Besides the above-mentioned sample of backbone anomaly detection mechanisms, there are numerous research and commercial systems designed to protect local networks. A typical representative of recently developed system is a SABER [6], which addresses not only threat detection, but attempts to actively protect the system by automatically generated patches. Technical perspective on many existing IDS systems, including SNORT [7] and other *signature matching* techniques that detect intrusions by detecting patterns specific to known attacks in network traffic, can be found in [8]. A good, even if slightly outdated review of classic research and systems in the domain is provided by [9].

3 Architecture

In our approach, we have decided not to develop a novel detection method, by rather to integrate several methods [1,2,3,4] with an extended trust models of a specialized agent. This combination allows us to correlate the results of the used methods and to combine them to improve their effectiveness. Most anomaly detection methods today are not fit for commercial deployment due to the high ratio of false positives (legitimate traffic classified as malicious) or false negatives (malicious traffic classified as legitimate). While their current level of performance is a valid scientific achievement, the costs associated with supervision of such systems are prohibitive for most organizations. Therefore, our main research goal is to combine the efficient low-level methods for traffic observation, with multi-agent detection process to detect the attacks with comparatively lower error rate see Table 2, and to provide the operator with efficient incident analysis layer presented in Section 3.3. This layer supports operator's decisions about detected anomalies by providing additional information from related data sources. The layer is also responsible for visualization of the anomalies and the detection layer status.

The architecture consists of several layers with varying requirements on on-line processing characteristics, level of reasoning and responsiveness. While the low-level layers need to be optimized to match the high wire-speed during the network traffic acquisition and preprocessing, the higher layers will use the preprocessed data to infer the conclusions regarding the degree of anomaly and consecutively also the maliciousness of the particular flow or a group of flows. Therefore, while the computation in the higher layers must be still reasonably efficient, the preprocessing by the lower layers allows us to deploy more sophisticated algorithms. System can be split into these layers, as shown in Figure 1:

• **Traffic Acquisition and Preprocessing Layer:** The components in this layer acquire the data from the network using the hardware accelerated NetFlow probes [10] and perform their preprocessing. This approach provides the real-time overview of all active unidirectional connections on the observed link. In order to speed-up the analysis of the data, the preprocessing layer will aggregate meaningful global and per-flow (or group of) characteristics and statistics.

• **Cooperative Threat Detection Layer:** This layer will principally consist of specialized, heterogeneous agents that would seek to identify the anomalies in the preprocessed traffic data by means of their extended trust models [11]. Their collective decision regarding the degree of maliciousness of a flow with certain characteristics shall use a reputation mechanism. The agents will run inside the A-globe agent platform [12] and will use its advanced features like agent migration and cloning to adapt the system to the traffic and relevant threats.

• **Operator and Analyst Interface Layer:** This layer is responsible for interaction with operator. The main component is the intelligent agent called Mycroft that will help the operator to analyze the output of the detection layer, by putting the anomaly information in context of other relevant information. When the detection layer detects suspicious behavior on the network it is reported to Mycroft. Mycroft opens the new case and retrieves relevant information from available data sources. Network operator can explore and evaluate the reported case subsequently. Another part of this layer is a set of lightweight, specialized visualization agents that will allow the operator to follow only the selected characteristics of the system.

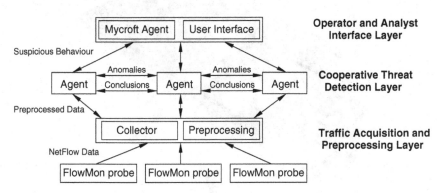

Fig. 1. System overview, with network probes, acquisition and preprocessing layer at the bottom, agent platform for anomalies detection in the middle and visualization on the top

3.1 Traffic Acquisition and Preprocessing Layer

The traffic acquisition and preprocessing layer is responsible for acquiring network traffic, preprocessing data and providing traffic characteristics to upper system layers. We use the flow characteristics, based on information from packet's headers.

In general, flows are a set of packets which share a common property. The simplest type of flow is a 5-tuple, with all its packets having the same source and destination IP addresses, port numbers and protocol. Flows are unidirectional and all their packets travel in the same direction. For the flow monitoring we use NetFlow protocol developed by Cisco Systems [13].

The amount of traffic in nowadays high-speed networks increases continuously and traffic characteristics change heavily in time (network throughput fluctuation due to time of day, server backups, DoS attacks, scanning attacks, etc.). Performance of network probes must be independent of such states and behave reliably in all possible cases. The quality of provided data significantly effects the upper layers and chances to detect traffic anomalies.

Therefore we use hardware accelerated NetFlow probes (see Figure 2). Flow-Mon probe is a passive network monitoring device based on the COMBO hardware [10], which provides high performance and accuracy. The FlowMon probe is preferred due to implemented features which contains packet/flow sampling, several sampling types, active/inactive timeouts, flow filtering, data anonymization, NetFlow protocol version 5 and 9 support. The FlowMon probe handles 1 Gb/s traffic at line rate in both directions and exports acquired NetFlow data to different collectors. Detailed evaluation of these crucial capabilities is described in Section 4.

Fig. 2. FlowMon - COMBO6X PCI-X 64/66MHz hardware accelerated card [14]

The collector stores incoming packets with NetFlow data from FlowMon probes into database. The collector provides interface to graphical and text representation of network traffic, flow filtration, aggregation and statistics evaluation, using source and destination IP addresses, ports and protocol.

To process acquired IP flows by upper system layers the preprocessing must be performed on several levels and in different manners. Packets can be sampled (random, deterministic or adaptive sampling) on input and the sampling information is added to NetFlow data. On the collector side the same flows are aggregated to reduce the amount of data without information loss and several statistic characteristics (average traffic values, entropy of flows) are computed.

Even after their deployment in monitored network, the probes can be reprogrammed to acquire new traffic characteristics. The system is fully reconfigurable and the probes can adapt their features and behavior to reflect the changes in the agent layer. As we can see in Section 4, presented solution can acquire unsampled flow data from even very fast links. The proposed traffic acquisition and preprocessing layer is able to provide real-time traffic characteristics to detect anomalies by upper system layers.

3.2 Cooperative Threat Detection Layer

Cooperative threat detection is based on the principles of trust modeling [15] that are an integral part of agent research. However, there are three important features [11] that must be added to trust modeling to cover our domain-specific requirements:

• **Uncertain Identity Modeling:** Baseline trust models evaluate the behavior of individual agents, whose identity is guaranteed (to an extent) by the multi-agent platform or similar computational environment. In the network domain, we have to evaluate the trustfulness of network flows, and while they can be distinguished as unique identities, this distinction is unpractical from the intrusion detection perspective. We represent the connections in a metric space, and use the associated distance function to assess the similarity of the flow representations in this space. All detection agents use the same NetFlow 5-tuple to construct the identity representations, but may obtain different results regarding the similarity due to the use of different distance functions. For example, one agent can emphasize the similarity of srcIP address (and likely host), while the others may concentrate on ports that are more application specific. This variability makes the agent perspective on the system multi-faceted, and the attacks are less likely to avoid multiple agents.

• **Context Modeling:** Merely representing the flow identities in the metric space and evaluating their trustfulness gives unsatisfactory results, as it ignores the most important information from the NetFlow data – the information about the other, similar flows in the current traffic sample. This information constitutes the context of the trusting decision [16], and together with the identity defines the Identity-Context metric space, where the detection agents assess the trustfulness of flow representations. Each of the agents uses its own particular context space, typically based on the existing anomaly detection methods. For instance, we can complement the information about the flow by the number of flows from the same srcIP and same dstPrt, or with an entropy of dstIP addresses computed over all flows with the same srcIP. The use of this information is twofold; each agent uses it to place the flow representations in the Identity-Context space of its trust model, and in the same time to provide the information about the degree of flow anomaly to other trusting agents.

• **Implicit Feedback Mechanism:** The principal input of classic trust models is a result of past cooperations with the partner: quality of service, degree of success, on-time delivery and other domain specific parameters. In our case, the system is deployed on backbone network and it is very difficult to obtain the feedback that can be associated with the current traffic on the network; cooperative IDS (like `dshield.org`) typically provide unsynchronized feedback, and not all the threats are Internet-wide. Obtaining the feedback from the connected operators or organizations is even more difficult: while the IETF has several working groups focusing on incident response interoperability, the bulk of the work is not suitable for real-time data processing, and concentrates on human-to-human interaction. Therefore, we use the information regarding the flow anomaly *as assessed by the other agents* to replace the direct feedback, therefore connecting the anomaly detection between diverse agents.

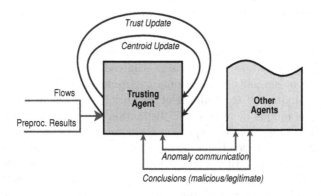

Fig. 3. Overview of detection (trusting) agent operations

While processing the information about the network flows (see Fig. 3), each trusting agent receives an identical copy of network flows list and associated pre-extracted statistics. Then, it uses its specific preprocessing to determine the anomaly of each flow (in most cases working only with already extracted statistics) and to communicate the list of anomalies to other agents. In its turn, the agent also receives the anomalies from the others, and starts the flow processing by its internal trust model. As we have implicitly suggested above, the trustfulness is not associated to individual flows, but rather to selected objects in the Identity-Context space. Individual flow is therefore represented by its identity (i.e. NetFlow 5-tuple) and the associated context is retrieved to determine its position in the Context subspace. Then, we retrieve the positions of nearby centroids from the current trust models and update their trustworthiness with an aggregated degree of flow anomaly as determined by the other agents. When there is no appropriate cluster in the vicinity of the observed flow, a new cluster is created. The details of the approach are presented in [11].

Performance of an isolated detection agent would be the same as a performance of the anomaly detection method it is based on. As we have suggested above, the agents base their evaluation of trustfulness not only on their local results, but also on the anomaly opinions of other agents (see Fig. 4). We argue that

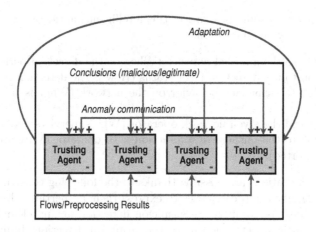

Fig. 4. Overview of collaboration between detection agents

this cross-correlation will help to filter-out most false positives on the level of individual agents, reducing the number of incidents to evaluate with other agents. In the second phase of evaluation, each agent selects the flows it considers as malicious and shares these flows with other agents. Agents then use a simple voting protocol to reach a collective conclusion regarding the estimated maliciousness of anomalous flows, further reducing the number of incidents reported. Collectively accepted flows are then sent to analyst interface layer for further filtering based on user preferences, and possibly assisted analysis by analyst.

The decision whether a given flow is trusted or untrusted depends on the typical degree of anomaly in the observed network traffic. This parameter varies widely with network type – the number of anomalies is typically low on corporate or government networks, but is significantly higher on public Internet backbone links, or in the university settings. To avoid the problems with manual tuning of the system, we use a fuzzy-inference process integrated with the trust model to decide whether the given flow is malicious, by computing its inference with Low and High trust values. These values are determined similarly to the trustfulness of individual flow representations, but are represented as two fuzzy intervals [17].

3.3 Operator and Analyst Interface Layer

The Cooperative Threat Detection Layer is coordinated by a super-agent Mycroft. Mycroft is again a multi-agent system. This system is constructed for context based inference over information synthesized from various data sources. The name of this system refers to the so called Mycroft problem well known from Doyle's Mycroft Holmes - a Sherlock Holmes' brother. Every detected suspicious behavior on the network is reported to the agent Mycroft by the detection layer. Mycroft opens the new case subsequently and retrieves relevant information from data sources to which it is connected. Then the network operator can explore and evaluate the reported case together with contextual information retrieved from connected data sources. Using of this intelligent multi-agent system brings the advantage of wide adaptation:

- adaptation to the guarded network (Where is the detection performed),
- adaptation to the detection layer status (How is the detection performed),
- adaptation to suspicious behavior on the network (What is the subject of detection),
- adaptation to the given network operator (Who supervises detection),
- adaptation to the purpose of interaction with operator (Why are detection results reported).

All these adaptations are possible thanks to the following construction: Individuals are classified into categories to express their properties. The individual can be any object relevant to the suspicion detection or any elementary relation between such objects relevant to the suspicion detection. Individuals are classified into categories with a measure from the interval $< -1, 1 >$. Value -1 stands for "certainly not in given category", value 1 stands for "certainly is in given category" and value 0 stands for "cannot decide". This classification is called the elementary fact and is realized always in some context. The context is nothing else than a set of elementary facts. All the information retrieved from data sources is disassembled into elementary facts. This approach allows data utilization in a way that fits actual situation. Presented construction is a straight extension of the so called diamond of focus presented in [18] for the first time. Formal definition of systems referred to as Knowledge and Information Robots is provided in [19]. Multi-agent system Mycroft is one of the representatives of Knowledge and Information Robots class.

• **Adaptation to the Guarded Network:** Mycroft uses available data sources to support network operator's evaluations and decisions. The system can be taught using quite formalized natural language what kind of information given data source contains, how to combine this information with all other information that Mycroft already knowns or is capable to acquire and how to access this data source. Mentioned teaching is to all intents and purposes done without programming. Summary of typical data sources is listed in Table 1. Communication with data sources is realized by the set of adapters for various types of data sources. This use-case saves operator's time and work in the decision making process.

• **Adaptation to the Detection Layer Status:** Multi-agent system can adapt to changes in the detection layer. It monitors the current status of the detection layer and of each agent present in the detection layer. It is also able to communicate with each agent. If the settings of detection layer change, the multi-agent system Mycroft can deal with it using rules or can be simply retaught to be ready for cooperation with new detection layer configuration. Interaction with the detection layer is shown in Figure 5.

• **Adaptation to Suspicious Behavior on the Network:** Multi-agent system can adapt to various types of suspicious behavior on the network. When suspicious behavior on the network is reported by the detection layer, Mycroft

Table 1. Table of data sources: first column contains data source description, second column contains importance for network operator (1 – most important, 3 – less important)

Topology		User information	
Routing tables of network hardware	1	Users allowed to log on target computers	1
Trace route utility	2	Users currently present in room with computer	2
Active network hardware configuration files	1	Users currently logged on target computer	1
SNMP information	2	Profiles of typical user behavior	3
Routing tables of computers	3		

Rules and policy		Available services	
Firewall configuration	1	Port scan utility	2
NAT configuration	1	OS detection utility	3

Address resolution		Administrative information	
DNS configuration	1	Computer configuration	1
DHCP address assignment	2	Ports vs. typical services database	2
ARP tables	2	Physical location of device	1
WHO IS service	2	Known attacks database	2

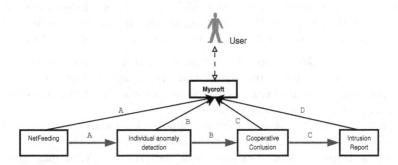

Fig. 5. Mycroft – detection layer interaction, Mycroft observes detection layer behavior **A** - NetFlow data and statistics, **B** - Each Agent's individual anomaly detection, **C** - Agent's negotiation and conclusions, **D** - Intrusion detection reports

asks connected data sources to get relevant information for given type of suspicious behavior. A pattern matching based transmutation is used to select the relevant information.

Using additional knowledge provided by these data sources Mycroft can perform basic evaluations of the traffic situation and present this evaluations to the network operator concurrently with the detection report. In some basic cases, using pre-learned transmutation patterns, system can even evaluate this

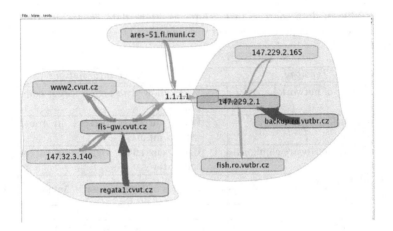

Fig. 6. Flow data visualization example. Oval nodes in graph represent network devices identified by IP addresses. Arrows show direction of the flows.

suspicion as false positive and drop it. Again this is possible by the means of the knowledge provided by suitable data sources.

If a new type of suspicious behavior on the network reveals, the system can deal with it using rules or can be retaught. There is no need of reprogramming.

• **Adaptation to Purpose of Interaction with the Operator:** There are various types of purposes of interaction with the operator (decision making request, notification, warning, etc.). Multi-agent system can adapt to these types of purposes and provide suitable interaction with the network operator.

• **Adaptation to Given Operator:** It means that multi-agent system is able to follow operator's procedures, habits and preferences. It learns during routine work. When operator is not satisfied he can ask the system for different presentation of given information.

Mycroft uses suitable visualization to support visual analytics and decision-making process. It makes the interaction between operator and the whole system efficient and helps operator to explore visual patterns while exploring the data. The visualization and interaction with operator approach is very close to Human-Centered Computing approach [20].

Mycroft uses various types of graphs as visual representation of network traffic situations (see Figure 6) with dynamic user control techniques:

- level of detail selection,
- filtering,
- detail on demand showing.

Level of detail selection allows network operator to explore given data on the level of subnets, particular IP addresses, ports or on the detailed level of individual flows. Filtering allows operator to filter out unimportant information. The operator can focus just on particular subnet or port and on the flows with given properties. Operator can also use detail on demand focus to get additional relevant information from available data sources.

4   System Evaluation and Performance

The performance is becoming a key concern of Network Intrusion Detection Systems (NIDS) in high-speed networks. The results of traffic acquisition and processing vary depending on the amount of acquired data. Numerous existing NIDS are based on commodity hardware with open-source software and very limited hardware acceleration.

The Figure 7 shows flow statistics for various IP packet's sizes transmitted on Ethernet at a rate of a gigabit per second. The tests were performed with the Spirent AX/4000 broadband test system [21]. The test system generated 5 flows with 500000 packets per flow for each packet size. The results of FlowMon probe correspond to the maximal theoretical throughput on gigabit Ethernet network. On the other hand the results of software NetFlow probe (nProbe [22]) are misrepresented for small IP packets. The nProbe was used on Linux OS with Intel PCI-X network interface card and default kernel configuration.

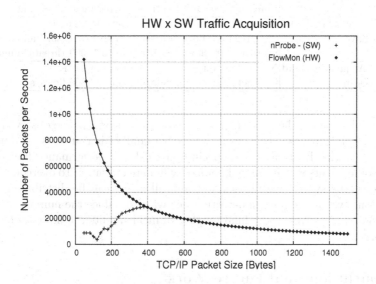

Fig. 7. Flow statistics acquired by SW and HW accelerated probe

Existing flow monitoring systems are mostly based on exporting flow data from routers. The routers are dedicated for routing the data in networks and enabling the flow export has often negative impacts on overall router performance especially during attacks. The exported flows are sampled or limited to maximal number of exported flows per second e.g. 5000 flows/s. In comparison with FlowMon probe, user defined extensions can't be added to the routers and the adaptability is very limited.

In our work we are focusing on the impact of packet sampling on anomaly detection. The articles [23,24] study whether existing sampling techniques distort traffic features that are critical for effective anomaly detection. They show that packet sampling methods (random packet sampling, random flow sampling, smart sampling, and sample-and hold sampling) introduce fundamental bias that degrades the performance of the anomaly detection algorithms. To avoid such a misbehavior the FlowMon probe provides non-sampled data, without packet loss at a line rate.

Table 2. Multi-agent system performance overview. The system process backbone link traffic with average load of 800 Mb/s.

Layer	Processed Data	
	Input	Output
Operator	Security Incidents - incidents at a certain priority levels	Incident Handling - resolving up to 10 high priority incidents per hour
Operator and Analyst Interface Layer	Network Anomalies - detected threats with additional network information	Detected Incidents - priority-based incidents up to 100 incidents/minute
Cooperative Threat Detection Layer	Network Traffic Statistics - aggregated flow statistics up to 100000 flows/minute	Detected Threats - network traffic anomalies up to 10000 threats/minute
Traffic Acquisition and Preprocessing Layer	Network Traffic - packets 125000 packets/s	Flow Statistics - flows 3800 flows/s

The Table 2 shows the performance of multi-agent system. The observed network traffic is processed by several layers to handle high amount of data in current networks. Each layer has specific physical and performance limits e.g. number of incidents which can be handled by human operator. To overcome such limitations, the system is fully scalable and all layers can be distributed. The multi-agent system adapts the detection behavior to reduce the number of false positives and negatives so the final number of incidents fits the limits of human operator.

5   Conclusions and Future Work

Our work presents a design of multi-agent system for network intrusion detection that is optimized for deployment on backbone networks. The designed system addresses two main limitations of existing intrusion detection systems – efficiency and effectiveness. Deployment on high-speed links implies the need to process the important quantity of data in near real-time, in order to prevent the spread of novel threats. Therefore, the individual agents do not acquire the data from the network directly, but receive the data already preprocessed, with the level of detail that is appropriate for anomaly-based intrusion detection. Each detection agent in the system is based on existing anomaly detection technique, which defines its perception of network flow identities in its trust model. Its private view of the data is complemented by the opinions of other agents regarding the anomaly of flows in the current traffic, therefore collaboratively improving the effectiveness of anomaly detection process. When the agents reach a conclusion regarding the untrustfulness of a particular subset of flows, they submit this observation to user-interface agent that automatically retrieves context information (DNS records, history, etc.) to allow rapid analysis by human supervisors.

At the time of this writing, the first preliminary version of the complete system is being integrated and the whole system is still under active development. Therefore, we don't present any definitive experimental results regarding its effectiveness of the complete system, but only the performance evaluations of critical components at the current integration stage. The data used for system testing are acquired on Masaryk University network, connected to the Czech national educational network (CESNET). In our future work, we will provide detailed experimental evaluation of system deployment, and also analyze its performance in countering a wide selection of currently observed malicious traffic.

Acknowledgment. This material is based upon work supported by the European Research Office of the US Army under Contract No. N62558-07-C-0001. Any opinions, findings and conclusions or recommendations expressed in this material are those of the author(s) and do not necessarily reflect the views of the European Research Office of the US Army. Also supported by Czech Ministry of Education grants 1M0567 and 6840770038.

References

1. Lakhina, A., Crovella, M., Diot, C.: Characterization of Network-Wide Anomalies in Traffic Flows. In: ACM SIGCOMM conference on Internet measurement IMC '04, pp. 201–206. ACM Press, New York (2004)
2. Xu, K., Zhang, Z.L., Bhattacharrya, S.: Reducing Unwanted Traffic in a Backbone Network. In: USENIX Workshop on Steps to Reduce Unwanted Traffic in the Internet (SRUTI), Boston, MA (2005)
3. Lakhina, A., Crovella, M., Diot, C.: Mining Anomalies using Traffic Feature Distributions. In: ACM SIGCOMM, Philadelphia, PA, pp. 217–228. ACM Press, New York (2005)
4. Ertoz, L., Eilertson, E., Lazarevic, A., Tan, P.N., Kumar, V., Srivastava, J., Dokas, P.: MINDS - Minnesota Intrusion Detection System. In: Next Generation Data Mining, MIT Press, Cambridge (2004)
5. Lakhina, A., Crovella, M., Diot, C.: Diagnosis Network-Wide Traffic Anomalies. In: ACM SIGCOMM '04, pp. 219–230. ACM Press, New York (2004)
6. Sidiroglou, S., Keromytis, A.D.: Countering network worms through automatic patch generation. IEEE Security & Privacy 3, 41–49 (2005)
7. Sourcefire, Inc.: Snort- Intrusion Prevention System (2007) Accessed in (January 2007), http://www.snort.org/
8. Northcutt, S., Novak, J.: Network Intrusion Detection: An Analyst's Handbook. New Riders Publishing, Thousand Oaks, CA, USA (2002)

9. Axelsson, S.: Intrusion detection systems: A survey and taxonomy. Technical Report 99-15, Chalmers Univ (2000)
10. CESNET, z. s. p. o.: Family of COMBO Cards. (2007), http://www.liberouter.org/hardware.php
11. Rehák, M., Pěchouček, M., Gregor, M.: Trust Modeling with Context Representation and Generalized Identities. Technical report, Gerstner Laboratory, CTU in Prague (2007)
12. Šišlák, D., Rehák, M., pěchouček, M., Rollo, M., Pavlíček, D.: A-globe: Agent development platform with inaccessibility and mobility support. In: Unland, R., Klusch, M., Calisti, M. (eds.) Software Agent-Based Applications, Platforms and Development Kits, pp. 21–46. Birkhauser Verlag, Berlin (2005)
13. Cisco Systems: Cisco IOS NetFlow (2007), http://www.cisco.com/go/netflow
14. Čeleda, P., Kováčik, M., Koníř, T., Krmíček, V., Špringl, P., Žádník, M.: FlowMon Probe. Technical Report 31/, CESNET, z. s. p. o (2006) (2006), http://www.cesnet.cz/doc/techzpravy/2006/flowmon-probe/
15. Sabater, J., Sierra, C.: Review on computational trust and reputation models. Artif. Intell. Rev. 24, 33–60 (2005)
16. Rehak, M., Gregor, M., Pechoucek, M., Bradshaw, J.M.: Representing context for multiagent trust modeling. In: IEEE/WIC/ACM International Conference on Intelligent Agent Technology (IAT 2006 Main Conference Proceedings) (IAT'06), pp. 737–746. IEEE Computer Society, Los Alamitos (2006)
17. Rehák, M., Foltýn, L., Pěchouček, M., Benda, P.: Trust Model for Open Ubiquitous Agent Systems. In: Intelligent Agent Technology, 2005 IEEE/WIC/ACM International Conference. Number PR2416, IEEE, Los Alamitos (2005)
18. Staníček, Z.: Universal Modeling and IS Construction. PhD thesis, Masaryk University, Brno (2003)
19. Procházka, F.: Universal Information Robots a way to the effective utilisation of cyberspace. PhD thesis, Masaryk University, Brno (2006)
20. Jaimes, A., Gatica-Perez, D., Sebe, N., Huang, T.S.: Human-centered computing: Toward a human revolution. Computer 40, 30–34 (2007)
21. Spirent, C.: Spirent AX/4000 Broadband Test System (2007), http://www.spirentcom.com/
22. Deri, L.: nProbe - An Extensible NetFlow v5/v9/IPFIX GPL Probe for IPv4/v6 (2007), http://www.ntop.org/nProbe.html
23. Mai, J., Chuah, C.N., Sridharan, A., Ye, T., Zang, H.: Is sampled data sufficient for anomaly detection? In: IMC '06: Proceedings of the 6th ACM SIGCOMM on Internet measurement, pp. 165–176. ACM Press, New York (2006)
24. Brauckhoff, D., Tellenbach, B., Wagner, A., May, M., Lakhina, A.: Impact of packet sampling on anomaly detection metrics. In: IMC '06: Proceedings of the 6th ACM SIGCOMM on Internet measurement, pp. 159–164. ACM Press, New York (2006)

A MultiAgent System for Physically Based Rendering Optimization

Carlos Gonzalez-Morcillo[1,2], Gerhard Weiss[2], Luis Jimenez[1],
David Vallejo[1], and Javier Albusac[1]

[1] Escuela Superior de Informatica,
University of Castilla-La Mancha, Spain
[2] Software Competence Center GmbH
Hagenberg, Austria
Carlos.Gonzalez@uclm.es, Gerhard.Weiss@scch.at, Luis.Jimenezo@uclm.es,
David.Vallejo@uclm.es, JavierAlonso.Albusac@uclm.es
http://oreto.inf-cr.ulcm.es, http://www.scch.at

Abstract. Physically based rendering is the process of generating a 2D image from the abstract description of a 3D Scene. Despite the development of various new techniques and algorithms, the computational requirements of generating photorealistic images still do not allow to render in real time. Moreover, the configuration of good render quality parameters is very difficult and often too complex to be done by non-expert users. This paper describes a novel approach called *MAgarRO* (standing for *"Multi-Agent AppRoach to Rendering Optimization"*) which utilizes principles and techniques known from the field of multi-agent systems to optimize the rendering process. Experimental results are presented which show the benefits of *MAgarRO* -based rendering optimization.

Keywords: MultiAgent, Rendering, Global Illumination, Optimization.

1 Introduction

The process of constructing an image from a 3D model comprises several phases such as modelling, setting materials and textures, placing the virtual light sources, and finally rendering. Rendering algorithms take a description of geometry, materials, textures, light sources and virtual cameras as input and produce an image or a sequence of images (in the case of an animation) as output. There are different rendering algorithms – ranging from simple and fast to more complex and accurate ones – which simulate the light behavior in a precise way. Such methods are normally classified in two main categories, namely, local and global illumination algorithms. High-quality photorealistic rendering of complex scenes is one of the key goals and challenges of computer graphics. Unfortunately this process is computationally intensive and may require a huge amount of time in some cases (especially when global illumination algorithms are used), and the generation of a single high quality image may take several hours up to several days, even on fast computers. As pointed out by Kajiya [1], all rendering algorithms aim to model the light behavior over various types of surfaces and try to

M. Klusch et al. (Eds.): CIA 2007, LNAI 4676, pp. 149–163, 2007.

solve the so-called rendering equation (which forms the mathematical basis for all rendering algorithms). Because of the huge amount of time it requires, *the rendering phase is often considered to be a crucial bottleneck in photorealistic projects.* In addition, the selection of the input parameters and variable values of the scene (number of samples per light, depth limit in ray tracing, etc.) is very complex. Typically a user of a 3D rendering engine tends to "over-optimize", that is, to choose values that increase the required rendering time considerably without affecting the perceptual quality of the resulting image.

This paper describes a novel optimization approach called *MAgarRO* based on principles, techniques and concepts known from the area of multi-agent systems. Specifically, *MAgarRO* is based on design principles of the FIPA standards (http://www.fipa.org), employs adaptation and auctioning, and utilizes expert knowledge. The key advantages of this approach are robustness, flexibility, scalability, decentralized control (autonomy of the involved agents), and the capacity to optimize locally.

The paper is structured as follows. The following section overviews the state of the art and the current main research lines in rendering optimization. Thereby the focus is on the most promising issues related to parallel and distributed rendering. This section also surveys approaches which aim at applying Artificial Intelligence methods to rendering optimization and, more specifically, it points to related work on rendering based on multi-agent technology. The next section describes *MAgarRO* in detail. Then, in the next section empirical results are shown that have been obtained for different numbers of agents and input variables. The final section offers a careful discussion and concluding remarks.

2   Related Work

There are a lot of rendering methods and algorithms with different characteristics and properties (e.g., [13,1,12]). Common to these algorithms is that different levels of realism of the rendering are always related in one way or another to the complexity and computation time required. Consequently, a key problem in realistic computer graphics is the time required for rendering due to the computational complexity of the related algorithms. Chalmers et al. [3] expose various research lines in the rendering optimization issues.

Optimization via Hardware. Some researchers use programmable GPUs (Graphics Processing Units) as massively parallel, powerful streaming processors than run specialized portions of code of a raytracer [6]. Other approaches are based on special-purpose hardware architectures which are designed to achieve maximum performance in a specific task [14]. These hardware-based approaches are very effective and even the costs are low if manufactured in large scale. The main problem is the lack of generality: the algorithms need to be designed specifically for each hardware architecture. Against that, *MAgarRO* works at a very high level of abstraction and runs on almost any existing rendering engine without changes.

Optimization using parallel/distributed computing. If the rendering task is divided into a number of smaller tasks (each of which solved on a separate processor), the time required to solve the full task may be reduced significantly. In order to have all processing elements fully utilized, a task scheduling strategy must be chosen. This task constitutes the elemental unit of computation of the parallel implementation approach [3], and its output is the application of the algorithm to a specified data item. There are many related approaches such as [4] which use Grid systems for rendering over the Internet. Compared to these parallel-computing approaches, *MAgarRO* uses dynamical in combination with non-centralized load balancing, and this makes *MAgarRO* more efficient especially when the number of nodes (thus the coordination overhead) increases.

Knowledge about the cost distribution across the scene (i.e., across the different parts of a partitioned scene) can significantly aid the allocation of resources when using a distributed approach to rendering. In fact, an estimated cost distribution is crucial in commercial rendering applications which otherwise could not be realized due to their enormous complexity. There are many approaches based on knowledge about cost distribution; a good example is [8].

Distributed Multi-Agent Optimization. The inherent distribution of multi-agent systems and their properties of intelligent interaction allow for an alternative view of rendering optimization. The work presented by Rangel-Kuoppa et al. [7] uses a JADE-based implementation of a multi-agent platform to distribute interactive rendering tasks (rasterization) across a network. The distribution of the tasks is realized in a centralized client-server style (the agents send the results of the rasterization of objects to a centralized server). Although this work employs the multi-agent metaphor, essentially it does not make use of multi-agent technology itself. In fact, the use of the JADE framework is only for the purpose of realizing communication between nodes, but this communication is not knowledge-driven and no "agent-typical" mechanism such as learning and negotiation is used.

The work on stroke-based rendering (a special method of Non Realistic Rendering) proposed by Schlechtweg et al. [9] makes use of a multi-agent system for rendering artistic styles such as stippling and hatching. The environment of the agents consists of a source image and a collection of buffers. Each agent represents one stroke and executes his painting function in the environment.

2.1 Comparison to *MAgarRO*

MAgarRO, the multi-agent approach to rendering proposed in this paper, significantly differs from all related work on rendering in its unique combination of the following key features:

- **Decentralized control.** *MAgarRO* realizes rendering in a decentralized way through a group of agents coordinated by a master, where the group can be formed dynamically and most services can be easily replicated. (As regards decentralized control, *MAgarRO* follows the principle of volunteer computing [2].)

- **Higher level of abstraction.** While other approaches typically realize parallel optimization at a low level of abstraction that is specific to a particular rendering method, *MAgarRO* works with *any* rendering method. All that is required by *MAgarRO* are the input parameters of the render engine to be used.
- **Use of expert knowledge.** *MAgarRO* employs Fuzzy Set Rules and their descriptive power [11] in order to enable easy modelling of expert knowledge about rendering and the rendering process.
- **Local optimization.** Each agent involved in rendering can employ different models of expert knowledge. In this way, and by using a fine-grained decomposition approach, *MAgarRO* allows for local optimization of each rendering (sub-)task.

In combining these features, *MAgarRO* exploits and integrates some ideas from related approaches to parallel rendering. For instance, *MAgarRO* 's cost prediction map (called *Importance Map* and described below) combines prediction principles described in [5,9] with elements of Volunteer Computing as proposed in [2] and demand driven auctioning known from agent-based task allocation.

3   The *MAgarRO* Approach

MAgarRO is a system which gets a 3D Scene as input and produces a resulting 2D image. From the point of view of the user the system works in the same way as local render engines do, but the rendering in fact is made by different agents spread over the Internet.

MAgarRO uses the ICE middleware (http://www.zeroc.com). The location service IceGrid is used to indicate in which computers the services reside. Glacier2 is used to solve the difficulties related with hostile network environments, making available agents connected through a router and a firewall.

The overall architecture of *MAgarRO* is based on the design principles of the FIPA standard. In figure 1 the general class diagram for the architecture is shown. There are some top-level general services (*Start Service, Agent Management System, Directory Facilitator* and *Agent Communication Channel*) available to all involved agents. On start-up, an agent is provided with a root service that describes or points to all the services available in the environment.

3.1   Architectural Overview

In *MAgarRO* a *StartService* offering two operations is available: an operation called *getServiceRoot* to obtain directly a description of all basic services, and another operation called *supplyBasicService* that enables the registration of a new basic service in the system. In accordance with FIPA, a basic service is defined by a unique Service Identifier (string in our case), at least one list of Transport Addresses and a description of Service Type.

The *Agent Management System* (AMS) is a general service that manages the events that occurs on the platform. This service also includes a naming service

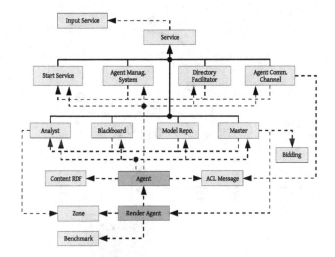

Fig. 1. *MAgarRO* general Class diagram

for *White Pages* which allow agents to find one another. The basic functionality of the AMS is to register, to modify a subscription, to unregister agents, and to search for agents.

A basic service called *Directory Facilitator* (DF) provides *Yellow Pages* for the agents. As suggested by the FIPA standard, the operations of this service are related to the services provided by an agent, the interaction protocols, the ontologies, the content languages used, the maximum live time of registration and visibility of the agent description in DF.

Finally, *MAgarRO* includes a basic service called *Agent Communication Channel* that receives and sends messages between agents. In *MAgarRO* only the receive functionality is employed, because the send operation is implemented as part of the agents. In accordance to FIPA standard, the data structure of each message is composed of two parts: the content of the message and the envelope (with information about the receiver and the sender of the message). The Agent Communication Language used in *MAgarRO* is based on XML and uses DTD as specified in the FIPA standard.

Each agent must implement a basic set of standard operations to be able to run in the *MAgarRO* environment. This operations are suspend, terminate, resume and receive a message.

In addition to the basic FIPA services described above, *MAgarRO* includes specific services related to Rendering Optimization. Specifically, a service called *Analyst* studies the scene in order to enable the division of the rendering task. A blackboard is used to represent some aspects of the common environment of the agents. The environmental models processed by the agents are managed by the *Model Repository Service*. Finally, a master service called (*Master*) handles dynamic groups of agents who cooperate by fulfilling subtasks. The Figure 2 illustrates this.

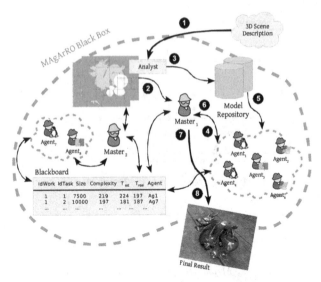

Fig. 2. General workflow and main architectural roles

Figure 2 also illustrates the basic workflow in *MAgarRO* (the circled numbers in this figure represent the following steps). **1** – The first step is the subscription of the agents to the system. This subscription can be done at any moment; the available agents are managed dynamically. When the system receives a new file to be rendered, it is delivered to the Analyst service. **2** – The Analyst analyzes the scene, making some partitions of the work and extracting a set of tasks. **3** – The Master is notified about the new scene which is sent to the Model Repository. **4** – Some of the agents available at this moment are managed by the Master and notified about the new scene. **5** – Each agent obtains the 3D model from the repository and an auction is started. **6** – The (sub-)tasks are executed by the agents and the results are sent to the Master. **7** – The final result is composed by the Master using the output of the tasks previously done. **8** – The Master sends the rendered image to the user. Key issues of this workflow are described in the following.

3.2   Agent Subscription

As shown in Figure 1, a *Render Agent* is a specialization of a standard *Agent*, so all the functionality and requirements related with FIPA are inherited in his implementation. There are two actions that could be done by an agent: to add a subscription or to unsubscribe from a group of rendering agents. These operations are related to a Master agent. The subscribe operation requires two parameters: the name of the agent and the *proxy* that represents the agent as a client. Using this proxy the Master could execute remote operations on the agent side. An agent can drop the association with a specific master by means of an unsubscribe operation.

The first time an agent subscribes to the system, he runs a benchmark to obtain an initial estimation of his computing capabilities. This initial value is adapted during rendering in order to obtain a more accurate prediction.

3.3 Analysis of the Scene Based on Importance Maps

MAgarRO employs the idea to estimate the complexity of the different tasks in order to achieve load-balanced partitining. Complexity analysis is done by the Analyst agent prior to (and independent of) all other rendering steps.

Fig. 3. Importance maps. *Left:* Blind partitioning (First Level). *Center:* Join zones with similar complexity (Second Level). *Right:* Balancing complexity/size ratio (Third Level).

The main objective in this partitioning process is to obtain tasks with similar complexity to avoid the delay in the final time caused by too complex tasks. This analysis may be done in a fast way independently of the final render process.

At the beginning, the Analyst makes a fast rasterization of the scene using an importance function to obtain a grey scale image. In this image (called *Importance Map*) the dark zones represents less complex areas and the white zones the more complex areas. In our current implementation a simple function is used (it only takes in account the recursion levels in mirror and transparent surfaces). As it is shown in Figure 3, the glass is more complex than the dragon because it has a higher number of ray interactions. The table is less complex because it does not have any of these properties. More advanced importance functions could be used in this grey scale image generation, using perception-based rendering algorithms (based on visual attention processes) to construct the importance map [10].

Once the importance map is generated, a partition is constructed to obtain a final set of tasks. These partitions are formed hierarchically at different levels, where at each level the partitioning results obtained at the previous level are used. At the first level, the partition is made taking care of the minimum size and the maximum complexity of each zone. With these two parameters, the *Analyst* makes a recursive division of the zones (see Figure 3). At the second level, neighbor zones with similar complexity are joined. Finally, at the third level the *Analyst* tries to obtain a balanced division where each zone has nearly the same complexity/size ratio. The idea behind this division is to obtain tasks that all require roughly the same rendering time. As shown below in the experimental results, the quality of this partitioning is highly correlated to the final rendering time.

3.4   Rendering Process

Once the scene is available in the *Model Repository*, the *Master* assigns agents to the individual tasks identified by the Analyst. These agents, in turn, apply in parallel a technique called profiling in order to get a more accurate estimation of the complexity of each task.[1] Specifically, the agents make a low resolution render (around 5% of the final number of rays) of each task and announce on the *Blackboard* the estimated time required to do the final rendering on the blackboard.

Blackboard service. The blackboard used by the agents to share their knowledge about the rendering task is implemented as a service offering a set of read and write operations. The basic blackboard data structure (as shown in Figure 2) has 7 fields labelled as follows. *IdWork* is a unique identifier for each scene, *IdZone* is a unique identifier for each task of each work, *Size* is the number of pixels of each task (width x height), *Complexity* is the estimated complexity of this task, calculated by means of the importance map, T_{est} is the Estimated Time calculated through profiling by the agents, T_{real} is the actual time required to finish a task, and *Agent* is the name of the agent who is responsible for rendering this task.

Adaptation. As said above, the estimated time is represented in the same way by all agents. More precisely, each agent has an internal variable that represents his relative computational power (V_{cp}). For example, assume that $Vcp = 1$ is chosen as a reference value. If a task T_A requires 5 minutes to be done by a particular agent A who has $V_{cp} = 0.5$, he annotates on the blackboard that the time required to complete T_A is 10 minutes (i.e., the agent takes into account that he runs on a *relatively* fast machine). On the other hand, if another agent B running on a slow machine ($V_{cp} = 2$) announces that she needs 2 minutes to complete task T_B, then agent A can infer from this information that it would take him 30 seconds to complete T_B on his machine.

During run time, each agent adapts the value of his variable V_{cp} to obtain a more accurate estimation of the required processing time as follows. Whenever there is a difference between the estimated time T_{est} and the actual completion time T of a task, then an agent updates his internal variable V_{cp} according to

$$V_{cp} = (1 - k) \times V_{cp} + k \times (T - T_{est}) \tag{1}$$

where k a constant. Small values of k assure a smooth adaptation. (k is set to 0.1 in the experiments reported below.) This mechanisms should be improved with

---

[1] Profiling is a technique which traces a small number of rays in a global illumination solution and uses the time taken to compute this few rays to predict the overall computation time.

a more complex learning method that takes in account the historical behavior and the intrinsic characteristics of the task (type of scene, rendering method, etc...).

Auctioning. At every moment during execution, all agents who are idle take part in an auction for available tasks. It is assumed that the agents try to obtain more complex tasks first. If two or more agents bid for the same task, the Master assigns it on the basis of the so called credits of these agents. The credit of an agent represents his the success and failure w.r.t. previous tasks: an agent is said to have success w.r.t. a task if he completes it in no more time than T_{est}, and otherwise he is said to fail w.r.t. this task. The amount added to, or subtracted from, an agent's credit is proportional to the time difference w.r.t. T_{est}.

Using Expert Knowledge. When a task is assigned to an agent, a fuzzy rule set is used in order to model the expert knowledge and optimize the rendering parameters for this task. Fuzzy rule sets are known to be well suited for expert knowledge modeling due to their descriptive power and easy exensibility. The output parameters (i.e., the consequent part of the rules) are configured so that the time required to complete rendering is reduced and the loss of quality is minimized. Each agent may model (or capture) different expert knowledge with a different set of fuzzy rules. In the following, the rule set we used for Pathtracing rendering is described. The output parameters of the rules are:

- **Recursion Level** [Rl], defined over the linguistic variables [15] {VS, S, N, B, VB}[2]. This parameter defines the global recursion level in raytracing (number of light bounces).
- **Light Samples** [Ls], defined over the linguistic variables {VS, S, N, B, VB}. This parameter defines the number of samples per light in the scene. The biggest, the more quality in the scene and the higher rendering time.
- **Interpolation Band Size** [Ibs], defined over the linguistic variables {VS, S, N, B, VB}. This parameter defines the size of the interpolation band in pixels, and it is used in the final composition of the image (as we will see in the next section).

The previous parameters have a strong dependency with the rendering method chosen (in this case Pathtracing). Against that, the following parameters, which are the antecedents of the rules, can be used for other rendering methods as well.

- **Complexity** [C], defined over the linguistic variables {VS, S, N, B, VB}. This parameter represents the complexity/size ratio of the task.
- **Neighbor Difference** [Nd], defined over the linguistic variables {VS, S, N, B, VB}. This parameter represents the difference of complexity of the current task in relation to its neighbor tasks.

---

[2] The notation used for the linguistic variables is typical in some works with Fuzzy Sets. This is the correspondence of the linguistic variables: VS is *Very Small*, S is *Small*, N is *Normal*, B is *Big* and finally VB is *Very Big*.

- **Size** [S], defined over the linguistic variables {S, N, B}. This parameter represents the size of the task in pixels (calculated as width x height).
- **Optimization Level** [Op], defined over the linguistic variables {VS, S, N, B, VB}. This parameter is selected by the user, and determines the level of optimization (more or less aggressive with initial parameters indicated by the user).

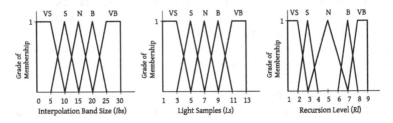

Fig. 4. Definition of the output variables

The definition of the fuzzy sets of input variables is done dynamically in that the intervals of this sets are calculated in runtime. For example, in a highly complex scene VS is higher than VB is in a simple scene. The partition of these variables is made by linear distribution. The same occurs with other parameters like Size and Neighbor difference. In the case of the Pathtracing method, the rule set is defined as follows (only two of 28 rules are shown, all rules have been designed by an expert in PathTracing):

- R_1: **If** C is {B,VB} $\wedge$ S is B,N $\wedge$ Op is VB
    then Ls is VS $\wedge$ Rl is VS
- R_{22}: **If** Nd is VB **then** Ibs is VB

The output variables have their own fuzzy sets; we use trapezoidal functions as shown in Figure 4.

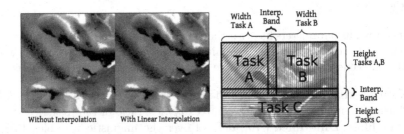

Fig. 5. Left: Without interpolation undesirable artefacts appear between neighboring parts. Linear interpolation solves this problem (at a price of slightly higher rendering costs). **Right:** Diagram of task decomposition and interpolation band situation.

3.5 Final Result Composition

With the results generated by the different agents, the *Master* composes the final image. A critical issue w.r.t. composition is that there may be slight differences (e.g., in coloring) between the neighboring parts obtained from different agents; these differences result from the random component which PathTracing contains as a Monte Carlo based method. Figure 5 (Left) illustrates this problem. For that reason, the *Master* smoothes the meeting faces of neighboring parts through a linear interpolation mask. More precisely, as shown in Figure 5 (Right) smoothing between neighboring parts is done in a zone called Interpolation Band.

In *MAgarRO* the size of the Interpolation Band is an output parameter of the rule set. In particular, the parameter gets a higher value if the difference between the quality of neighboring zones is important. To reduce rendering time, this parameter should be kept as small as possible to avoid unnecessary "'double work"' done by different agents. This is particularly important if the zone is very complex, as this also implies high costs for rendering the interpolation band. (In our applications the amount of time required by *MAgarRO* for interpolation was between 2% and 5% of the overall rendering time).

4 Experimental Results

The results reported in this section have been generated with the implementation of *MAgarRO* [3]. Moreover, these results are based on the following computer and parameter setting: eight identical computers were connected to run in parallel (Pentium Intel Centrino 2 Ghz, 1GB RAM and Debian GNU/Linux); as a rendering method Pathtracing (*Yafray* 0.0.9 render engine (http://www.yafray.org)) was used; eight oversampling levels; eight recursion levels in global configuration of raytracing; and 1024 Light samples by default. The scene to be rendered contains more than 100.000 faces, 5 levels of recursion in mirror surfaces and 6 levels in transparent surfaces (the glass). With this configuration, rendering on a single machine without any optimization took 121 minutes and 17 seconds (121:17 for short, below the amount of time needed for rendering is sometimes expressed in the format *Minutes:Seconds*).

Table 1 shows the time required using different partitioning levels. These times ha have been obtained using the N (*Normal*) optimization level (Figure 6 Left). Using a simple first-level partitioning, a good render time can be obtained with just a few agents in comparison to third-level partitioning. The time required in the third partitioning level is larger because more partitions in areas having higher complexity (i.e., in the glasses) are needed. This higher partition level requires the use of interpolation bands and as an effect some complex parts of the image are rendered twice. For example, the rendering time with one agent is *105* minutes in the third level and *93* minutes in first level. However, when the number of agents grow, the overall performance of the system increases because

---

[3] Download at (http://code.google.com/p/masyro06/) under GPL Free Software License.

Table 1. Different partitioning with *Normal* optimization level

Agents	1st Level	2nd Level	3rd Level
1	92:46	82:36	105:02
2	47:21	41:13	52:41
4	26:23	23:33	26:32
8	26:25	23:31	16:52

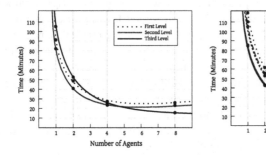

Fig. 6. Left: First/second/third level of partitioning with the N (*Normal*) optimization level. **Right:** Different optimization levels (all with third level of partitioning).

Table 2. Third level of partitioning with different Number of Agents and level of optimization

Agents	VS	S	N	B	VB
1	125:02	110:50	105:02	85:50	85:06
2	62:36	55:54	52:41	42:55	42:40
4	31:10	27:11	26:32	22:50	22:40
8	23:43	20:54	16:52	16:18	15:58

the differences in the complexity of the tasks are relatively small. In first- and second-level partitioning, there are complex tasks that slow down the whole rendering process even if the number of agents is increased (the time required with four or eight agents is essentially the same). On the other hand, the third partitioning level works better with a higher number of agents.

Table 2 shows the time required to render the scene using different levels of optimization but always third-level partitioning (Figure 6 Right). By simply using a *Small* level of optimization results are obtained that are better than the results for rendering without optimization. The time required with *Very Small* optimization is exceeds the time to render the original scene. This is because additional time is required for communication and composition.

Excellent results are also obtained when only four agents are used. For instance, in the case of Normal level optimization, the time required to render the scene is just about 26 minutes (whereas the original rendering time is 120 minutes). Figure 7 shows the results of rendering obtained for different configurations.

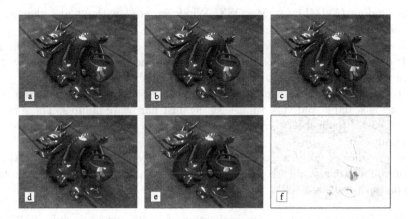

Fig. 7. Result of the rendering using different optimization levels. (a) No optimization and render in one machine. (b) *Very Small* (c) *Small* (d) *Normal* (e) *Very Big* (f) Difference between *(a)* and *(e)* (the lighter colour, the smaller difference).

As a final remark, note that optimization may result in different quality levels for different areas of the overall scene. This is because more aggressive optimization levels (i.e., Big or Very Big) may result in a loss of details. For example, in Figure 7.e, the reflections on the glass are not so detailed as in Figure 7.a.

The difference between the optimal render and the most aggresive optimization level (Figure 7.f) is minimal[4].

5   Discussion and Conclusion

The media industry is demanding high fidelity images for their 3D scenes. The computational requirements of full global illumination are such that it is practically impossible to achieve this kind of rendering in reasonable time on a single computer. *MAgarRO* has been developed in response to this challenge.

The experimental results show that *MAgarRO* achieves excellent optimization results. The use of the importance map assures an initial time estimation that minimizes the latency of the latest task. In particular, as a result of optimization *MAgarRO* achieves overall rendering times that are below the time required by one CPU divided by the number of agents. Most important, *MAgarRO* is a novel multi-agent rendering approach that offers several desirable features which together make it unique and of highest practical value. In particular:

- It is FIPA-compliant.
- Due to the use of ICE Grid middleware layer, *MAgarRO* can be used in heterogeneous hardware platforms and under different operating systems

---

[4] In this example, the difference between the optimal render (Figure 7.a) and the image obtained with the *Very Big* optimization level (see Figure 7.e) is 4.6% including the implicit noise typical to monte carlo methods (around 1.2% in this example).

(including GNU/Linux, MacOSX, Windows, etc.) without any changes in the implementation.

- It enables importance-driven rendering through its use of importance maps.
- It employs effective auctioning and parameter adaptation, and it allows the application of expert knowledge in form of flexible fuzzy rules.
- It applies the principles of decentralized control and local optimization, and thus is scalable and very robust e.g. against hardware failures. The services are easily replicable, thus possible bottlenecks in the final deploy can be minimized.
- In presence of failures (e.g. an agent does not complete an assigned task or obtain a wrong result), it is easy to apply the typical techniques of volunteer computing systems [2].

MAgarRO is pioneering in its application of multi-agent principles to 3D realistic rendering optimization. This opens several interesting research avenues. Specifically, we think it is very promising to extend *MAgarRO* toward more sophisticated adaptation and machine learning techniques. The open technology used in the development of *MAgarRO* allow the agents to reside and run on different machines around the world. Such a Grid version is a very effective answer to the challenge of handling the enormous complexity of real-world rendering applications.

In our current work, we concentrate on two research lines. First, the combination of different rendering techniques within the *MAgarRO* framework. Due to the high abstraction level of *MAgarRO*, in princple different render engines can be combined to jointly generate an image, using complex techniques only if needed. Second, we are exploring the possibilities to equip *MAgarRO* with agent-agent real-time coordination schemes that are more flexible than auctioning.

Acknowledgments

This work has been funded by the Junta de Comunidades de Castilla-La Mancha under Research Project PBC06-0064.

References

1. Kajiya, J.T.: The Rendering Equation. Computer Graphics. Proceedings of SIG-GRAPH 20(4), 143–150 (1986)
2. Anderson, D.P., Fedak, G.: The computational and storage potential of volunteer computing. In: Sixth IEEE International Symposium on Cluster Computing and the Grid (CCGRID'06), pp. 73–80. IEEE Computer Society Press, Los Alamitos (2006)
3. Chalmers, A., Davis, T., Reinhard, E.: Practical Parallel Rendering. In: Peters, A.K. (ed.) (2002)
4. Fernandez-Sorribes, J.A., Gonzalez-Morcillo, C., Jimenez-Linares, L.: Grid architecture for distributed rendering. Proc. SIACG '06, 141–148 (2006)

5. Gillibrand, R., Debattista, K., Chalmers, A.: Cost prediction maps for global illumination. In: Proceedings of Theory and Practice of Computer Graphics, pp. 97–104 (2005)
6. Hachisuka, T.: GPU Gems 2: Programming Techniques for High Performance Graphics and General-Purpose Computa tion. Addison-Wesley Professional, London, UK (2005)
7. Kuoppa, R.R., Cruz, C.A., Mould, D.: Distributed 3D rendering system in a multi-agent platform. In: Proc. ENC'03, vol. 8 (2003)
8. Reinhard, E., Kok, A.J.F., Jansen, F.W.: Cost Prediction in Ray Tracing. In: Proceedings of the Seventh Eurographics Workshop on Rendering, pp. 41–50 (1996)
9. Schlechtweg, S., Germer, T., Strothotte, T.: Renderbots multiagent systems for direct image generation. Computer Graphics Forum 24, 137–148 (2005)
10. Sundstedt, V., Debattista, K., Longhurst, P., Chalmers, A.: Visual attention for efficient high-fidelity graphics. In: Spring Conference on Computer Graphics (SCCG 2005), pp. 162–168 (2005)
11. Tanaka, K.: An introduction to fuzzy logic for practical applications. Springer, Heidelberg (1998)
12. Veach, E., Guibas, L.J: Metropolis light transport. In: SIGGRAPH '97, pp. 65–76. New York (1997)
13. Whitted, T.: An improved illumination model for shaded display. In: SIGGRAPH '79, vol. 14, New York (1979)
14. Woop, S., Schmittler, J., Slusallek, P.: Rpu: a programmable ray processing unit for realtime ray tracing. In: ACM SIGGRAPH, pp. 434–444. New York (2005)
15. Zadeh, L.A.: The concept of a linguistic variable and its applications to approximate reasoning. part i, ii, iii. Information Science (1975)

Neural Network Based Multiagent System for Simulation of Investing Strategies

Darius Plikynas[1] and Rimvydas Aleksiejūnas[2]

[1] Informatics department, Vilnius Management Academy,
Basanaviciaus 29a, Vilnius, Lithuania
dariusplikynas@vva.lt
[2] Informatics department, Vilnius Management Academy,
Basanaviciaus 29a, Vilnius, Lithuania
a_rimvydas@yahoo.com

Abstract. Recent years of empirical research have collected enough evidences that for efficient markets the process of lower-wealth accumulation by capital investment is approximated by log-normal and high-wealth range by Pareto wealth distribution. This research aims to construct a simple neural network (NN) based multiagent system of heterogeneous agents' targeted to get on the efficiency frontier by combining investments to the real life index funds and nonrisky financial assets, diversifying the risk and maximizing the profits. Each agent is represented by the different stock trading strategy according to his portfolio, saving and risk aversion preferences. The goal is, following empirical evidences from the real investment markets, to find enough proofs that NN-based multiagent system, in principle, has the same fundamental properties of real investment markets described by the log-normal, Pareto wealth and Levy stock returns distributions and can be used further to simulate even more complex social phenomena.

Keywords: Social Agents, Neural Networks, Investment Strategies.

1 Introduction

One of the most fruitful trends in empirical macroeconomics over the last fifteen years has been the effort to construct rigorous micro foundations for macroeconomic models. Broadly speaking the goal was to find empirically sensible models for the behavior of individual agents (people, firms, banks), which can then be aggregated to derive implications about macroeconomic dynamics. Separately, but in the similar spirit, researches at the Santa Fe Institute, the CESD, and elsewhere have been exploring "agent-based" models that examine the complex behavior that can sometimes emerge from the interactions between collections of simple agents [1].

One of the questions is how to describe outliers, phenomena that lie outside of patterns of statistical regularity. Collaborative work joining economists and physicist has begun to lead to modest progress in answering questions of interest to both economists and physicists. Following statistical physics, it is becoming clear that almost any system comprised of the large number of interacting units has the potential of displaying power-law behavior. Since economic systems are comprised of a large

M. Klusch et al. (Eds.): CIA 2007, LNAI 4676, pp. 164–180, 2007.
© Springer-Verlag Berlin Heidelberg 2007

number of interacting units, it is perhaps not unreasonable to examine economic phenomena within the conceptual framework of scaling and universality [2].

Recent attempts to make models that reproduce the empirical scaling relationships suggest that significant progress on understanding firm growth may be well underway, leading to the hope of ultimately developing a clear and coherent "theory of the firm". One utility of the recent empirical work is that now any acceptable theory must respect the fact that power laws hold over typically six orders of magnitude. As Axtell put the matter rather graphically: "the power-law distribution is an unambiguous target that any empirically accurate theory of the firm must hit [1]. If one takes seriously power laws that appear to characterize financial and economic fluctuations, then he will have to face question what sort of understanding could eventually develop out of it.

Recently, it has come to be appreciated that many such systems consisting of a large number of interacting subunits obey universal laws that are independent of the microscopic details. The finding, in physical systems, of universal properties that do not depend on the specific form of the interactions gives rise to the intriguing hypothesis that universal laws or results may also be present in economic and social systems. If so, then our simplified social model of interacting agents should exhibit the same laws too.

Recent extensive analysis of short-term returns has shown that price differences, log-returns, and rates of return are described extremely well by a truncated Levy distribution [1]. In this respect, proposed research is potentially significant, since it provides a simulation framework within which empirical laws like power-law and Levy distribution can be observed and studied within the parameters space of the simulated model. Specifically, the model fulfils these requirements for such a basic microscopic model of the stock market. It is founded on the basic realistic features of the stock market, and reflects the view that market participants (agents) have of the functioning of the market, as well as the main determinants of their trading behavior[3].

Following rationality assumption, our agent is a subjective expected utility maximizer. We hold this assumption true than designing our model. The level of savings, the portfolio and risk aversion that agents choose may be affected by these individual factors. One way to think of this is to think of agents who have the same objective but different expectations, and, therefore, hold different portfolios. This will cause wealth to flow between the agents over time [1].

When agents have identical discount factors, those who survive best in terms of owned assets are those whose forecasts are closest in relative entropy to the truth. The agent with rational expectations survives, and most of agents who do not have rational expectations vanish in terms of possessed economic wealth. In the economy in which agents have heterogeneous beliefs or heterogeneous learning rules, structure on long-run prices arises from the forces of market selection. It has long been assumed that those with better beliefs will make better decisions, driving out those with worse beliefs, and thus determining long run assets prices and their own welfare.

Bearing in mind financial implications for rational agents, a number of theoretical and empirical studies have suggested that efficient market hypothesis is true at least in a weak form [4]. A natural outcome of a belief in efficient market is to employ some type of passive strategy in owning and managing common stocks. If the market is totally efficient, no active strategy should be able to beat the market in a risk-adjusted basis [4]. Passive strategies do not seek to outperform the market but simply to do as well as the market. The emphasis is on minimizing transaction costs and time spent in managing the portfolio because any expected benefits from active trading or analysis are likely to be less than the costs[1].

The discussion above shows us that there are different approaches, often contradictory to each other, which are employed by investors as different investment strategies. In our approach, we hold that a very effective way to employ a passive strategy with common stocks is to invest in an indexed portfolio[2]. An increasing amount of mutual fund and pensions fund assets can be described as passive equity investments. Using index funds[3], these asset pools are designed to duplicate as precisely as possible the performance of some market index.

In our model, we give to investors even more freedom as they are free to choose their portfolios composed from different index funds. In this way, they achieve higher diversification and individual stocks related risk aversion, which is based on the weighted performance of the portfolio of indices. Therefore, the proposed approach is semi-passive as our agents invest to the whole market having ability to switch from one index fund to another if better profit opportunities are forecasted.

Regarding the methodical aspect of our novel technique (see next sections), reinforcement learning seems to be close match as it is a sub-area of machine learning concerned with how an agent ought to take actions in an environment so as to maximize some notion of long-term reward. Reinforcement learning algorithms attempt to find a policy that maps states of the world to the actions the agent ought to take in those states.

This is exactly what our supervised learning approach based on NN learning and simulating is aimed to do, but it differs essentially from the reinforcement learning algorithms (like Q-learning) in a way how correct input/output pairs are presented and sub-optimal actions are explicitly corrected.

The paper is organized as follows. Section 2, below, introduces with research framework, i.e. here we formulate the major assumptions and model setup for the multiagent system of agents'. Section 3 introduces with the experimental setup. Section 4 offers an overview of operational procedures, i.e. how agent operates in the investment environment. Section 5 summarizes the findings and discusses the further implications towards the more complicated system of interacting social agents giving future direction for the related research.

---

[1] Evidence to support this view comes from a study by Odean and Barber, who examined 60000 investors. They found the average investor earned 15.3 percent over the period 1991-1996 whereas the most active traders (turning over about 10 percent of their holdings each month) averaged only 10 percent [4]. An investor can simply follow a buy-and-hold strategy for whatever portfolio of stocks is owned.

[2] Index funds arose in response to the large body of evidence concerning the efficiency of the market, and they have grown as evidence of inability of mutual funds to consistently, or even very often, outperform the market continues to accumulate. The available evidence indicates that many investment companies have failed to match the performance of broad market indexes. For example, for the period 1986-1995, 78 percent of general equity mutual funds were outperformed by the S&P500 Index, and 68 percent of international stock funds were outperformed by their comparative index.

[3] The largest selection of index funds in the industry are offered by Vanguard Group of Investment Companies, e.g. Index Trust 500 Portfolio (had over $100 billion in assets in 2001, placing it in the top two mutual funds in terms of assets in US), Extended Market Portfolio, Total Stock Market Portfolio, the Value Portfolio etc.

2 Research Framework

In our model, each investor is represented as an agent, where the total number of agents varies from 100 to 1000. We assume that the whole market is sufficiently big, i.e. the total number of agents' N>>1000, where our subsystem of agents composes just a small fraction in the whole market. Therefore, it doesn't significantly influence the overall market behavior. In fact, at this stage we want to create large enough subsystem of investing artificial agents, but we don't want it to disturb the real market behavior and be accounted for the feedback estimates.

The empirical study described below seeks to create self-organizing system of investing agents. Following well-established definition [9], investors simulated as NN-based economic agents have properties of distributed/concurrent systems resembling computational information processing entities, i.e. computational agents[4]. We are modeling a reactive system that cannot adequately be described by the relational or functional view. Our modeled system of economic agents maintains one-way interaction with environment (financial market), and therefore must be described and specified in terms of their on-going behavior. This is because each individual module in a concurrent system is a reactive subsystem, interacting with its own environment which consists of other modules [11].

Agents technically are modeled using multilayer perceptron [MLP] backpropagation[5] neural networks [NN] having fast Levenberg-Marquardt [LM] learning algorithm[6], using Matlab NN Toolbox and Simulink simulation environment. All real market data is preprocessed (missing values eliminated, data sets time adjusted etc), but not detrended (using e.g. nonlinear detrending techniques like Hodrick-Prescot [HP]) as we want all periodic and nonperiodic fluctuation be present for NNs to learn. As a standard procedure, each NN goes through the initial training and validation stage. Only then they are ready for the simulation stage, i.e. forecasting. NN learns to forecast the market behavior (index fluctuations, trade volume, market timing moments[7], specific patterns described by technical analyses etc).

In fact, each investment asset is analyzed by the separate NN. Meaning that an agent is a modular entity composed by a number of different NNs, where different NNs are dealing with different market data analyses and forecasting, and one more

---

[4] An agent is a computer system that is situated in some environment, and that is capabale of autonomous action in this environment in order to meet its design objectives [10].

[5] Properly traines backpropagation networks tend to give reasonable answers when presented with inputs that they have never seen. Typically, a new input leads to an output similar to the correct output for input vectors used in training that are similar to the new input being presented. A network can be trained for function approximation (nonlinear regression), pattern association, or pattern classification. The training process requires a set of examples of proper network behavior - network inputs p and target outputs t. During training the weights and biases of the network are iteratively adjusted to minimize the mean square error mse - the average squared error between the network outputs a and the target outputs t.

[6] Like the quasi-Newton methods, the Levenberg-Marquardt algorithm was designed to approach second-order training speed without having to compute the Hessian matrix.

[7] Some of our agents attempt to earn excess returns by varying the percentage of portfolio assets in equity securities. One has to observe a chart of stock prices over time to appreciate the profit potential of being in the index fund at the right times and being out of the index fund at the bad times.

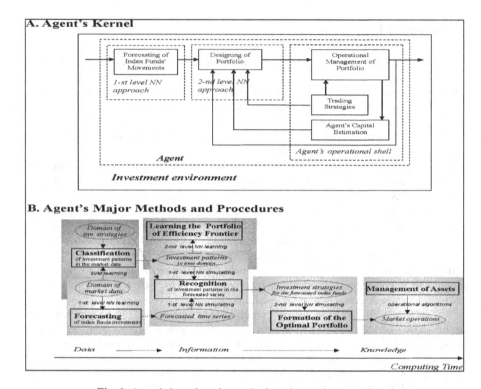

Fig. 1. Agent's kernel, major methods and procedures employed

NN is responsible for the formation of portfolio and operational decisions (see Fig. 3). Whereas, management of agent's actions and capital in the market is further conducted by the algorithmic shell of procedures (see Fig. 1 and 4).

As a result, for the strategy S7 we need two learning phases. Why not take 1 NN divided over several layers and components, where the layers and components take care of the frontline, and the end-layers take care of the combination? The answer: we would loose application-free flexibility, modularity and functionality as it would create cumbersome neural network with a very poor performance. However, we also shed light on the fact that proposed modular structure gives us very promising opportunities to implement information exchange between net of agents in the next research stage. We will only need to add another NN for the information interchange and interpretation. In this way, we simulate intelligence following numerous examples from the natural nets of neural networks reported from the neuroscience [8].

Hence, profitable decision making is not solely based on the 1-st level NN performance alone, see Fig.1. Therefore, an agent has the 2-nd NN level too, which is designed for portfolio weights estimations and for assistance to the procedural shell, which maintains market interventions in operational sense, feedback, and agent's capital estimates.

As it is shown in Fig. 1, our agent in consequent stages is learning and later simulating some procedures based on NN and algorithmic methods, which are single out in the Table 1. Consistency of applied methods plays a key role. The logic is based on a very natural perception of how real agents make decisions. First, they learn how to discriminate the environment. Afterwards, they make some forecasting and, according to the projected future, they try to recognize the future states of the environment for making appropriate decisions to meet those coming perspectives (strategic decisions).

Table 1. Applied methods and involved procedures in consecutive time steps

Time Stages	1	2	3	4	5
Methods	Discrimination of trends*	NN1 learning	NN1 simulating NN2 learning	NN2 simulating	Oper. algorithms
Procedures	Classification	Forecasting	Recognition; Learning the Eff. Frontier	Optimal Portfolio Formation	Management of Assets

* we employ a set of technical indicators (see Table 3) and with the use of different algorithmic approaches recognize and discriminate different investment triggers present in the time series; combinations of those triggers compose investment strategies, which are used by agents' for investing decisions.

Modularity of an agent efficiently helps to represent investors as having the same goal (to minimize forecasting error and gain the maximum of investment capital having lowest risk by placing timely bids/asks[8] in the market represented by the real SE data) but with different expectations (different strategies how to achieve it).

At this stage, we are not using any kind of genetic algorithms for selection of the best agents, nor social field approaches for information and energy [Plikynas, Kyoto] distribution. We simplify our research to the simplest framework. So we could start from the scratch and further develop the system for more sophisticated tasks.

The market is represented here using real data set from two index funds, namely, Index Trust 500 Portfolio, Extended Market Portfolio, and US Treasury Bond (value, open/close values, trade volumes, volatility etc). Each agent, according to his investment strategy[9] (see below), is optimized (trained by the NN, see above) for the best possible forecasting of index fund movements in order to place profitable and risk adjusted decisions. His primary goal is to find optimal risk-profit on the efficient frontier according to Markowitz theory [5].

Let's talk about agents' investment strategies now. Having formed capital market expectations (i.e. than NN-based agent, according to historical data, generates his own forecasting), we next construct the portfolio based on the CAPM policy statements[10].

---

[8] Transaction costs (ranging 0.1-1% of stock value) as a cost function is regulating an agent from too frequent market interventions.

[9] Heterogeneity of investing strategies is achieved having different NN topology, learning algorithms, other NN parameters and wrapped software parameters etc.

[10] Investment policy statement according to CAPM (Capital Asset Pricing Model derived by Sharpe, Lintner and Mossin) is concerned with the equilibrium relationship between the risk and the expected return on risky assets. Included here are such issues as asset allocation, portfolio optimization, and security (or fund) selection.

Because of the complexity of the real world, following CAPM, additional assumptions are made to make agents more alike: 1. All agents have identical probability distributions for future rates of return; they have homogeneous expectations with respect to the three inputs of the portfolio model, i.e. expected returns, the variance of returns, and the correlation matrix. Therefore, given a set of index values and a risk-free rate, all agents use the same information to generate an efficient frontier; 2. All agents have the same one-period time horizon; 3. There are no or small transaction costs; 4. There are no personal income taxes – agents are indifferent between capital gains and dividends.

The procedure of building of a portfolio of financial assets is following certain steps. Specifically it identifies optimal risk-return combinations available from the set of index funds and nonrisky assets. We employ Markowitz efficient frontier analysis. Each agent is assumed to diversify his portfolio according to the Markowitz model, choosing a location on the efficient frontier that matches his return-risk preferences. This step uses the expected returns, variances and covariance for chosen portfolio of financial assets.

Markowitz portfolio theory is normative, meaning that it tells agents how they should act to diversify optimally. Markowitz approach to portfolio selection is that an agent should evaluate portfolios on the basis of their expected returns and risk as measured by the standard deviation [5].

Approximately half of the riskness of the average stock can be eliminated by holding a well-diversified portfolio. This means that part of the risk of the average stock can be eliminated and part cannot. In our case, though, agents are investing not to the stocks but to the indexes themselves. It means that we are not dealing with the stocks related risk but with the whole market's systemic risk. To reduce it we further diversifying risk associated with different markets (represented by different indexes).

In the proposed model, each agent is forming his portfolio composed from the three investment options (two index funds and Treasury bond, see above).

Each agent enters investment market with the same initial capital K_0, which is further allocated according to individual investment preferences. In short, each agent has to perform (i) stock portfolio selection in order to diversify his investment risk, which is equally important as return maximization [5]; (ii) to distribute available capital so as to get on the efficiency frontier line (maximum return for the chosen risk level, or minimum risk level for the chosen return rate). Agent utilizes optimization procedure; (iii) to choose the final portfolio (i.e. determine the proper portfolio weights), consisting of the risk-free asset and the optimal portfolio of risky assets, based on an investor's (agent's) preferences.

The first step is a priori determined in the model, i.e. there are only three different investment options: two investment funds and the nonrisky asset investment (see above). Agent's portfolio rate of return R^t_p for the time moment t

$$R^t_p = x^t_1 * R^t_A + x^t_2 * R^t_B + x^t_3 * R^t_C, \tag{1}$$

$$R^t_{A,B,C} = (K^t_{A,B,C} - K^{t-1}_{A,B,C})/K^{t-1}_{A,B,C},$$

$$\sum K^t_{A,B,C} = K_0, \text{ than } t=0; \ K_t - K_0 <> 0 \text{ than } t \neq 0,$$

$$\sum x_i = 1.$$

here multipliers $x^t_{1,2,3}$ represent weighting coefficients; $R^t_{A,B,C}$ – rate of return of the A, B or C investment assets; $K^t_{A,B,C}$ and $K^{t-1}_{A,B,C}$ – total financial capital in two subsequent time periods for the appropriate investment asset. Agents can redistribute their portfolio as market intervention timing is another factor in the model.

Portfolio rate of return, as investment factor, is only one side of the Markowitz optimal portfolio selection. The other side is dealing with the associated risk measured by the mean square error. For three assets portfolio

$$\sigma_p^2(t) = \overline{(R_p^t - \overline{R}_p^t)^2} = x_1^2(t)\overline{(R_A^t - \overline{R}_A^t)^2} + x_2^2(t)\overline{(R_B^t - \overline{R}_B^t)^2} + x_3^2(t)\overline{(R_C^t - \overline{R}_C^t)^2} + \\ +2x_1^t x_2^t \overline{(R_A^t - \overline{R}_A^t)(R_B^t - \overline{R}_B^t)} + 2x_1^t x_3^t \overline{(R_A^t - \overline{R}_A^t)(R_C^t - \overline{R}_C^t)} + 2x_2^t x_3^t \overline{(R_B^t - \overline{R}_B^t)(R_C^t - \overline{R}_C^t)}$$ (2)

Let denote $\sigma_i^2(t)$ as investment return dispersion of the assets A (i=1), B (i=2) or C (i=3) respectively and $\sigma_{ij}(t)$ as covariance between returns of those assets, then

$$\sigma_p^2(t) = \sum_{i=1}^{3} x_i^2(t)\sigma_i^2(t) + 2\sum_{i=1}^{3}\sum_{j>1}^{3} x_i(t)x_j(t)\sigma_{ij}(t), \sum_{i=1}^{3} x_i = 1.$$ (3)

In fact, Eq. (3) can be further reduced having in mind that we can choose nonrisky C asset in such a way as to diminish associated risk with it to the minimum (e.g. investing to the treasury bonds with low rate of returns, but almost zero dispersion)

$$\sigma_p^2(t) = x_A^2(t)\sigma_A^2(t) + x_B^2(t)\sigma_B^2(t) + 2x_A(t)x_B(t)\sigma_{AB}(t).$$ (4)

All resulting portfolios are restricted in the area of the shape depicted in Fig. 2. The area contains investment portfolios with all possible return-risk combinations. The best choices, though, are on the efficient frontier (see e.g. AC curve in Fig. 2) because it gives for the chosen return level the minimum associated risk.

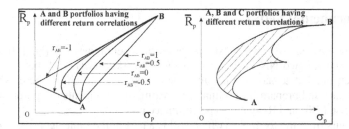

Fig. 2. Distribution of risk-returns (for two and three assets) portfolios due to various asset returns mutual correlations

In terms of wealth, initially homogenous agents are gradually collecting different wealth depending on chance and performance of the investment strategies they employ. To our knowledge, there are no experiments in the field that focuses on the intelligent agents, simulating empirically tested distributions of wealth or stock returns.

Pareto discovered that at the high-wealth range, wealth (and also income) are distributed according to a power-law distribution. The parameters of this distribution may change across societies, but regardless of the social or political conditions, taxation, and so on, Pareto claimed that the wealth distribution obeys this general distribution law, which became known as the Pareto distribution or Pareto law. The Pareto distribution is given by the following probability density function

$$P(W) = CW^{-(1+\alpha)} \quad for \ W \geq W_0,$$ (5)

where W stands for wealth, P(W) is the density function, W0 is the lower of the high-wealth range, C is a normalization constant, and α is known as the Pareto constant. Pareto finding has been shown in numerous studies to provide an excellent fit to the empirical wealth distribution in various countries [1].

Pareto distribution is related with wealth distribution, which is acquired by cumulative process of price differences. Meanwhile, distribution of stock returns, especially when calculated for short time intervals, are not fitted well by the normal distribution. Rather, the distributions of stock returns are leptokurtic or "fat tailed". Mandelbrot [6] has suggested that log-returns are distributed according to the symmetrical Levy probability distribution defined by

$$L_{\alpha_L}^{\gamma}(x) \equiv \frac{1}{\pi} \int_0^{\infty} \exp(-\gamma \Delta t q^{\alpha_L}) \cos(qx) dq,$$ (6)

where $L_{\alpha_L}^{\gamma}(x)$ is the Levy probability density function at x; α_L is the chatacteristic exponent of the distribution; $\gamma \Delta t$ is a general scale factor; γ is the scale factor for $\Delta t=1$, and q is an integration variable [2]. Recent analysis of returns has shown that price differences, log-returns, and rates of return are described extremely well by a truncated Levy distribution [6].

3 Experimental Setup

For each investment asset we construct appropriate neural network [NN]. In our case, it gives three (or two NN in the case of one nonrisky asset) MLP backpropagation NN, which are but a part of the same agent. We teach each of these NN on the historical data and prepare for optimal forecasting of movements of appropriate index fund (we use MSE and R-square as performance criteria).

The inputs from the forth NN (which still is a part of the same agent) are connected to the outputs of the former three (or two) NN. In this way we are building an optimal investment portfolio, i.e. determine the proper portfolio weights depicted from the fourth NN's input weight matrix [7].

For the effective implementation of portfolio management each agent has to be enrolled in the operational framework, which we are going to discuss below. First of all, functionality of an agent is described by the major decision concerning his targets in terms of profit and risk. In this sense, proposed model is tailored to depict major possible strategies explicitly applied in the real markets, see Table 2.

The initial experimental setup is designed so as to represent each trading strategy by allocating equal number of agents following them. Agents are with slightly different modifications inside each group, but they can't jump from one strategy to another. Investment strategies are given beforehand. Each agent randomly picks one strategy which forms six groups of agents with different investment strategies.

Table 2. Major algorithmic investing strategies simulated by different groups of agents

#	Strategy	Functionality	Remarks
S_1	*"Buy and hold"*	Naïve or conservative approach, where initial portfolio is randomly selected and kept fixed, i.e. keeping up with the portfolio initial diversification. Agent gets his return-risk which is directly related with market's tendencies.	The idea is that the market itself knows best, therefore, there is no need to intervene; just follow the aggregated trend. Minimum transaction costs are involved.
S_2	*"Bull" trading*	Whenever appears growing trend in the subsequent periods of 50 or 200 business days, proportional investing intervention takes place, and vice versa than decreasing tendency shows up. Frequent changes in portfolio risk-profit diversification are observed. Agent wants to keep with growing "bull" assets having returns as high as possible in expense of risk and transaction costs.	The idea: to keep up with the growing trend. Dynamic approach. Maximum transaction costs are involved.
S_3	*Support&Resistance trading*	An agent is looking for the high volume trends. either of selling or buying, which are influenced by the demand and supply. This is so called price search of support or resistance level, i.e. if price is high, but trade volume is very low – sell the asset, and vice versa if price is low and volume is high – buy the asset.	The threshold values of high and low volume of market trade and "firing" ratios of current price vs. market trend price have to be defined in advance.
S_4	*Contrary trading*	The idea is to trade contrary to most investors, who supposedly almost always lose – in other words to go against the crowd. Under this scenario, when index fund liquidity is low (high price and low trade volume) contrarians would consider this is a good time to buy and otherwise when liquidity is high.	This is an old idea on Wall Street, and over the years technicians have developed several measures designed to capitalize on this concept.
S_5	*Risk averse trading*	The agent is extremely risk averse in expense of his profits. It holds with low risk index if it meats certain dispersion criteria, otherwise, it invests to nonrisky assets.	Diversification is taking place only if indexes exhibit low volatility fluctuations.
S_6	*Filter rule*	A well known technical trading rule is the so-called filter rule, which specifies a breakpoint for a market average, and trades are made when the index fund price change is greater than this filter. For example, buy an asset if the price moves up 10% from some established base, such as previous low (or moving average), hold it until it declines 5% from its new high, and then sell it.	There are two algorithms applied, where one gives "green" light for investing and another "red" light for withdrawal of capital.
S_7	*NN portfolio optimization method*	This strategy is solely based on NN_{1-3} performance, where optimal portfolio weights are estimated using NN_3 input weights, see Fig. 4. Portfolio return-risk is recalculated using a trendline for every 20 business day period.	We apply a system composed by the first and second level NN, which learn/simulate, classify/ recognize and optimize the agent's performance

Note: the model uses daily and weekly time series, all inputs are normalized

Consequently, we were looking if a model can be found that could relate the domain space (real investment market data) with solutions space (simulated agents' investment decisions). We have proposed a modular approach for simulation of complex strategies existing in the investment markets. The major brick of it is composed by MLP (multilayer perceptron) error backpropagation NN (neural network), which is aimed to give (i) forecasting of index values, (ii) make market timing decisions, (iii) generate decisions for the technical investment, (iv) make volatility estimates as depicted in Fig. 3.

In more formal terms, let us assume that the financial problem domain $\Omega(\Omega^{in}; \Omega^{out})$ is characterized by 1) sub domain Ω^{in}, which consists of the set of input data space vectors $\{I_n^i\}$ (where n denotes the input space dimensionality and i=[1,.. k] indicates the input data vector); 2) sub domain Ω^{out}, which consists of the set of output data space vectors $\{O_m^i\}$ with appropriate output dimensionality m.

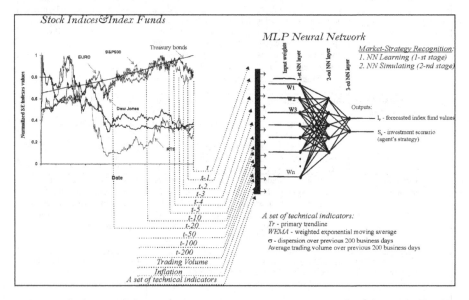

Fig. 3. Investment market data (autoregression time series) and technical preprocessed data projected as inputs to the forecasting MLP NN

NN is used for mapping a given input space onto the desirable output space (NN decisions). Our goal consists of investigating the mapping function Φ

$$\Phi(\{I_{in}\}) \rightarrow \{O_{im}\} \tag{7}$$

A multilayer perceptron (MLP) network serves as a universal approximator, which learns how to relate the set of the input space vectors $\{Iin\}$ to the set of output (solutions) space vectors $\{O_{im}\}$. The transformation function Φ is then characterized by the MLP structural parameters like weights' matrix W, biases B, number of neurons N, topology structure T, learning parameters L

$$\Phi = \Phi(W, B, N, T, L) \tag{8}$$

This is supervised learning. MLP gains experience by learning how to relate the input space vectors $\{I_{in}\}$ to the known output vectors $\{O_{im}\}$. Now we parameterize Eq. (7)

$$\Phi_{W, B, N, T, L}(\{I_{in}\}) \rightarrow \{O_{im}\} \tag{9}$$

Eq. (9) gives us needed relations in terms of agent's capability to recognize markets behavior and also strategic data like market timing, technical data and volatility (see Fig. 3 and Table 2), which are later employed for portfolio management operations. For the effective implementation of MLP networks, we have to investigate the major NN structural parameters like W, B, N, T and L in a decomposition manner [7]. Whereas, the output vectors $\{O_{im}\}$ are described by a set of technical indicators, see Table 3.

We have to bare in mind that regarding portfolio setup of three assets (composed of two SE based index funds and one nonrisky asset like Treasury bond), the scheme depicted above applies for each risky asset separately composing the system of two neural networks.

4 Management of Agents' Operational Procedures

This section discusses research issues related with how investment strategies $\{S_i\}$ are going to be implemented modus operandi in operational terms. As we see from the Table 3 above, only the first strategy S_1 doesn't requires dynamic management of the portfolio (agent's risk-profit performance depends on initially selected portfolio in the beginning of simulation). This is so called naïve investment approach, where portfolio selection is taking place only once without any regard to the changing market infrastructure. We have chosen it as a benchmark to compare performance of the simplest "buy and hold" strategy against other dynamic strategies S_{2-7}.

The other strategies, though, are more sophisticated as they are not hard wired to the initial portfolio and reflect dynamic approach to the changing investment environment. Following previous discussion, we are going to investigate how remaining six strategies are classified in our multiagent model. First of all, see Fig. 1, all six strategies are wrapped in the same principal simulating scheme, which is further detailed in Fig. 4. But there are major differences concerning operational and portfolio management:

1. Strategies S_{2-6} operate using forecasted information from different assets. For operations management they use market intervention signals (see Table 3) and are not bound to optimize their portfolio according to the Markowitz portfolio optimization theory. So these are mostly short-term opportunistic-speculative investors, which follow well known (on Wall Street) investment scenarios based on technical indicators.

2. Strategy S_7 is also short-term and opportunistic, but it follows portfolio optimization theory using a novel nonlinear portfolio optimization technique, which is sustained by NN_3. Operational and portfolio management issues are solely controlled by the second level NN_3, which learns how to optimize the portfolio and get portfolio weights out from the input weights.

Our main purpose by doing so is to construct heterogeneous investment market where we could compare performance of different investment strategies (represented by appropriate agents) against the newly constructed holistic investment method (investment strategy S_7). We also expect to get Pareto wealth and Levy returns distribution [1,6].

The most important notation about Fig. 4 is concerning a special way how proposed NN learning and simulating scheme automatically depicts the portfolio weights in nonlinear manner just by getting the input weights from the NN_3, where each NN_3 input neuron is connected with only one input node [7].

In fact, our NN-based system is not dealing with the classification task (see Fig. 1), which is left outside the NN scope. We shifted classification of initial data to the preprocessing stage than a set of known technical indicators (see Table 3) and procedures are applied in order to find market intervention signals and investment timing. We have based our technical analysis approach on a number of theoretical and empirical studies [4], which have suggested that technical analysis remains popular with many investors. So we had to follow the reality on the hand and design the appropriate model.

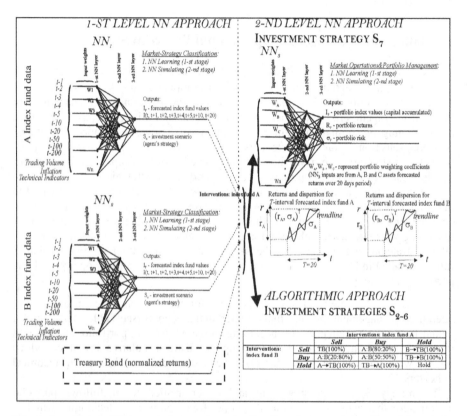

Fig. 4. Agent's two NN layers scheme, where the first layer learns and classifies, the second – makes investment decisions

In this paper, we have proposed novel ways in which each investing strategy can be facilitated by a set of appropriate technical indicators and procedures, which once satisfied give incentives for the market interventions, see Table 3. We have chosen aggregate indicators instead of charts analysis because we are interested in longer term portfolio investments, which are better described by aggregated measures. For instance, the most basic measure of an index direction is the trend line T_r, which simply shows the direction the index is moving.

Table 3. Estimation of technical indicators for classification and recognition of investing scenarios (i.e. agents' investment strategies, see Table 2)

Inv. Strat.	Interv. Signals	Technical Indicators					
		T_r	WEMA	V_o	F	σ	T
S_1		-	-	-	-	-	-
S_2	Buy	$a_{50}>0.001$	OR $I>WEMA_{50}$ AND $\alpha=0.05$	AND $V>\bar{V}$	-	-	$T>10$
	Hold*	$0.0001<a_{50}<0.001$	-	-	-	-	$T<10$
	Sell	$a_{50}<0.0001$	OR $I<WEMA_{50}$ AND $\alpha=0.05$	AND $V_t<\bar{V}$	-	-	$T>10$
S_3	Buy	$T_r>1.05*I_t$	-	AND $1.1*V_t<\bar{V}$	-	-	$T>20$
	Hold	-	-	-	-	-	$T<20$
	Sell	$T_r<0.95*I_t$	-	AND $1.1*V_t>\bar{V}$	-	-	$T>20$
S_4	*Buy*	$T_r<0.90*I_t$	-	AND $1.1*V_t>\bar{V}$	-	-	$T>20$
	Hold	-	-	-	-	-	$T<20$
	Sell	$T_r>1.1*I_t$	-	AND $1.1*V_t<\bar{V}$	-	-	$T>20$
S_5	*Buy*	-	AND $I_t>WEMA_{200}$ AND $\alpha>0.01$	-	-	$\sigma_t<\sigma_k$	$T>50$
	Hold	-	-	-	-	$\sigma_t<\sigma_0$	$T<50$
	Sell	-	OR $I_t<WEMA_{200}$ OR $\alpha<0.01$	-	-	$\sigma_t>\sigma_0$	$T>50$
S_6	*Buy*	-	-	-	$I_t>1.1*I_{t-1}$	-	$T>10$
	Hold	-	-	-	Treasury bonds	-	$T<10$
	Sell	-	-	-	$I_{t+n}<0.95*I_{t+n-1}$	-	$T>10$

* - "Holding" scenario is always applied (for all trading strategies) if appropriate technical indicators don't give signals of buying or selling.
T_r – index fund primary trendline; it is represented by linear approximation I=at+b, where I – normalized [0,1] index value, a - stands for the slope coefficient (a>0); t – is measured in business days, x=50 (Tr_{50}) and x=200 (Tr_{200}); or we could use the secondary trendline represented by the nonlinear polynomial, HP or other approximation.
WEMA – normalized weighted exponential moving average, over 50 ($WEMA_{50}$) and 200 ($WEMA_{200}$) business days; V_o – normalized volume of trade ($\bar{V}$ - the average volume of trade for consecutive 200 business days), F - filter rule, σ - dispersion ($\sigma_k<\sigma_0<\sigma_t$, where threshold values of σ_k and σ_0 are defined in advance), T- market timing (for the most of the cases this is market intervention frequency).

The leading experts in the field [4] have published a number of papers providing support for two basic technical indicators, moving average and support and resistance. Following their advice, we have employed weighted exponential moving average

WEMA, where a comparison of a current market price to the moving average produces a buy or sell signal. The general buy signal is generated when actual prices I_t rise through the moving average on high volume, with the opposite applying to a sell signal[11]. We also included volume V_o, which is an accepted part of technical analysis too[12]. Besides, we have included one example from trading rules, i.e. filtering rule F "buy a stock if the price moves up 10 percent from some established base, such as a previous low, hold it until it declines 10 percent from its new high, and then sell it". We also have included the estimates of dispersion σ, which is very important for some investing strategies mostly executed by risk averse institutional market players, see Table 3. Market timing T is of high importance too as different players have different attitudes concerning time scale and transaction costs.

In fact, the initial pool of agents will be equally divided among six groups, which are described in Table 3 and follow correspondent trading strategies. Among each group, though, we can randomly select slightly different values of technical indicators having nonidentical but justifiable (in the certain limits) variance in the investing behavior. This gives better exploration of feature space and better represents the real market.

5   Results and Discussion

Hence, our main concern is related with the simulation of the heterogeneous investment market, where a big number of different agents (represented by a set of well known investment strategies) play short-term opportunistic games of optimal capital allocation in the stock market. In the model, agents are using forecasting for prediction of short and long-term trends. It gives incentives to adjust appropriately their portfolios by doing market interventions whenever technical indicators are signaling the presents of algorithmically described "firing" (buying or selling) moments.

The ultimate long-term goal for each agent differs, but in general each of them is looking for the initial capital maximization. We have to admit, though, that it is almost impossible to model the complexity we face in the real markets, where investors behave irrationally following rumors, subjective judgments and believes. Therefore, we have enriched the model with some heuristics based on the nonlinear properties and generalization features of neural networks, which, as numerous studies have shown [7,8], have systematically outperformed the traditional regression methods so popular in finance.

The modular structure of the system of specialized NN gives us an edge over the straightforward algorithmic approaches as it performs better forecasting, it is more flexible, and, being nondeterministic, mimics the complexity in its own terms. This

---

[11] A variation of the moving average popular on some Web sites is the moving average convergence-divergence, or MACD. This involves a longer moving average (such as the 200 day) and a shorter moving average (such as the 50 day). As a stock price rises, a bullish signal is generated if the short-term average consistently is greater than the long-term average. A warning signal is generated when the short-term average falls below the long-term average.

[12] High trading volume, other things being equal, is generally regarded as a bullish sign. Heavy volume combined with rising prises is even more bullish.

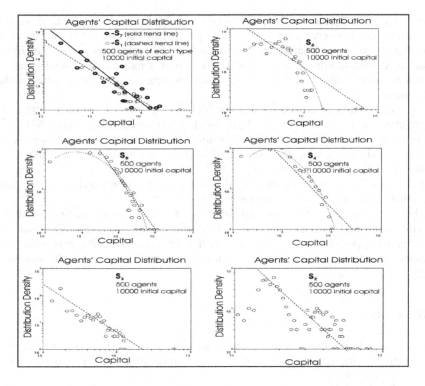

Fig. 5. Comparison of Pareto wealth distribution for investment strategies S1-7

paper provides insights and an overview for creating a society of artificial investing agents, where each agent can classify, recognize, learn and forecast in order to make best use of available information. It is impossible to cover the full range of our empirical work here, but to give a flavour of it a few details of current research are included. Below we plotted some charts of obtained Pareto distribution of wealth for different agent strategies S_{1-7}, see Fig. 5.

The experimental setup below includes 500 NN-based agents for each strategy simulation, 10000 initial capital allocated for each agent, trading costs per transaction 0.5%. Values for technical indicators were depicted as it is shown in the Table 3. Though, we have to admit that parameter space for the investment strategies S_{1-6} has to be explored and estimated much more thoroughly as it gives huge difference for the final results. Besides, for the prospective experimentation we have to enlarge the total number of agents, which would give much better estimates of agents' distribution density in accumulated wealth terms.

Unfortunately, the limited space available doesn't allow us to deliver more empirical results here. In short, S_7 gives Pareto wealth distribution, which is very close to S_1 performance (see Fig. 5), meaning that NN trading strategy works as good as buy-and-hold approach, which has no transaction cost involved. Therefore, NN in equal terms is very competitive, well represents the real market situation and fits for financial market simulation. Meanwhile, strategies S_2 ("Bull" trading) and S_6 (Filter trading) are not well fitted with Pareto empirical law. In case of Bull trading strategy, the lost market position and high dispersion of results are mostly related with (i) frequent market interventions, i.e. very high transaction costs, (ii) not stable or overreacted activity. Meanwhile, filtering rule is failing mostly because of biases associated with filtering rule thresholds. These results indicate that sound statistical estimations are still needed for much wider experimentation setup having richer parameter space.

We are at pains to emphasize that our findings represent an initial first 'take' on simulation of Pareto wealth and Levy stock returns distribution using NN-based multiagent system of heterogeneous investors. This work, though, gives some clear outlines and their explanatory sources. Moreover, we have spent a considerable amount of this paper examining the methodological novelty of our work. Subsequent papers will focus more on the empirical aspects of our research once we have established the methodological importance of our work.

Future research will focus on experimenting with more representative investment portfolios as well as search for better fitted algorithmic parameters. Proposed NN modularity approach let us envisage a system of networking social agents, which will interact among each other creating behavioral patterns and social phenomena in general.

References

1. Blume, L.E., Durlauf, S.N (eds.): The Economy as an Evolving Complex System III. A Volume in the Santa Fe Institute in the sciences of complexity. Oxford University Press, New York (2006)
2. Mantegna, R.N., Staney, H.E.: An Introduction to Econophysics: Correlations and Complexity in Finance. Cambridge University Press, Cambridge (2000)
3. McCauley, J.L.: Dynamics of Markets: Econophysics and Finance. Cambridge University Press, Cambridge (2004)
4. Jones, C.P.: Investments: Analysis and Management, 8th edn. John Wiley & Sons, Chichester (2002)
5. Markowitz, H.M.: Portfolio Selection:Efficient diversification of investments. Blackwell Publishers, Malden (1991)
6. Levy, M., Solomon, S.: New Evidence for the Power-Law Distribution of Wealth. Physica A 242 (1997)
7. Plikynas, D., Akbar, Y.H.: Application of modified MLP input weights matrices: An analysis of sectorial investment distribution in the emerging markets. Neural Comput & Applic 15, 183–196 (2006)
8. Haykin, S.: Neural Networks: A Comprehensive Foundation, 2nd edn. Prentice-Hall, New Jersey (1999)
9. Wooldridge, M.: An Introduction to Multiagent Systems. John Wiley&Sons, West Sussex, England (2002)
10. Wooldridge, M., Jennings, N.R.: Intelligent Agents: Theory and Practise. The Knowledge Engineering Review 10(2), 115–152 (1995)
11. Pnueli, A.: Specification and development of reactive systems. Information Processing 86, 845–858 (1986)

Business Ecosystem Modelling: Combining Natural Ecosystems and Multi-Agent Systems

César A. Marín*, Iain Stalker, and Nikolay Mehandjiev

School of Informatics
University of Manchester
PO Box 88, Sackville St
Manchester M60 1QD, UK
Cesar.Marin@postgrad.manchester.ac.uk,
Iain.Stalker@manchester.ac.uk,
Nikolay.Mehandjiev@manchester.ac.uk

Abstract. The increasing popularity of the "business ecosystem" concept in (business) strategy reflects that it is seen as one way to cope with increasingly dynamic and complex business environments. Nevertheless, the lack of a convincing model of a business ecosystem has led to the development of software which only give organisations a partial aid whilst neglecting their need for adaptation. Research in Multi-Agent Systems has proved to be suitable for modelling interactions among disparate sort of entities such as organisations. On the other hand, natural ecosystems continue to adapt themselves to changes in their dynamic and complex environments. In this paper, we present the Dynamic Agent-based Ecosystem Model. It combines ideas from natural ecosystems and multi-agent systems for business interactions.

1 Introduction

Increasing technological complexity means that market demands are often best satisfied through networks of collaborating organisations. Indeed, to achieve a sustainable growth and improve competitiveness in increasingly complex and dynamic environments, many businesses actively seek partnerships and alliances. Moreover, the volatility of many contemporary markets mean that such partnerships form opportunistically and are short-lived. To flourish in such environments, businesses must continually adapt and evolve. This requires that a business engage in an ongoing dialogue with its environment and with others with which it shares this environment.

The notion of a "business ecosystem" is a strategic planning concept, introduced by Moore [1,2], that supports the perspective stated above. Moore defined a "business ecosystem" as a collection of companies which co-evolve, developing capabilities in response to new, wide-ranging innovations; companies both

---

* César A. Marín thanks the support provided by Consejo Nacional de Ciencia y Tecnología (CONACyT) through sponsorship No. 197297/218103.

M. Klusch et al. (Eds.): CIA 2007, LNAI 4676, pp. 181–195, 2007.

cooperate and compete, as appropriate, as they vie for survival and dominance. Moreover, Moore remarks that a business ecosystem is obtained when a set of randomly interacting companies develops into a more structured community; and it typically supports competition at a higher level, i.e., competition among business ecosystems.

Natural ecosystems do indeed offer a clarifying metaphor: they encompass species competing for the same resources and interacting to create complex networks, such as food webs, and capable to adapt to different situations. A natural ecosystem has been characterised as a complex adaptive system (CAS) [3] because it comprises sets of individuals exhibiting "emergent" behaviours not apprehended by any one individual. Such systems continuously adapt to changes in their extremely dynamic environment. Four essential properties describe a natural ecosystem (cf. CAS): aggregation, diversity, flow, and non-linearity [3]. These properties have been studied by two comunities: biology (e.g. [4,5]) and multi-agent systems (e.g. [6,7]) and are widely accepted.

A business environment (cf. markets) has also been characterised as a CAS, yet there remains the lack of a convincing model through which to explore the metaphor of a business ecosystem. In this paper, we present the Dynamic Agent-based Ecosystem Model (DAEM): a synthesis of ideas from natural ecosystems and multi-agent systems which provides a framework to leverage the strategic concept of a business ecosystem. We approach business ecosystems from a holistic point of view, in particular, by studying the propagating influence of local relationships on the ecosystem as a whole. That is, we subscribe to the view that interactions are fundamental to the creation of ecosystems [8] and the development of adaptive behaviours [9]. Thus, here we abstract from the details of a particular trade or transaction between two organisations and instead develop a model which assumes the presence of only three elements in a business interaction, namely: services being offered; services being consumed and evaluated; and feedback of service evaluations.

We present the fundamentals of our approach in Sect. 2 in an incremental way: We begin with a clarification of what is meant by the notion of service in the model and how we characterise an agent; We introduce notions of environment and interactions, especially random interactions to support opportunism, which motivate extensions of our model of an agent; We then add appropriate capabilities which allow our agents to forage for services and customers as they move around their environment interacting; We conclude the exposition of the model by showing that it naturally admits supply networks; We then present related work in Sect. 3; We close the paper with a few remarks.

2   Dynamic Agent-Based Ecosystem Model (DAEM)

We begin with a brief overview of DAEM where we mainly focus on a supply chain domain. We then proceed to incrementally add more details to it.

In DAEM, an agent represents an organisation and function as both a supplier of one service and a customer (cf. consumer) of another. Thus, an agent offers his services to potential customers and analyses service offers from potential suppliers. Service offers are evaluated to determine who is more convenient to buy from, and therefore, to flag him as a preferred supplier. Service evaluations are sent back to the potential suppliers for them to know how good their services are evaluated and whether they are considered as preferred suppliers. Likewise, suppliers determine who their preferred customers are according to the received evaluations. These evaluations depend on past experience and the service added value each particular agent prefers, such as price, response time, warranty, etc. An agent knowing he is the preferred supplier of his preferred customer will seek to have more frequent interactions with him; and vice versa. Preferred supplier and preferred customer together become partners and constitute a link in a supply chain.

Competition is essential to a business ecosystem [1]. Evaluation and feedback mechanisms encourage competition among agents who provide the same service, i.e. *intra-class* competition. Service evaluations are relative to the highest evaluation recently made, i.e. the highest evaluation is used to assess other service offers. Therefore, a relative evaluation functions as a measure for the minimum relative improvement required in a service offer for an agent to become a preferred supplier. Service improvements are considered out-of-scope for the current paper; we simply note that improvements occur and they are noticed when new evaluations are made. Improvements in services cause changes in partner preferences, for instance, by raising standards and thus expectations; and in turn they trigger more changes across all agents.

Supply chains arise when a service provided by an agent is preferred by another agent whose service is preferred by another and so on. Supply chains resemble food chains in nature. When an agent offers and requires more than one service, supply networks are formed. These resemble food webs in nature. The whole collection of agents offering and requiring different services forms the multi-agent business ecosystem. Adaptation globally emerges when agent preferences change and agents change their interactions accordingly, i.e. ecosystem nodes are replaced but the ecosystem itself remains.

2.1   Services and Agents

We begin our more detailed account by clarifying services and agents. For convenience, we write $\prod_{j=1}^{M} a_j$ to signify an ordered list. Similarly, we use $\prod_{i=1}^{N} (a_i, b_i,$ $\ldots, c_i)_i$ to signify a table containing the i-th tuple as the i-th row. Additionally, we use subscripts to distinguish multiple uses of terms.

Definition 1. A service is a pair (sv, adv) where sv is a service description and adv is its added value. The service description specifies the basic details of the service, i.e. what the service is about. The added value specifies aspects such as price, response time, warranty policy, etc.

The internal structure of both service description and added value are not discussed here. We simply assume that an agent is able to use them to make evaluations and decisions. Additionally, we assume that the added value can vary for the same service description, e.g., according to the agent offering the service.

Definition 2. The tuple $(aid, isv, osv, A, ms, mls, mlc)$ is used to describe an agent ag and related interactions as follows:

- aid is the agent's unique identifier,
- isv is a service description specifying what service is required,
- osv is a service description specifying what service this agent can offer, where $osv \neq isv$,
- $A = \{sendof, sendev\}$ is a set of actions available to the agent:
  - $sendof(aid, aid_c, osv, adv)$ signifies that agent aid offers service (osv, adv) to agent aid_c,
  - $sendev(aid, aid_s, sv_s, adv_s, ev_s)$ signifies that agent aid informs supplier agent aid_s that the evaluation of the service (sv_s, adv_s) is ev_s.
- $(aid, adv, ev)_{aid_i}$ is an interaction tuple recording an interaction that took place with agent aid_i, specifying his identifier aid_i, added value adv of service offered or received and the received or given evaluation, accordingly. Typically, we abuse notation for compactness when referring to agent aid_i by simply writing agent i, e.g. $(aid, adv, ev)_i$.
- ms is the interaction memory size, i.e. the number of recent interactions in memory,
- $mls = \prod_{r=0}^{ms}(aid, adv, ev)_r$ is a table containing the last ms interactions this agent had with supplier agents for the corresponding service described by isv.
- $mlc = \prod_{r=0}^{ms}(aid, adv, ev)_r$ is a table containing the last ms interactions this agent had with customer agents for the corresponding service described by osv.

An agent identifier aid is unique to an agent. Actions $sendof$ and $sendev$ are used to send service offers and evaluation, respectively. For example, when agent i sends an offer to agent j, denoted $sendof(aid_i, aid_j, osv_i, adv_i)$, agent j immediately verifies whether the description of the offered service corresponds to its required service isv_j. In such case, agent j evaluates it according to the given added value adv_i and returns a positive evaluation ev, denoted $sendev(aid_j, aid_i, sv_i, adv_i, ev_i)$, where $sv_i = osv_i = isv_j$. Otherwise, a negative value will be sent meaning that agent j does not require (or is not interested on) that service as a supply. The evaluation ev_i sent back to the supplier is relative to the highest evaluation agent j has given in the last ms interactions. That is, the maximum evaluation ev found in mls. Likewise, agents store in mlc the last ms evaluations received from customers. The limited memory agents have gives them the forgetting mechanism which has been proposed as a principle for engineering multi-agent systems as natural systems [7].

The value of ev_i sent to a supplier agent i depends on the particular service evaluation and agent j's past recent experience; it is calculated as follows:

$$ev_i = \begin{cases} 1 & \text{if } eval(sv_i, adv_i) \geqq ev_{\max} \\ \frac{eval(sv_i, adv_i)}{ev_{\max}} & \text{otherwise} \end{cases} \qquad (1)$$

Where $(aid, adv, ev_{\max}) \in mls_j$.

Function *eval* evaluates services according to agent's preferences. Thus, a service from the same supplier can be evaluated differently by two separate customers. Evaluation *ev* received through other agent's action *sendev* and stored in *mlc* is always within the range $[0, 1]$ because it is a relative evaluation. In contrast, evaluations in *mls* have an open range $[0, \infty)$ because they directly come from function *eval*; they are not relativised. In summary, relative evaluations reflect the relative difference in the perceived quality of a service offered by different suppliers.

For instance, consider an agent i who grants x as the evaluation of a service offered by agent j. Say x turns out to be the highest evaluation i has recently made, i.e. $ev_{max} = x$. Then j receives a relative evaluation of $ev = 1$ after (1) and now j knows he is the preferred supplier for i. Now, consider a similar situation with agent h. He sends a service offer to i who concedes y as the corresponding evaluation. Say y turns out to be greater than ev_{max}, i.e. $y > x$. Then ev_{max} new value is y and h receives $ev = 1$ as well as the relative evaluation. Likewise, h now knows he is i's preferred supplier. However, when j returns to interact again with i and the latter compares his service with the new value of $ev_{max} = y$, j will obtain a relative evaluation of $ev = x/y$ where $x/y < 1$ as mentioned above. Finally, at this point j will know he has be replaced as a preferred supplier.

For an ecosystem to develop, it is necessary to have a diverse set of individuals [3,7,10,11]. This diversity is allowed to exist in our model, but it depends on the specific case one is modelling. Here, diversity is not seen as species or classes of agents, but rather as the market they are targeting by the services they offer. That is, one might say that agents offering the same service *osv* belong to the same "class". And as markets dictate, agents offering the same service become competitors whereas an agent requiring what other agent offers effectively becomes his partner (in the context of a supply chain).

Interactions are important because they are a factor to form ecosystems [8] and to develop adaptive behaviours [9]. These interactions consist of offering services, evaluating them, and reporting on such evaluations. Moreover, these interactions encourage competition specially because of service relative evaluations. Competition is necessary for a business ecosystem to function [1]. And it allows agents to have an idea of how much improvement they need to achieve in order to be a top supplier and survive in their environment. This improvement can be seen as individual adaptation which is out of the scope of the model presented here. However, its effect can be noticed through differences in service evaluations once adaptation is achieved.

2.2 Environment and Random Interactions

An environment is a fundamental element of both business and natural ecosystems. Roughly, an environment is defined as a provider of conditions under which an agent exists [12]. For our purposes, it is a virtually observable surface where inhabitants wander across and encounter others in order to interact. It is a common environment [13]. Its importance arises because it supports capabilities such as a sense of positioning and displacement, proximity-based interactions, and surrounding awareness. These capabilities permit an agent to orient himself and follow energy flows or a (notional) gradient on the environment [7,11]. Moreover, it allows spatial diversity and the creation of niches [11] where only some services can be found. An environment also permits the most important species, i.e. keystone species [8], to cluster food webs around themselves, which at the same time comply with the view business ecosystem [1,2].

Definition 3. An environment *env* is a toroidal grid[1] characterised using the tuple (es, Ads, itr) where

- es is the environment size, i.e. $es \times es$, where each side goes from 0 to $(es-1)$,
- $(eid, epos, eisv, eosv)_i$ is an agent description tuple specifying agent i's identifier eid, position $epos = (x, y)$ on the environment, and services descriptions $eisv$ and $eosv$,
- Let Ag denotes the set of all existing agents. $Ads = \prod_{i=0}^{|Ag|}(eid, epos, eisv, eosv)_i$ is a table where all agent positions and services are described. The number of description tuples depends on the number of existing agents $|Ag|$.
- itr specifies an agent interaction radius, where $itr \geqq 1$.

The environment size es depends on the number of agents $|Ag|$ inhabiting it. And it must be sufficiently large for agents to avoid to come across one another too frequently if they move entirely randomly on it. Ads is a table containing descriptive information for each agent on the environment. Such information includes agent identifier eid, current position $epos$ on the environment, and two service descriptions $eisv$ and $eosv$. The interaction radius, itr, identifies the maximum distance at which two agents may interact in this particular environment. From Def. 2 and Def. 3, we observe that are elements in common, namely, aid, isv, osv and $eid, eisv, eosv$, respectively. Elements from the environment can be viewed as reflections from the agent elements onto the environment itself, i.e. agent elements are exposed. In this way, agents can be detected by others and can detect others on the environment as well, i.e. we are creating a common environment [13].

Definition 4. Reflection is an environment property which denotes that some values are exposed from the agent onto the environment in such a way that if a change occurs to the original one, it will be immediately updated on the reflected part. We represent reflection as relation ρ and say that the first argument is reflected on the second one.

Agent identifier aid and agent services isv and osv are reflected on the environment as $eid, eisv, eosv$, respectively. Thus, we have the following reflections—using an infix notation—between agents and the environment (read $\leftarrow$ as "gets"):

$$aid_i \; \rho \; eid_i := eid_i \leftarrow aid_i. \tag{2}$$

$$isv_i \; \rho \; eisv_i := eisv_i \leftarrow isv_i. \tag{3}$$

$$osv_i \; \rho \; eosv_i := eosv_i \leftarrow osv_i. \tag{4}$$

---

[1] Consider an $n \times n$ grid ($n \in \mathbb{N}$) in the plane: both x and y run from 0 to n, where lines $y = 0$ and $y = n$ overlap as well as lines $x = 0$ and $x = n$. This results in a torus necessary to ensure that an agent can wander continuously, cf. Sect. 2.3.

The last 3 equations apply for all agents $i \in Ag$ such that $(eid, epos, eisv, eosv)_i \in Ads$, where Ads is part of the environment env and aid_i, isv_i, osv_i are associated with agent i.

We can now expand Def. 2 to make use of the environment (additional or revised items in bold for the purpose of emphasising).

Definition 5. The tuple $(aid, \boldsymbol{apos}, isv, osv, \boldsymbol{A}, ms, \boldsymbol{mls}, \boldsymbol{mlc})$ is used to describe an agent ag and related interactions; where we extend Def. 2 as follows:

- $apos = (x, y)$ is the agent's current position on the environment,
- $A = \{sendof, sendev, \boldsymbol{walk}\}$ is the (revised) set of actions available to the agent, where
  - $walk(ang)$ signifies that the agent changes its current position on the environment in the direction specified by ang.
- $(\boldsymbol{apos}, aid_i, sv_i, adv_i, ev_i)$ is an interaction tuple recording an interaction that took place with agent i. It specifies agent i's identifier aid, location $apos$ where this agent was when interaction took place, service (sv, adv) that was offered to or by i and the received or given evaluation, accordingly.
- $mls = \prod_{r=0}^{ms}(apos, aid, sv, adv, ev)_r$ is a table containing the last ms interactions this agent had with supplier agents for service described by isv,
- $mlc = \prod_{r=0}^{ms}(apos, aid, sv, adv, ev)_r$ is a table containing the last ms interactions this agent had with customer agents for service described by osv.

Interaction tuple $(apos_j, aid_i, adv_i, ev_i)$ now contains the element $apos$ which allows agent j to record into memories mls and mlc the location where j was when an interaction took place on the environment with agent i. Agents need to know their current location to have a sense of positioning on the environment. $apos$ denotes the agent current position and it is reflected from the environment value $epos$; thus analogous to previous reflections for an agent $i \in Ag$ we have

$$epos_i \ \rho \ apos_i := apos_i \leftarrow epos_i. \tag{5}$$

To express that agents move on the environment, we have added action $walk(ang)$. This action transforms the environment every time it is performed because. It re-locates the agent to a position adjacent to its current one towards direction ang, giving the agent the sense of displacement. Thus, we represent this transformation as

$$\omega(i, epos_i, ang) := epos_i \leftarrow epos_i + ang. \tag{6}$$

Parameter ang is a direction-vector which, since the environment is a grid, gives an unambiguous indicator of an agent's next position. Because the environment is a toroidal grid and we allow multiple occupancy of a given node, there are 8 possible directions to move towards from any position. But the direction depends on the final destination $dpos$ the agent would decide to move to. Then we first define $dpos$ as follows:

$$dpos = (\lfloor rand() \cdot es \rfloor, \lfloor rand() \cdot es \rfloor). \tag{7}$$

Where $rand()$ is a function returning a random number within the range $[0,1]$. Thus ang is defined as

$$ang = \left(\left\{ \begin{array}{ll} -1 & \text{if } |p - x| > es/2 \\ 1 & \text{if } |p - x| < es/2 \\ 0 & \text{otherwise} \end{array} \right\}, \left\{ \begin{array}{ll} -1 & \text{if } |q - y| > es/2 \\ 1 & \text{if } |q - y| < es/2 \\ 0 & \text{otherwise} \end{array} \right\} \right). \quad (8)$$

Having defined $dpos$ and ang, the new agent position $epos$, according to (6), then is simply

$$epos_i \leftarrow epos_i + ang = ((x + u + es) \bmod es, (y + v + es) \bmod es). \quad (9)$$

For instance, if an agent i is located at $epos = (0,6)$ on an environment of size $es = 100$, and walks ($walk(ang)$) with e.g $ang = (-1,1)$ randomly selected, then the new position will be $epos = (99,7)$ because of the toroidal grid. Then the new position will be immediately reflected on the agent because of (5).

Agents can now walk randomly on the environment and because of that, agent interactions have a tendency to occur randomly as well. A business ecosystem starts its interactions randomly and as time elapses it gains structure as the system develops [1]. To acquire such structure every two agents have to come close enough to each other for interacting. Thus, to identify what determines "close enough" for interacting, an interaction radius itr is used to delimit a circle of influence around each agent.

Definition 6. An agent j senses the proximity of another agent i within a distance itr by receiving a proximity table T containing agent i's identifier, position and service descriptions:

$$T = \{ \forall_{|epos_j - epos_i| \leq itr} i \mid (eid, epos, eisv, eosv)_i \}. \quad (10)$$

Where $|epos_j - epos_i|$ is an Euclidean distance.

The proximity table T is transient: it is updated with every change regarding agent positioning occurring on the environment. It is used by agents to decide with whom to interact, according to services offered and required by agents in their surrounding. In other words, receiving a proximity table T is the mechanism all agents have for starting interactions with potential customers nearby. Moreover, the environment limits agent action $sendof$ in such a way that it is only enabled under the condition that agent i's details are in agent j's proximity table. It is important to remark that whenever a potential customer is in T, agents offer their services whether that customer is the preferred one or not. However, service evaluations are sent back regardless the supplier agent being in customer agent's proximity table.

The environment plays an important role in adaptation because the latter's properties are based on the existence of a geographical space (cf. a surface): aggregation or agent spatial clustering occur [11]; spatial diversity influence the creation of service niches [11]; flows allow agent to orient themselves [7,14]; and non-linearity occurs when the combination of interactions and environment changes makes difficult for the system to go to a previous state, i.e. it creates history [14,15].

2.3   Foraging for Services

Moore mentions in [1] that a business ecosystem is obtained when a set of randomly interacting companies develops into a more structured community. Up to this point, in our model agents can offer their services to whoever they randomly encounter on the environment, and make evaluations accordingly. However, they can hardly increment the interaction frequency with their preferred supplier and preferred customer. Thus, a structured community such as a supply chain cannot surface. Therefore, we look again into nature in order to get more ideas.

In natural ecosystems, individuals move across their environment in their search for food by following trails of potential preys. In order to do this, they perceive their environment beyond their reach by using various "senses", e.g. sight, smell, etc. That is, individuals use a notion of gradient on the environment to guide their exploratory behaviour [7,14]. By adding these capabilities to our agents we provide them with mechanisms to guide their exploratory behaviour as well towards where it seems to be services of their interest. In particular, agents might (i) "see" beyond their interaction space to identify others and their services; and (ii) "follow a trail" left by other on the environment. We now extend Def. 3 to equip our agents with "senses".

Definition 7. An environment env is a toroidal grid described using the tuple $(es, Ads, \mathbf{Mks}, R)$ where the new elements are defined as follows:

- $(epos, eisv, eosv, ods)$ is a mark description tuple specifying that an agent left a mark at position $epos = (x, y)$, with service descriptions $eisv$ and $eosv$, and the mark has an odour strength ods,
- $Mks = \prod_{m=0}^{|Ads| \cdot ods_{\max}} (epos, eisv, eosv, ods)_m$ is a table where all marks are described. The number of mark description tuples depends on the number of existing agents exposed on the environment $|Ads|$ and the maximum value ods can take. Duplicates can exist.
- $\{itr, \mathbf{sgr}, \mathbf{smr}\} \in R$ is the perceptions radii set where itr is the interaction radius, sgr is the sight radius, and smr is the smell radius; where $smr > sgr > itr \geqq 1$.

Table Mks contains information about all marks on the environment. Notice that this information does not include any agent identifier, but only the position $epos$ where an agent passed by, service descriptions $eosv$ and $eisv$ he offers and requires, accordingly, and an odour strength ods. The latter is a diminishing value specifying how long ago the agent was there, i.e. the greater the odour, the shorter time has elapsed since the agent was there. This value helps agents to find a foraging direction by simply following the mark with the greater odour strength. The sight radius sgr allows an agent to identify others within a greater space than itr allows, but not to make any sort of contact. The agent would need to approach closer in order to know the added value those agents can offer or be interested in. The smell radius smr permits an agent to perceive beyond sgr permits. It allows agents to sense marks left on the environment and follow such smells in case the agent gets interested in services described by the marks. The smell radius and marks do not allow an agent to identify others, but only to give him a hint of where to look for.

Whilst mark locations depend on where agents are positioned, the agents do not decide whether or not to leave a mark. It is a property of the environment: all agents leave marks. We represent it as environment property $\kappa(Ads, Mks)$ signifying that agent information contained in Ads is projected as new marks in Mks and left on the environment:

$$\kappa(Ads, Mks) := Mks \leftarrow \prod_{i=0}^{|Ads|} (epos, eisv, eosv, \mathrm{ods_{max}})_i. \tag{11}$$

Where $epos, eisv$, and $eosv$ come from tuples in Ads; $ods_{\max}$ is the maximum odour strength.

Moreover, mark odours vanish as time elapses until the mark disappears. This implies another transformation to reduce the odour strength from existing marks on the environment. We represent this as environment function $\tau(Mks)$ saying that existing marks vanish as time elapses

$$\tau(Mks) := Mks \leftarrow TMks \text{ where}$$
$$\{\forall\, m \in Mks \mid TMks \leftarrow (epos, eisv, eosv, ods-1)_m\} \tag{12}$$
$$\{\forall\, t \in TMks \mid TMks \leftarrow TMks \setminus (epos, eisv, eosv, 0)_t\}.$$

Thus, the total number of marks on the environment at any time is $|Ads| \cdot ods_{\max}$ as specified in Def. 7. Now, we add "senses" to our agents:

Definition 8. An agent j sees another agent i beyond a distance itr and up to sgr by receiving a sight table G containing agent i's identifier, position and service descriptions:

$$G = \{\forall_{itr \,<\, |epos_j - epos_i| \,\leqq\, sgr}\, i \mid (eid, epos, eisv, eosv)_i\}. \tag{13}$$

The sight table G is transient: it is updated with every change regarding agent positioning occurring on the environment, i.e. transformation ω. All agents described in such table cannot receive service offers until they come to at most a distance of itr. Table G can be used by agents to analyse service descriptions that other agents offer and require. But mainly, for identifying agents at a distance, i.e. it can be used to decide whether to approach an agent or not due to he being a preferred supplier according to recent interactions. Receiving the sight table G is the mechanism all agents have for identifying agents at a distance and decide whether to come closer or not for interacting.

Definition 9. An agent j smells mark m beyond a distance sgr and up to smr by receiving a smell table L containing mark m's position, service descriptions and odour strength:

$$L = \{\forall_{sgr \,<\, |epos_j - epos_m| \,\leqq\, smr}\, m \mid (epos, eisv, eosv, ods)_m\}. \tag{14}$$

Likewise, the smell table L is transient: it is updated every time a transformation κ and τ occurs on the environment. Marks perceived through the table only contain service descriptions. For identifying agents, it is necessary to come to at most a distance sgr. The list is used to track down other agents according to the services they offer and require. Receiving a smell table L is the mechanism all agents have for perceiving a notional gradient on the environment and guide their exploratory behaviour towards where it seems to be something of their interest. Seeing agents and smelling services permit agents to perceive their environment and orient themselves on it. Thus, we have already set the elements for allowing agent to have more control on their exploratory behaviour, i.e. to forage for services on the environment. Foraging requires an objective to follow which can easily be the preferred customer and preferred supplier regarding to the service evaluations received and sent, accordingly. Therefore, extending Def. 5 we have

Definition 10. We use the tuple $(aid, apos, \boldsymbol{svp}, isv, osv, A, ms, mls, mlc)$ to characterise and agent ag and its interactions; where svp is a pointer to the service description the agent is currently foraging for. Agents have two main objectives when foraging: to find the preferred customer and the preferred supplier. The preferred supplier is that with the highest evaluation ev according to mls. The preferred customer is that who sent the highest evaluation ev according to mlc. And according to that, destination point $dpos$ on the environment is selected.

Agents have two foraging behaviours: (i) directed and (ii) semi-random. The directed foraging behaviour starts when, e.g. $svp \rightarrow isv$, agent j tries to reach the location where the interaction with the preferred supplier took place. If that agent is seen, then agent j will simply follow him until interacting either with him or with a better supplier found in the way. Then, svp will switch to the other service, i.e. $svp \rightarrow osv$, and the process starts over. However, when agent j arrives at the final position and the supplier agent is not found there or at sight, then the semi-random foraging behaviour (ii) is activated. At this point, agent j tries to smell the service svp he is foraging for and to track down its source. If the preferred supplier is not found at sight, svp will switch to the other service, i.e. $svp \rightarrow osv$. Otherwise, the directed behaviour is re-activated. However, if isv is not smelled, agent j will forage randomly on the environment until isv is smelled.

Agents forage for services required and offered by those who have given the best evaluation and have provided the best service, accordingly. Agents interact with whoever come across their way whilst foraging. When an agent i is the preferred supplier for agent j and at the same time j has given the highest evaluation i has received, they will forage for each other. If no other agent with a better service comes across, i and j will eventually reduce the distance between them, and creating and strengthening in this way a link in a supply chain. Thus, a supply chain is a collection of links going from the supplier in the bottom up to the final customer in the top. Because this links are formed by preferred suppliers and preferred customers, they will either pull competitors closer whilst they keep trying to form part of the supply chain, or create a separate, similar, competing supply chain. This depends on how close they are on the environment. Thus, this behaviour resembles what occurs in business ecosystems [1,2].

2.4 From Supply Chains to Supply Webs

In previous sections, we have presented how agents interact by offering services and sending back an evaluation feedback; agents move across an environment whilst they forage for services and form part of a supply chain. However, in business ecosystems, companies typically participate in more than one supply chain. They consume services from more than one supplier and at the same time they have more than one service to offer. This situation resembles food webs in natural ecosystems where one species acquires nutrients from more than one (other) species. Conversely, more than one species typically depends on a single species. The more species depending on one particular species, the more important that species becomes for the survival of the entire ecosystem. This species is called *keystone* [5]. Analogous situations arise in business domains, where an organisation is well positioned in the market in such a way that it sets the rules and standards for others to adopt [1]. To introduce this feature into our model, we augment Def. 10

Definition 11. The tuple $(aid, apos, \boldsymbol{S}, A, \boldsymbol{M})$ is used to characterise an agent ag; where

- $\{svp, ISV, OSV\} \in S$ is the service description set, where
  - svp is a pointer to the service description the agent is currently foraging for,
  - $ISV = \prod_{i=1}^{g} isv_i$ is a list with all the services this agent requires,
  - $OSV = \prod_{i=1}^{h} osv_i$ is a list with all the services this agent offers,
- $M = \{ms, MLS, MLC\}$ is the memory set, where
  - ms is the interaction memory size,
  - $MLS = \prod_{i=1}^{g} mls_i$ is a list with all the interaction memories corresponding one-to-one to service $isv_g \in ISV$,
  - $MLC = \prod_{i=1}^{h} mlc_i$ is a list with all the interaction memories corresponding one-to-one to service $osv_h \in OSV$.

Set S holds lists OSV and ISV which describe services the agent offers and requires, respectively. svp still functions as a pointer to the service the agent is currently foraging for. But now it switches from an element in ISV to one in OSV and then back to another element in ISV and so on. This extension adds more dynamism to the ecosystem because agents can not only participate as a link in a supply chain but as a node in a supply web, resembling food webs in natural ecosystems. In addition, set M contains lists MLS and MLC. These describe the last ms interaction the agent has had with suppliers and customers for the different services the agent requires and offers, respectively. There is a one-to-one relation from a service to an interaction memory: $isv-mls$, $osv-mlc$.

Definition 12. An environment env is a toroidal grid expressed through the tuple $(es, \boldsymbol{Ads}, \boldsymbol{Mks}, R)$ where the extended elements are defined as follows

- $(eid, epos, EISV, EOSV)_i$ is an agent description tuple specifying agent i's identifier eid, position $epos = (x, y)$ on the environment, and lists of services descriptions $EISV$ and $EOSV$,
- $Ads = \prod_{i=0}^{|Ag|} (eid, epos, EISV, EOSV)_i$ is a table where all agent positions and services are described. The number of description tuples depends on the number of existing agents $|Ag|$.
- $(epos, EISV, EISV, ods)$ is a mark description tuple specifying that an agent left a mark at position $epos = (x, y)$, with service description lists $EISV$ and $EOSV$, and the mark has an odour strength ods,
- $Mks = \prod_{m=0}^{|Ads| \cdot ods_{\max}} (epos, EISV, EOSV, ods)_m$ is a table where all marks are described. The number of mark description tuples depends on the number of existing agents exposed on the environment $|Ads|$ and the maximum value ods can take. Duplicates can exist.

Since extensions have been made to the number of services an agent offers and requires, corresponding extensions have to be made to their reflections onto the environment, i.e. extensions to (3) and (4). Thus, $\forall\, i \in Ag$ we have

$$ISV \;\rho\; EISV := EISV \leftarrow ISV. \tag{15}$$

$$OSV \;\rho\; EOSV := EOSV \leftarrow OSV. \tag{16}$$

Because Ads and Mks have been extended according to Def. 12, perception tables T, G, and L are affected accordingly.

Agents now can forage for more services and give preference to the best suppliers found for each service. Moreover, they also can give preference to several customers giving them the best evaluations for each service. Since agents now participate in more than one supply chain, they became nodes in supply webs. Agent foraging for more services will pull, in the long run, more than one supply chain to it and increase competition. If an agent improves its service, offers a new services or stops offering a service, it will be noticed in service evaluations and preferences will change. Replacing a supplier for a better one or preferring one customer over another. All these changes in interactions affect directly the agents, i.e the individual level. However, at system level supply webs are expected to remain and adapt to changes occurring at the individual level. When the system develops forward like this, and makes it difficult to assign states to the system because there is no way to make the whole system go back; it is called *transformational evolution* [5]. Moreover, it complies with the non-linearity adaptation property [14].

3   Related Work

The concept of business ecosystem has been around for more than a decade and naturally there has been much research in connection with it. Indeed to do justice to the wealth of material available would require many volumes and certainly occupy much more space than remains available to us. Accordingly, we confine our attention to three works of particular relevance.

(i) The New and Emergent World models Through Individual, Evolutionary, and Social learning (NEW TIES) [16] is also inspired by nature but they pursue a more generic goal: to develop autonomously created societies, cf. business ecosystems, within virtual environments. They primarily focus on the emergence of generic societies through evolutionary learning, agent cooperation, and agent interactions whereas our purpose is, more specifically, to help businesses to survive in their dynamically changing environment.

(ii) The Digital Business Ecosystem (DBE) [17] mainly makes extensive use of evolutionary processes, i.e. using genetic algorithms to find the optimal composition of suppliers for a specific service request. However, they focus on one-shot requests where supplier compositions only last until the service is fulfilled. Their main objective is optimisation of one-shot service requests whereas we focus on the uninterrupted search of long-lasting partnerships.

And (iii) the Open Negotiation Environment (ONE) [18] study adaptation and spontaneous composition of disparate services by means of Dynamic Electronic Institutions. They basically focus on short-term associations (cf. temporary electronic institutions) whose members align their norms and objectives "on the fly" whereas we concentrate on the opportunistic discovery of potential long-lasting partnerships such as those in a supply chain.

4   Concluding Remarks

We have focused above on an incremental exposition of the theoretical aspects of DAEM. However, it is worth noting (i) how the model can be used, (ii) how it can be applied, and (iii) what is the expected impact:

(i) We are currently developing a DAEM prototype. With it, we will go through an iterative cycle to identify a set of conditions/parameters under which a business ecosystem can be maintained. For example, the environment size; amount of service offers required for the agents to gain global structure; number of agents necessary to maintain a stable system; etc. None of the projects mentioned in Sect. 3, focus on this aspect.

(ii) Using such conditions/parameters it would be possible to feed the model with existing business ecosystem information in order to discover new and (possibly) profitable interactions for organisations. This would help them to survive in their business ecosystem.

Finally, (iii) findings can both enhance the theoretical underpinnings of DAEM and provide valuable input into new theoretical models of complex business ecosystems.

The above are matters for future work. Thus, in conclusion, we have presented the Dynamic Agent-based Ecosystem Model: a synthesis of ideas from natural ecosystems and multi-agent systems. We believe this to be a novel and promising framework for modelling business ecosystems, especially at the level of interactions so fundamental to their creation and development.

References

1. Moore, J.F.: Predators and prey: a new ecology of competition. Harvard Business Review 71(3), 75–86 (1993)
2. Moore, J.F.: The Death of Competition: Leadership and Strategy in the Age of Business Ecosystems. HarperBusiness (1996)
3. Holland, J.: Hidden Order: How Adaptation Builds Complexity. Helix books. Addison-Wesley, London, UK (1995)
4. Olson, R.L., Sequeira, R.A.: Emergent computation and the modeling and management of ecological systems. Computers and Electronics in Agriculture 12(3) (1995)
5. Levin, S.A.: Ecosystems and the biosphere as complex adaptive systems. Ecosystems 1(5), 431–436 (1998)
6. Maes, P.: Modeling adaptive autonomous agents. Artificial Life 1, 1–2 (1994)
7. Parunak, H.V.D.: Go to the ant: Engineering principles from natural mutli-agent systems. Annals of Operation Research 75, 69–101 (1997)
8. Green, D.G., Sadedin, S.: Interactions matter–complexity in landscapes and ecosystems. Ecological Complexity 2(2), 117–130 (2005)
9. Marín, C.A., Mehandjiev, N.: A classification framework of adaptation in muti-agent systems. In: Klusch, M., Rovatsos, M., Payne, T.R. (eds.) CIA 2006. LNCS (LNAI), vol. 4149, Springer, Heidelberg (2006)
10. Allen, P.M., Strathern, M.: Evolution, emergence, and learning in complex systems. Emergence 5(4), 8–33 (2003)
11. Maurer, B.A.: Statistical mechanics of complex ecological aggregates. Ecological Complexity 2(1), 71–85 (2005)
12. Odell, J.J., Parunak, H.V.D., Fleischer, M., Brueckner, S.: Modeling agents and their environment. In: Giunchiglia, F., Odell, J.J., Weiss, G. (eds.) AOSE 2002. LNCS, vol. 2585, pp. 16–31. Springer, Heidelberg (2003)
13. Omicini, A., Ricci, A., Viroli, M., Castelfranchi, C., Tummolini, L.: A conceptual framework for self-organising MAS. In: AI*IA/TABOO Joint Workshop Dagli oggetti agli agenti: sistemi complessi e agenti razionali, WOA, Turin, Italy (30 November – 1 December 2004), pp. 100–109. Pitagora Editrice, Bologna (2004)
14. Jørgensen, S.E., Fath, B.D.: Application of thermodynamic principle in ecology. Ecological Complexity 1(4), 267–280 (2004)
15. Ulanowicz, R.E.: On the nature of ecodynamics. Ecological Complexity 1(4) (2004)
16. Griffioen, A., Schut, M., Eiben, A., Bontovics, A., Hévízi, G.: New ties agent. In: Socially Inspired Computing Joint Symposium AISB, SSAISB (2005)
17. Briscoe, G., De Wilde, P.: Digital ecosystems: Evolving service-orientated architectures. In: First International Conference on Bio Inspired mOdels of NETwork, Information and Computing Systems (BIONETICS), IEEE, Los Alamitos (2006)
18. Muntaner Perich, E., de la Rosa Esteva, J.L: Using dynamic electronic institutions to enable digital business ecosystems. In: Muntaner Perich, E. (ed.) International Workshop on COIN, ECAI 2006. LNCS (LNAI), vol. 4386, Springer, Heidelberg (to appear, 2007)

From Local Search to Global Behavior: Ad Hoc Network Example

Osher Yadgar

SRI International, 333 Ravenswood Avenue Menlo Park, CA 94025-3493, USA
yadgar@ai.sri.com

Abstract. We introduce the *Consensual N-Player Prisoner's Dilemma* as a large-scale dilemma. We then present a framework for cooperative consensus formation in large-scale MAS under the *N-Person Prisoner's Dilemma*. Forming consensus is performed by demonstrating the applicability of a low-complexity physics-oriented approach to a large-scale ad hoc network problem. The framework is based on modeling cooperative MAS by a physics percolation theory. According to the model, agent-systems inherit physical properties, and therefore the evolution of the computational systems is similar to the evolution of physical systems. Specifically, we focus on the percolation theory, the emergence of self-organized criticality, and the exploitation of phase transitions. We provide a detailed low-ordered algorithm to be used by a single agent and implement this algorithm in our simulations. Via these approaches we demonstrate effective message delivery in a large-scale ad hoc network that consists of thousands of agents.

1 Introduction

The *N-Person Prisoner's Dilemma*, *NPD*, is an evolvement of the prisoner's dilemma. More than two players are involved in *NPD*, and therefore a group cooperation issue arises (Akiyama & Kaneko 2002) (Lindgren & Johansson 2001). In this paper we scale up the number of players to thousands, and thus consider large-scale problems. As in the case of the prisoner's dilemma, *NPD* offers each player two choices. The first choice is to *cooperate* and the second is to *defect*. While having no information about other players, the dominant strategy for the dilemma is to defect. Following the same strategy, the equilibrium is for all players to defect. As in the prisoner's dilemma, this equilibrium is a Pareto-suboptimal solution since cooperation between all agents leads to a better solution.

Consensus is a cooperative decision-making process that seeks general agreement over one solution. According to traditional game theory, consensus formation techniques such as voting require central control or coordination via communication among all agents (Ephrati 1993). The famous prisoner's dilemma is one example of a well-studied process of the benefits that arise from reaching consensus and from adopting the general decision. We define a *Consensual*

M. Klusch et al. (Eds.): CIA 2007, LNAI 4676, pp. 196–208, 2007.

N-Person Prisoner's Dilemma, Con-NPD, as a generalized form of *NPD* and present an example for emergent consensus in a large-scale environment.

In large agent communities, maintaining centralized control or direct connection between all agents is too costly. Therefore, when the number of agents increases, the complexity of most of the cooperation methods becomes unbearable. To resolve the scale-up computational explosion of cooperation mechanisms in a large multiagent system we present a different approach. We apply a model based on methods from percolation theory to model large-scale agent communities. The physics-oriented methods are used to construct a beneficial cooperative consensus formation algorithm to be used by the single agent within the system. We show via simulations that using the physics-oriented approach enables effective emergence of consensus formation in very large agent-systems.

Many problems arise in large-scale MAS research. In this paper we concentrate on investigating the consensus emergence in large-scale MAS. More specifically, we consider cases in which cooperative autonomous agents form a large-scale ad hoc network. We describe a model that allows for the dynamic consensus emergence and is appropriate for a large-scale MAS and test it. The latter is performed by simulating a dynamic agent-system that follows our suggested mechanisms and consists of thousands of agents.

The main contributions of this paper are the following. We introduce the *Con-NPD* for the first time and show how it relates to consensus problems in large-scale MAS. We use percolation theory for emergence of consensus in large autonomous agent communities. Our approach is fully distributed and results in very low complexity. By exploiting phase transition phenomena that are typical of percolation systems we show how agents can predict the utility of the solution and how they should choose their strategy. The framework we present provides a solution to problems that were not addressed previously in large-scale MAS, and may be the basis for future solutions for a larger class of problem domains. We show here applicability to one domain. A demonstration of the presented configuration approaches was presented (Yadgar 2007); however, more research is necessary to determine applicability to additional domains.

1.1 Assumptions

We assume that the agents with which we deal have the ability to perceive the virtual displacement in the problem-space, can change their virtual displacement gradually, and can perceive the properties of other adjacent agents. This may be done by sensors integrated into the agents. This assumption is necessary since one agent is expected to propagate from state to state within the problem-space according to the properties of the surrounding agents. To enable such propagation, some knowledge regarding neighbors is necessary. We assume that each agent has a performance capability that can be measured using standard measurement units. The standard measurement will be used as a quantitative way of measuring the agents' success in becoming part of the consensus while fulfilling their goals. In addition, we assume that a scaling method is used to represent the displacements of the agents in the problem-space and to evaluate the mutual distances between goals and agents within this space. This assumption is necessary since virtual and physical distances are a significant factor in the model we present.

1.2   Percolation and Physics Notation

Our model relies on the percolation theory (Grimmett 1999), specifically on the site-percolation model. According to the percolation model a geometrical construction of sites and bonds between them is referred to as a lattice. A lattice may be considered a graph of vertices and edges, respectively. An edge may exist only between two sites considered as closest neighbors; that is, the distance between them is smaller than a predefined range. The percolation model may be characterized as two types, the bond and the site. In a bond percolation model a probability p represents the probability for an edge to be open. In a site percolation model a probability p represents the probability for a vertex to be open. An open edge or vertex means that current, such as water or electricity, can pass along that edge or vertex. Site percolation is more general since every bond percolation may be reformatted to a site percolation while not all site percolation can be represented by bond percolation. In this paper we focus on the site percolation model.

We say that two occupied vertices belong to the same cluster if there is a path of edges leading from one vertex to the other. As we increase p we may see that the number of clusters is reduced while the average cluster size is increased. Percolation theory focuses on the behavior of an infinite lattice and studies the condition for existence of an infinite cluster. Such a cluster spans from one side of the lattice to the other, and current can pass from one side of the lattice to the other. Studies show that the probability for the existence of such a cluster is either zero or one, whereas for small p the probability is zero and for large p the probability is one. Given that, there is a critical probability p_c such that if $p > p_c$ almost surely, there is an infinite cluster for that lattice, while for $p < p_c$ almost surely, there is no infinite cluster. We refer to the probabilities below p_c as the subcritical phase, to the probabilities above p_c as the supercritical phase, and to p_c as the phase transition.

Large-scale random distribution of nodes has a very distinguished self-similarity property (Willinger et al. 1996). Benefiting from our physics metaphor, we may use scale-invariant statistical field theory to deduce that phase transition is not dependent on the scale of the system as long as the system is big enough. The same theory proves that the important property is the particle density rather than the scale. That is, if we increase the size of the area and the number of agents by the same factor we will witness the same global utility. This means that if we keep the same agent density, it takes the same evolution time for the model to reach the same utility regardless of the scale of the problem.

Moreover, following scale invariance laws, if an agent senses other agents in a big enough area it may use the same conclusions about the whole system. In that case, an agent may find the probability to achieve consensus and act upon that knowledge. If it finds that a consensus could emerge, it would act to achieve it. If it finds that a consensus could not emerge, it could save its resources by not trying to cooperate in vain. In our study we validated these deductions when applying our model to the large-scale ad hoc network example.

Using a lattice and playing the prisoner's dilemma against all neighboring agents is not a new concept. Feldman and Nagel (Feldman & Nagel 1993) and Outkin (Outkin 2003) arrange agents on a lattice, apply a cellular automaton mechanism, and let them play with their neighbors. Our framework is different since we use physics metaphors and insights from percolation and scale-invariant statistical field theories to motivate agents and deduce the performance of the system.

Krishnamachari et al. (Krishnamachari et al. 2001) use random graph theories and a fixed radius model to discuss a similar phase transition phenomenon in ad hoc networks; however, they do not suggest ways to leverage this phenomenon and ways to enhance it. In this paper we demonstrate how to use the phase transition phenomenon to predict the likelihood for consensus emergence and suggest a novel algorithm to achieve this phase transition with fewer resources.

In the next sections we define the *Con-NPD* and present the large-scale ad hoc network as an example to that dilemma. We then detail the algorithm for the autonomous agent and demonstrate our solution with simulations. We conclude with a detailed discussion of our results and the conclusions arising from this framework.

1.3 Emergent Behavior and Self-organization

Emergent behavior and self-organization are well-studied fields. Camazine et al. (Camazine et al. 2001) define self-organization as a process in which a pattern at the global level of a system emerges solely from numerous interactions among the lower-level components of the system. Much of the work that has been done in these fields is inspired by biology and animal behavior such as ants, bacteria, and other insects (Vitorino & Ajith 2005). In their paper, Conradt and Roper (Conradt & Roper 2005) study consensus decision making in animals. They explain the motivation of social species for jointly made decisions by their need to stay together and review empirical and theoretical studies of consensus decision making in animals. They conclude that consensus decision making is common in nonhuman animals, and that cooperation between group members in the decision-making process is likely to be the norm, even when the decision involves significant conflict of interest. Understanding the motivations and the behavior of social species is very important but could take a long time to converge. We, on the other hand, present a general dilemma in which the equilibrium strategy is for agents not to cooperate; however, this equilibrium is a Pareto-suboptimal solution since cooperation between all agents leads to a better solution. We use percolation theory rather than biological theories. We apply our solution to an ad hoc network formation problem, demonstrate how fast a consensus can emerge, and draw guidelines to deduce the likelihood for a consensus to form.

2   Consensual N-Person Prisoner's Dilemma

We focus on a specific set of problems of consensus decision making. We consider the problems in which it is beneficial to become a part of a consensus, but partial information disrupts distinguishing between vast majority and very small minority groups.

Definition 1. *Consensus group*
Let A be a set of agents and S a set of solutions to the problem P.
  A group $C \subseteq A$ is said to be a consensus group if for each $c \in C$ there is a solution function $f(c) \rightarrow s$ where $s \in S$ and $|A| - |C| = \epsilon$ such that $\epsilon \rightarrow 0$.

In an agent community, a *consensus group* is a subset of that community such that every member in that group agrees with all members of the group on the solution to the problem, and the size of the group is very close to the size of the whole community.

Let us consider the following case of a large-scale version of the prisoner's dilemma, in which the authorities tend to accept the vast majority opinion. King Solomon and the Bees is a variant to the King Solomon and the Bee tale by H. N. Bialik.

One day while King Solomon was napping under his fig tree, a group of bees passed the guards and stung his nose. Jumping up with pain, he knew that his nose was stung by one or more bees. Angrily, he summoned all the bees in his garden to find and punish those who were responsible for the crime. Knowing that the bees knew which ones were responsible, he cried, "I'll give each one of you two choices, to keep silent or to tell me who are responsible for this." Since King Solomon liked honey and did not want the bees to stop producing it, he decided not to severely punish the entire bee community. Being fair and just, he added, "and here is the way I will punish the guilty ones."

Table 1. Consensual N-Person Prisoner's Dilemma

	Consensus Stays Silent	Consensus Betrays	No Consensus
Bee Stays Silent	0.5	10	10
Bee Betrays	Free	2	Free

Following Table 1, a bee serves six months in prison if she and the general community keep silent. She serves two years in prison if she and the general community betray each other. She is sent free if she betrays and there is no general behavior for the rest of the community, and finally, she serves ten years in prison if she chooses to be silent and there is not common behavior among the other bees.

The intelligent bee will have to carefully consider her best move and take into account the other bees' reactions. If the vast majority stays silent, her best move is to betray and gain freedom. If the vast majority betrays, her best move also would be to betray since she will serve two years instead of ten in prison. Finally, if there is no consensus among the bees, betraying is also the best move that will grant her freedom. Following the same strategy, all bees arrive at the same conclusion to betray and therefore will serve two years in prison. However, had the bees cooperated they could have achieved a better result and served only six months in prison. Moreover, even if the bees know that cooperation is beneficial they will still need to find out how to form a consensus in a large-scale community since traditional mechanisms are not applicable.

In the next section we present the large-scale ad hoc network problem as an example to the *Con-NPD*.

3   Large-Scale Ad Hoc Network Problem

In a large-scale disaster such as an earthquake or a flood, one of the first things to collapse is the communication infrastructure. Communication is an essential means for rescue operations, and survivability of victims of such disasters decreases sharply with every minute that passes. Therefore, rapidly reviving the communication infrastructure in a large-scale disaster zone is crucial to saving many lives.

Nodes of an ad hoc network should support both local and global requirements. The local requirement is to allow nearby users of phones, computers, and PDAs to receive service, while the global requirement is to maintain network connectivity. Agencies responsible for establishing such networks wish to balance between the need to service as many users as possible and the maintenance of network connectivity. Formally, this balance could be defined by the agencies as a utility function of the system. These agencies wish to maximize the utility of the system under the limitation of a given budget.

Reflecting recent technology advances, we propose to use thousands of small, short-range, inaccurate, low-cost, nodes to support the requirements. Following the spatial behavior of the network we show a phase transition in the utility of the system as a function of the number of nodes. That is, (i) for a few nodes, there is no applicable deployment while (ii) for many nodes any random deployment achieves the maximum utility, and (iii) the transition between zero utility and maximum utility is sharp.

Considering the phase transition phenomenon, we show how to determine the number of agents needed to maintain such ad hoc networks and demonstrate an algorithm to drastically reduce the number of nodes required for the phase transition. We propose to apply intelligent behavior to every node. This behavior considers very limited partial information obtained by the node. To reduce resource consumption, a node does not use communication with other nodes for implicit cooperation but only for acquiring the locations of its neighboring nodes.

We may consider the balancing between local and global requirements as a *Consensual N-Person Prisoner's Dilemma* if we use autonomous agents as network nodes. In the ad hoc network problem each node has two options: to stay in its original location and thus be at the optimal location to communicate with its users or to spend resources to improve network connectivity. These two options are equivalent to *defect* and *cooperate*, respectively, as described in the previous section. Spending resources to communicate with neighboring agents might not be beneficial since these neighbors might not be connected to the rest of the nodes.

Following Definition 1 we say that *consensus* in the ad hoc network domain problem is a conclusion of the vast majority of the agents to cooperate and *consensus group* is the group of those agents. We suggest using the phase transition phenomenon to infer whether a *consensus group* is likely to exist. The strategy of a single agent would be then to *cooperate* only if there is such likelihood and otherwise to *defect*.

4   Physics-Oriented Agent

Using only partial information, each agent is motivated to increase the global utility while following its own utility function. To do that we built a utility function for an agent that is combined out of two elements, the local motivating the agent to *defect* and the global motivating the agent to *cooperate*. Based on our previous work (Shehory et al. 1999) and (Yadgar et al. 2003), we represent the global considerations by a potential field that drives the agent toward other agents but rejects it from them if it is too close to them. The influence of an adjunct agent to the potential is represented in a grid around the agent, and the size of the grid depends on an *awareness range* defined for the agent. The potential contribution of each adjunct agent may reach up to an *interaction range* and its value is between -1 and 1 while negative potential values represent attraction and positive values represent rejection.

To represent the local motivations we used an exponentially decaying function. This function depends on the distance of the agent from its origin and aims to motivate the agent to remain in its original location. We used the same grid that was used to represent the potential field to incorporate the local considerations. The contribution of the local consideration to every cell in the grid may vary between 1, for maximum contribution, and 0, for minimum contribution.

To model the overall agent utility we prefer to use a model that does not require long-range interactions. In such a model we may take into account only local interaction and reduce the computational complexity. Since we want the local consideration to be dominant we represent the utility as a multiplication between the local and the global motivations. We use the grid to represent the range of possibilities to gain utility if the agent were in a certain location on the grid. We find the best places that result in the highest utility value and randomly choose one of them. The agent then moves in the direction of this cell.

As we stated above, the density of agents is the dominant property. Therefore, if the density is too high we may reduce the *awareness range* such that for any number of agents the awareness area will include approximately the same number of agents.

To describe the algorithm that follows the above guidelines we first present our notation. Then we follow the steps that each agent should autonomously repeat every second.

Notation:

- a : current agent
- $< a.x, a.y >$: current location of the agent
- $< a.o.x, a.o.y >$: original location of the agent
- $< a.x + i, a.y + j, R >$: circular area defined by its center and radius
- R_A : range of awareness
- R_I : range of interaction
- V : potential
- C_c, C_d : *cooperate* and *defect* coefficients

Algorithm:

```
first = true                       // Forming a grid of possible
for (i=-R_A;i<R_A;i++) {           // utilities around the agent:
  for (j=-R_A;j<R_A;j++) {         // Looking for neighboring
    d_2 = i^2+j^2                  // agents within the
    if (d <= R_A_2) {             // awareness perimeter
      SA = agents in <a.x+i,a.y+j,R_I>
      V = 0                        // Computing the contribution
      for each sa in SA {          // of the global considerations
        dsa_2=(a.x+i-sa.x)^2+(a.y+j-sa.y)^2
        V -= 2*dsa_2/R_I_2-1
      }
      cooperate = -C_c * V

      D_o_2 = a.o.x^2+a.o.y^2       // Computing the contribution
      defect = exp(-C_d*D_o_2)     // of the local considerations

      u = cooperate*defect         // Integrating global and local
                                   // contributions
      if (first || u > maxU) {     // Remembering the location
        first = false              // of the best utility
        maxU = u
        maxUset.clear();
      }
      if (u == maxU)
        maxUset.add(<a.x+i,a.y+j>)
    }
  }
}
k = (int)(random()*maxUset.size())        // Moving towards
move 1 unit in direction maxUset.get(k) // the best utility
```

Since the number of sensed agents is approximately constant and since the *awareness range* may be reduced if the agent density is high but not increased if it is low, the time complexity of this algorithm is $O(1)$.

5   Simulation and Results

To examine our model and show its applicability to real problems, we performed a set of simulations. Via these we demonstrate effective consensus emergence in a dynamic MAS that consists of thousands of agents. The problem domain for which the simulations were performed is as follows. We simulate a large-scale disaster zone in which thousands of sites each demand means to communicate with any other site. Such problems in real environments are commonly solved by dictating a central decision and using a few stationary, well-equipped communication nodes. We simulate by using many low-cost mobile nodes instead to provide a communication service such that the nodes will move around to form a network but stay close enough to the users to provide better reception.

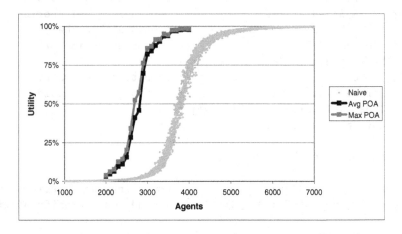

Fig. 1. Bringing phase transition forward

We consider a large-scale area in the size of 1,000,000 x 1,000,000 meters. We assume that users are randomly spread across the simulated zone, and nodes are in the best places to gain reception. Each node has an *interaction range* of 20,000 meters and an *awareness range* of 100,000 meters. That is, each node may interact with other nodes residing within its interaction area that covers 0.13% of the overall area. Each node may also be aware of the location of other nodes residing within its awareness area that covers 3.14% of the overall area. While in their positions, nodes may not have the capability to interact with other nodes since they might be out of each other's *interaction range*. We consider this deployment as the best for reception and hence the best deployment to service local requirements.

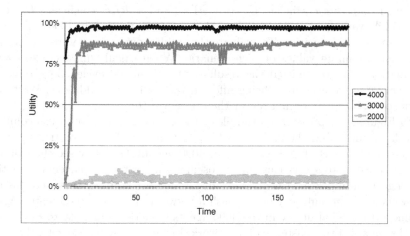

Fig. 2. Evolution of deployment of 2000, 3000, and 4000 agents

To study the emergence of a network in such an environment we started by looking for the probability that two nodes would be able to communicate. In an examined scenario we randomly placed N nodes in our tested large-scale area. In this initial study, nodes stayed stationary. We refer to this as a *naive* deployment and compare the deployment achieved by applying the percolation theory to it. For the *naive* deployment we examined 8,000 different scenarios such that in each scenario we used a different number of nodes and we varied the number of nodes from 1000 to 9000. For each scenario we connected each node to its adjacent nodes if they were inside its *interaction range*. Following that procedure a set of separate clusters was formed. Each cluster represents a group of nodes that may exchange messages.

We used the following metrics to evaluate connectivity: we randomly picked two nodes and marked a success if the two belonged to the same cluster and a failure if not. We repeated choosing pairs of nodes for a large number of pairs (ten times the number of nodes) and calculated the percentage of successfully delivered messages. We decided not to check all the possible pairs, which would be computationally expensive. Checking all the possible pairs has the time complexity of $O(N^2)$ while our procedure has the time complexity of $O(N)$; N is the size of the cluster. Since we randomly picked up a large number of pairs ($>> N$) our pair population is a good representation of the whole possible population. We tested each scenario 20 times. Each time, we randomly redeployed the same number of nodes and reported the average value.

As we anticipated, we discovered a phase transition point (see Figure 1). We saw that while randomly spreading less than 3200 nodes, only 10% of the messages arrived at their destination, while to have 90% of the messages successfuly arrive at their destinations we need 4300 nodes.

We then attached to each node a Physics-Oriented Agent that was allowed, autonomously, to move according to the detailed algorithm. In our settings we assigned *cooperate* coefficient C_c to be 1E-7, and a *defect* coefficient D_d to be 0.5. While varying the number of agents from 2000 to 4000 in 100-agent steps, we examined 21 scenarios. To evaluate the results, our metrics were the average and the maximum values of each scenario. When calculating averages over a simulation run, we omitted the results of the first 100 cycles. This omission prevents the averages from being affected by the initial conditions. We tested each scenario ten times and report the average of the two metrics.

In Figure 2 we present an example of three scenarios of consensus evolution. The figure describes the global utility as a function of the evolution cycle for large-scale systems of 2000, 3000, and 4000 agents. One may see that all scenarios converge to a very close utility oscillation. For the 2000 agents formation the algorithm improved the global utility from 0.6% to 4.7%. However, this improvement is not sufficient to establish a network. The Physics-Oriented Agent bounced the global utility in the 3000 agents scenerio from 2.9% to 87.4% and by that enabled the existence of a network. In the case of 4000 agents, the naive approach formed a network; however, the Physics-Oriented Agent improved the utility from 79% to 97.2%, thus resulting in a much more reliable network.

Analyzing Figure 2 we may also notice that the converged value is not the maximal value and that the maximum value for scenarios with a low number of agents is achieved at a very early stage. There may be formations that produce higher utility but are less stable and robust. This explanation is supported by our findings showing that the average degree of a node in those scenarios increases as the network evolves and converges when the utility converges.

The maximum utility formation may be useful if the deployment is simulated or if the agents have backtracking capabilities. In these cases, a decision must be made about whether to have a more robust network or a better-performing network. We will show next that adding a small number of agents may compensate for the difference between the utility gained by the network at the stable state and its maximum value.

Figure 1 presents the global utility as a function of the number of agents for the *naive* model and for the Physics-Oriented Agent model. As shown, following the *naive* approach we will need at least 4300 nodes to establish a network with a utility that is over 90%. To achieve the same utility while using the proposed Physics-Oriented Agent model we will need only 2700 nodes. This reduction of about 40% can be used to establish a more efficient network. We can also notice that the maximum utility value curve is a little ahead of the average. This means that we may backtrack to the formation that resulted in the maximum utility if the utility converged value is lower, or we may compensate by adding more agents to have the utility of that maximum value.

Through the scale invariance feature of the system, an agent may sense other agents in its *awareness range* and deduce about the prospects for a network to emerge. If the density of the agents in its awareness area is equal to or greater than the global density of the agents right after phase transition, a network would emerge. If the density of the agents in its awareness area is equal to or smaller than the global density of the agents right before phase transition, a

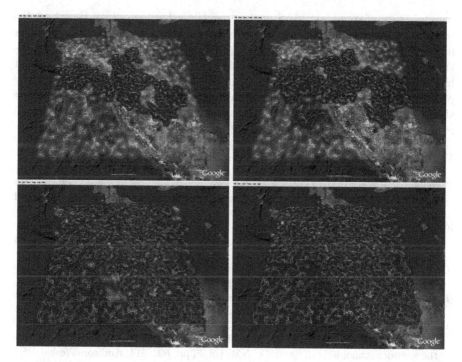

Fig. 3. Emergence of an ad hoc network of 3000 nodes over Indonesia: after 0 (naive), 1, 10 and 238 cycles

network would not emerge. More specifically, in our settings, if the local density is larger than $2700/1,000,000^2$, that is if there are more than 84 agents in an *awareness range* of 100,000 meters, a network would emerge. If the local density is smaller than $2300/1,000,000^2$, that is, if there are fewer than 73 agents in an *awareness range* of 100,000 meters, a network would not emerge. An agent may then act upon that knowledge and save energy and resources if it concludes that there is no chance for a network to emerge, or follow the algorithm described above to form a network if it concludes that a network would emerge.

Figure 3 illustrates the emergence of an ad hoc network over Indonesia. This network combines 3000 nodes following the Physics-Oriented Agent model. In the first image we see the naive solution that did not produce a reliable network since its utility is only 9%. The naive distribution is also referred to as distribution in evolution cycle 0. After one cycle the model formed a network that produced a utility of 16%. After only ten evolution cycles a relible network emerged and gained a utility of 93%. Keeping the network evolving results in a more effective solution with a utility of 99% after 238 cycles.

6   Conclusion

We introduced the *Consensual N-Person Prisoner's Dilemma* and showed how it relates to consensus problems in a large-scale multiagent system. Achieving consensus in a large-scale MAS environment is a hard problem. Based on the percolation theory we presented a novel approach for emergence of consensus in large autonomous agent communities. Our presented approach is fully distributed and results in very low complexity on the part of a single agent that has the order of $O(1)$. Following our model, each agent may also predict the prospects for network emergence and act efficiently upon this knowledge. We showed that by applying our model we can achieve an early phase transition and exploit that to dramatically improve the system's utility and resource usage.

References

[Akiyama & Kaneko 2002]Akiyama, E., Kaneko, K.: Dynamical Systems Game Theory - A New Approach to the problem of Social Dilemma. Physica D 167(1–2), 36–71 (2002)

[Camazine et al. 2001]Camazine, S., Franks, N., Sneyd, J., Bonabeau, E., Deneubourg, J., Theraula, G.: Self-Organization in Biological Systems. Princeton University Press, Princeton, NJ, USA (2001)

[Conradt & Roper 2005]Conradt, L., Roper, T.: Consensus Decision Making in Animals. Trends in Ecology and Evolution 20, 449–456 (2005)

[Ephrati 1993]Ephrati, E.: Planning and Consensus among Autonomous Agents. Ph.D. Dissertation, The Hebrew University of Jerusalem (1993)

[Feldman & Nagel 1993]Feldman, B., Nagel, K.: Lattice Games With Strategic Takeover. In: Stein, D., Nadel, L. (eds.) Lectures in Complex Systems, Papers from the summer school held in Santa Fe, NM, vol. 5, pp. 603–614. Addison-Wesley, USA (1993)

[Grimmett 1999]Grimmett, G.: Percolation. In: Zwiers, J. (ed.) Compositionality, Concurrency, and Partial Correctness. LNCS, vol. 321, Springer, Heidelberg (1989)

[Krishnamachari et al. 2001]Krishnamachari, B., Wicker, S., Bejar, R.: Phase Transition Phenomena in Wireless Ad-Hoc Networks. In: Proc. of the Symp. on Ad-Hoc Wireless Networks (2001)

[Lindgren & Johansson 2001]Lindgren, K., Johansson, J.: Coevolution of strategies in n-person Prisoner's Dilemma. LNCS. Addison-Wesley, London, UK (2001)

[Outkin 2003]Outkin, A.: Cooperation and Local Interactions in the Prisoners' Dilemma Game. Journal of Economic Behavior & Organization 52(4), 481–503 (2003)

[Vitorino & Ajith 2005]Vitorino, R., Ajith, A.: ANTIDS: Self Organized Ant-Based Clustering Model for Intrusion Detection System. Swarm Intelligence and Patterns special session, 977–986 (2005)

[Willinger et al. 1996]Willinger, W., Taqqu, M., Erramilli, A.: A Bibliographical Guide to Self-Similar Traffic and Performance Modeling for Modern High-Speed Networks. Oxford University Press, Oxford (1996)

[Yadgar et al. 2003]Yadgar, O., Kraus, S., Ortiz, C.: Cooperative Large Scale Mobile Agent Systems. In: Proc. of the Autonomous Intelligent Networks and Systems Symposium (2003)

[Yadgar 2007]Yadgar, O.: Emergent Ad Hoc Sensor Network Connectivity In Large-Scale Disaster Zones. In: Proc. of the Sixth International Joint Conference on Autonomous Agents and Multi Agent Systems (AAMAS07) (2007)

[Shehory et al. 1999]Shehory, O., Kraus, S., Yadgar, O.: Emergent Cooperative Goal-Satisfaction in Large Scale Automated-Agent Systems. Artificial Intelligence 110(1) (1999)

A Generic Framework for Argumentation-Based Negotiation

Markus M. Geipel[1] and Gerhard Weiss[2]

[1] Chair of Systems Design, ETH Zurich, Kreuzplatz 5,
8032 Zurich, Switzerland
mgeipel@ethz.ch
[2] Software Competence Center Hagenberg GmbH, Softwarepark 21,
4232 Hagenberg, Austria
gerhard.weiss@scch.at

Abstract. Past years have witnessed a growing interest in automated negotiation as a coordination mechanism for interacting agents. This paper presents a generic, problem- and domain-independent framework for argumentation-based negotiation that covers both essential agent-internal and external components relevant to automated negotiation. This framework, called Negotiation Situation Specification Scheme (N3S), is both suited as a guideline for implementing negotiation scenarios as well as integrating available approaches that address selective aspects of negotiation. In particular, N3S contributes to the state of the art in automated negotiation by identifying and relating basic argument types and negotiation stages in a structured and formal way.

1 Introduction

Agents are autonomous entities that are situated in an environment and capable of flexible action [30,31]. Agents typically have conflicting interests, incompatible goals and limited capabilities, and so there is a need for principles and mechanisms that enable agents to coordinate themselves. In the multi-agent area various such principles and mechanisms have been proposed, including partial global planning [3], contracting [1,29], commitments and conventions [6], GPGP/TAEMS [10], auctioning, voting, and many others. Recent years have witnessed a growing interest in automated negotiation as a coordination mechanism especially for complex applications [11,14,19].

Negotiation is a communication-based, knowledge-intensive process during which agents try to come to mutually acceptable agreements by exchanging appropriate arguments that influence – convince, persuade, etc. – the respective other agents [13,21]. Many approaches to automated negotiation have been described in the literature. In [20] these approaches are divided into three groups: the more traditional game-theoretic and heuristics-based approaches, and argumentation-based approaches. Compared to each other, the argumentation-based approaches are better suited for applications in which agents have incomplete or inconsistent information about each other and the

M. Klusch et al. (Eds.): CIA 2007, LNAI 4676, pp. 209–223, 2007.

environment in which they act. Argumentation-based negotiation (ABN), however, has many facets and is a quite complex mechanism. As a consequence, and as has been already argued in [20], there is a strong need for a framework that identifies the essential components that are needed to conduct automated negotiation. This paper presents a generic argumentation framework, called Negotiation Situation Specification Scheme (N3S), that we developed in response to this need. N3S, which in part is inspired by the real-world negotiation method proposed in [5], covers negotiation both from an agent-internal perspective and an external perspective. While the former perspective focuses on the individual agents' cognitive and societal abilities needed for negotiation, the latter perspective concentrates on the communication and argumentation processes in which the agents are involved during negotiation. Specifically, N3S addresses the open issue of structuring the overall negotiation and argumentation process by identifying argument types and relating them to different negotiation stages.

The paper is structured as follows. Section 2 describes selected related work. Sections 3 and 4 introduce the basic agent-internal elements and the external perspective of N3S, respectively. Section 5 illustrates the overall flow of argumentation induced by N3S. Finally, Section 6 summarizes the key features of N3S, compares N3S to related approaches, and shows important research issues raised by N3S.

2   Related Work

Several frameworks for ABN have been proposed in the literature. Four of the most prominent representatives of these frameworks are overviewed below; others are, for instance, described in [15,16,26,27].

The interest-based negotiation framework [17] builds on the observation that in human negotiation an agreement is often easier to achieve through argumentation about interests (i.e., desires and goals adopted to fulfill the desires) rather than positions – while positions may be fully incompatible, some of the interests underlying incompatible positions may even be fully shared by the negotiating parties. As a consequence, the concept of conflicts is essential to this framework, and different conflicts (namely, conflicts w.r.t. resources, goals, desires and beliefs) and conflict resolution methods (namely, concession making, exploration of alternative offers/goals, and persuasion) are distinguished. Specifically, three types of arguments are identified, including arguments for beliefs, arguments for desires, and arguments for plans. Other key elements of the framework are a communication language and protocol and a methodology for negotiation strategy design.

The trust/persuasion-based negotiation framework [22] concentrates on the reduction of multiple sources of uncertainty a negotiating agent typically is confronted with. Specifically, the framework includes two main components. First, an interaction-based trust model, called CREDIT, that allows to reduce uncertainty regarding the reliability and honesty of negotiation partners. CREDIT aims to assess the trustworthiness of other agents by taking into account own experience from direct interactions and experience gathered by other agents. Second, a model of persuasive negotiation (i.e., negotiation based on rewards that are either given or asked from one agent to another) that allows to reduce uncertainty regarding the preferences and action sets of negotiation partners during bargaining. This model is complemented by a protocol for persuasive argumentation and a strategic reasoning mechanism for generating persuasive arguments. The framework distinguishes two broad classes of illocutions to be used in persuasive negotiation, namely, negotiation illocutions (*propose* and *accept*) and persuasive illocutions (*reward* and *ask_reward*).

The logic-based negotiation framework proposed in [9] distinguishes six argument types known from human negotiation: threats to produce goal adoption or abandonment; promises of future reward; appeal to past reward; appeal to precedents as counterexamples; appeal to prevailing practice; and appeal to self-interest. In the case multiple arguments are available to an agent, he applies them successively in a predefined order (which appears to be plausible but not necessarily optimal). The framework emphasizes agent-internal elements, and does not provide explicit communication protocols.

The formal framework suggested in [28] is intended to cover negotiation scenarios in which an agent tries to convince another agent to execute a particular task on his behalf. Negotiation, considered as a sequence of offers and counter offers containing values for the issues under negotiation, is achieved through the exchange of illocutions in a shared communication language. Two types of negotiation-relevant illocutionary acts are distinguished: acts used to make offers (*offer, request, accept, reject, withdraw*), and acts used for argumentative persuasion (*appeal, threaten, reward*). The framework allows to specify attack and support relationships among arguments as well as authority relationships between agents. Based on these relationships, arguments are evaluated and generated.

While the available ABN frameworks provide guidance regarding the types of arguments and illocutions needed during negotiation, they do not structure the overall argumentation process per se and it remains unclear whether and how the different arguments and illocutions correspond to specific phases of negotiation. Such a correspondence, however, is essential to the realization of complex automated negotiation scenarios.

3 Agent-Internal Elements of N3S

N3S distinguishes five agent-internal elements relevant for automated negotiation: options, constraints, interests, utilities and arguments. These are described below. In the following, let $A = \{a_1, \ldots, a_n\}$ denote the set of agents participating in the negotiation process.

3.1 Options

N3S assumes that each agent a has his own set of options which he deems feasible. This set is denoted by $O^{(a)}_{feasible} \subseteq \mathbb{O}$ where $\mathbb{O}$ is the set of all possible options.

According to N3S, an option defines a world state, no matter which modeling perspective is applied. In the case of a task-oriented perspective of the world, each option $o \in O_{feasible}^{(a)}$ defines an ordered set of task allocations, that is, a collection of tasks $T = \langle T_1, \ldots, T_n \rangle$ where $\forall i\ T_i$ is the task assigned to $a_i \in A$.

Example 1. Two agents, a_1 and a_2, negotiate the assignment of the tasks "collect data from database" ($T_{collect}$)and "analyze data" ($T_{analyze}$). An option of a_1 may be to take care of data collection while a_2 takes care of data analysis, i.e., $\langle \{T_{analyse}\}, \{T_{report}\} \rangle$.

In value-oriented domains, each option $o \in O_{feasible}^{(a)}$ defines a set of values of parameters describing the world or a part of it.

Example 2. An agent a_b wants to buy a car from a sales agent a_s, where one of his buying options specifies the following car attribute values ("AC" stands for "Air Conditioning"): $(price \leq 50000, brand \in \{\texttt{Mercedes}, \texttt{BMW}\}, hasAC = \texttt{true}, topSpeed \geq 200)$ (assuming prices in € and speed in km/h throughout this paper).

An important subclass of options distinguished by N3S are *non-negotiated options*, that is, options that can be achieved by an agent a even *without* agreement of any of the agents in $A \setminus a$. Note that each non-negotiated option, $o_{nn}^{(a)}$, of an agent a is feasible by definition, that is, $o_{nn}^{(a)} \in O_{feasible}^{(a)}$. In this paper, $O_{nn}^{(a)}$ denotes the set of all non-negotiated options of agent a. (The utility of the best $o_{nn}^{(a)}$ defines the acceptance bottom line for the agent (see 4.2), and non-negotiated options play an important role in the argumentation process (see 3.4)).

Example 3. Consider the car dealer scenario from Example 2. If a_s is not the only sales agent out there, a_b can also buy a car from another sales agent a_{s2}. The options a_{s2} proposes to a_b are a_b's non-negotiated options in his negotiation with a_s.

Negotiation is a chance for the agent to ameliorate his situation by cooperation, but not every negotiation necessarily generates acceptable options. If agents agree upon an option, they commit themselves to achieving a specific world state. In the sequel, o^* denotes the option upon which the agents finally agree. A negotiation is said to fail, if $o^* = \epsilon$ (i.e., o^* is empty).

3.2   Constraints

A constraint defines an option or a set of options which an agent considers as impossible to realize (under current circumstances) and thus as not negotiable. More specifically, because options define world states, a constraint defines world states which an agent does consider as *not* feasible. N3S assumes that each agent a has his own set of constraints $C^{(a)} \subseteq \mathbb{C}$ where $\mathbb{C}$ denotes the set of all possible constraints. The union of all individual constraint sets $C = \bigcup_{a \in A} C^{(a)}$ rules out the options that are out of question for mutual acceptance. The constraint set C is not necessarily congruent with the set of constraints C_{real} that reality imposes. In the case that the agents settle on an option $o^* \in (C_{real} \setminus C)$, the outcome of the negotiation is void and from that it follows that $o^* = \epsilon$.

Example 4. The constraints for the sales agent a_s from Example 2 may include that she does not sell BMWs going faster than 250 km/h: $(\neg((brand = \text{BMW}) \wedge (topSpeed > 250))) \in C^{(a_s)}$. The constraints for the buyer a_b may include that he cannot pay more than 60K: $(\neg(price > 60000)) \in C^{(a_b)}$.

3.3   Interests and Utilities

It is up to an agent to decide whether to accept or reject an option proposed by another agent. To make this decision in a rational way, N3S suggests to take into account the utility of an option (hence of the world state it defines). Thereby, N3S assumes that interests, especially conflicting ones, are central to calculate the utility of options and, with that, to decide on acceptance and rejection: the better the interests are addressed by an option, the higher is the utility of the option. In particular, N3S exploits interests to calculate the utility of an option in two steps, namely, in a first step from option to interest and in a second step from interest to utility. This crucial separation splits the debatable part of an agent's reasoning from the undebatable part. Clearly the mapping from an option to an interest follows rational principles (in the case of rational agents), while the mapping from interest to utility does not. For instance, while it is possible to argue rationally whether air conditioning (part of the option) adds to comfort (interest), it is *not* possible to argue rationally how much comfort it adds to an agent's personal utility. In other words, an agent's utility function is private, but an agent's mapping from options to interests is not. (An agent, as a rational entity, may not only keep track of his own mind state, but may also try to model the mind states – including the interests – of his negotiation partners.)

Formally, there is a mapping from the option $o \in \mathbb{O}$ to the degree of the interest $i \in I$ in o: $\mathbb{O} \times \mathbb{I} \mapsto \mathbb{R}$. This mapping is called *interest-satisfaction-mapping (ISM)*. Let I_{a_1} be the set of interests of agent a_1. As this set is finite, *ISM* can be formulated as follows: $\mathbb{O} \mapsto \mathbb{R}^{|I_a|}$. The result of this formulation is a vector called *interest-satisfaction-vector is*. With that, the utility function can be written as $u : \mathbb{R}^{|I_a|} \mapsto \mathbb{R}$.

Example 5. Consider the car dealer scenario from Example 2 and assume that a_b is interested in some comfort $i_{comfort} \in I_{a_b}$. The availability of air conditioning (AC) fulfills this interest, while not having AC leaves $i_{comfort}$ unaddressed. $hasAC \Rightarrow i_{comfort}$ would then be part of the *ISM*. A utility function could, for example, calculate the utility of an option as the weighted average of the degrees of interest fulfillments: $u(o) = \sum_{i \in I} \omega_i ISM(o, i)$. If the existence of an AC makes up 50 percent of a car's utility for the agent, $\omega_{i_{comfort}}$ would be 0.5 in that case.

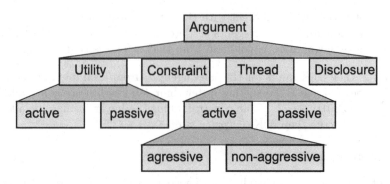

Fig. 1. The taxonomy of arguments (see the text for formal definitions)

3.4 Arguments and Argument Taxonomy

A rational agent always aims to maximize the utility he gets from negotiation. This means he will try to get the best of his options to which the other agent's will agree. According to N3S, argumentation is used to set the stage for a final agreement (negotiation stages distinguished by N3S are described in 4.3). Specifically, N3S allows to distinguish different types of arguments, depending on the way in which they aim to influence the opposing agents' "mind". These types, which are summarized in Figure 1, are described below.

Constraint-Arguments. Constraint-Arguments, referred to as *c-arguments*, aim at influencing the other agents' constraint sets. The idea is to widen $O^{(a)}_{feasible}$ where a is the opponent or to narrow $O^{(a)}_{feasible}$ where a is the agent itself. Note that for an effective negotiation it should be in the interest of all parties to achieve a view of the constraint set which is as coherent as possible.

Example 6. In the car dealer scenario from Example 2, a c-argument could be "I'm sorry Sir, but we don't sell BMWs that go faster than 250":

$$(\neg(brand = \text{BMW} \wedge speed > 250)) \in C.$$

Thread-Arguments. Threads target the agents' non-negotiated options repertoire. Arguments of this type that aim to decrease the utility of the other agent's $O^{(a)}_{nn}$ are called *active threads*. Moreover, active threads are called *aggressive* if the agent announces personal involvement in downgrading the other agent's $O^{(a)}_{nn}$. Non-negotiated-option arguments, shortly referred to as *n-arguments*, that aim to increase the utility of the agent's own $O^{(a)}_{nn}$ are *passive threads* from the perspective of the opponents. A rational agent a should be aware of the fact that passive threads only make sense if his $O^{(a)}_{nn}$ is better that the other agents expect it to be. (From an ethical point of view, threads are legitimate as long as they are not aggressive.)

Example 7. In the car dealer scenario from Example 2, a passive thread is to tell the sales agent that one got an offer for a Mercedes with AC and a top speed of

300 km/h for 51,000 € from another car dealer: $(51000, \text{Mercedes}, \text{true}, 300) \in O_{nn}^{(a_c)}$.

Disclosure-Argument. These arguments, referred to as *d-arguments*, do not change the option spaces. They rather support the other agents in their joint problem solving by disclosing own interests *i* and by mapping them to a utility measure.

Example 8. A customer could reveal to the sales agent that she is interested in comfort in the car: $(i_{comfort} \in I_a, hasAC \Rightarrow i_{comfort})$. The first part states that she is interested in comfort, and the second part states that AC needs to be true to fulfill the interest.

Utility-Arguments. Utility-Arguments, or *u-arguments* for short, aim at influencing the other agent's utility function. The idea is to widen $O_{agreeable}^{(a)}$ where *a* is the opponent (*active* utility-argument) or to narrow $O_{agreeable}^{(a)}$ where *a* is the agent itself (*passive* utility-argument).

Example 9. Consider Example 8. The sales agent could argue that an AC does not necessarily lead to higher comfort. If the customer intends to use the car in Greenland, she won't need an AC to keep the car cold: $(\neg AC \rightarrow (\text{Temperature}(car) = \text{Temperature}(country)), \text{Temperature}(country) < 23, \text{Temperature}(car) < 23 \rightarrow comfort)$. So, the sales agent is saying that without an AC the car will have the same temperature as the country and this temperature is blow 23 degrees.

4   External Elements of N3S

4.1   Communication

N3S suggests that negotiation essentially is the communication of suggestions (including both options and arguments). Communication starts by suggesting to negotiate, and is continued by the agents through rejecting and accepting suggestions. Communication ends if all agents accept the same suggestion or if one or several agents decide to not longer participate in the negotiation process. Figure 2 depicts the general scheme.

Starting and Terminating a Negotiation. Negotiation is a conversation that has a well defined beginning and ending. Before the real negotiation takes place, it has to be clear which agents participate and what the option space is. Thus one agent will suggest a negotiation be sending a message with the following information:

1. a specification of the option space (*what* is negotiated?);
2. a list of known participating agents (*who* negotiates?);
3. a unique identifier of the (intended) negotiation.

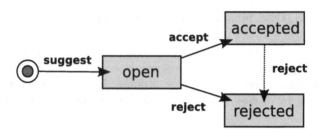

Fig. 2. The logic of suggesting, rejecting and accepting

The option space may be defined by an ontology accessible to each participating agent in order to ensure an efficient negotiation. (Note that N3S does not prescribe the provision of an ontology by the user. Rather, it would be possible as well that agents themselves choose or "meta-negotiate" the ontology they want to use).

Negotiation starts if all participating agents accept the offer to run a negotiation. An ongoing negotiation ends if one or several partners refuse to continue the negotiation process or if all partners agree on (accept) the same suggestion.

Making Proposals. An option is always suggested by an agent. N3S requires that each of the participating agents has the choice between sending a reject or an accept message once an option has been suggested. In the case of the acceptance of an option by all agents, the negotiation ends.

Providing Arguments. N3S assumes that arguments, just like options, are subject to the suggest-accept-reject scheme. This makes sense because the participating agents can now track whether their negotiation partners accept an argument. In particular, it allows them to adjust their own reasoning and to model the reasoning of the other agents. As another means for increasing negotiation efficiency, a negotiation protocol could be applied that allows to propose only arguments that are (logically) consistent with previously accepted arguments.

4.2 Option Spaces

N3S structures the option space from the perspective of rationally acting agents. An obvious option class being relevant to rational agents is the class of options deemed feasible by all agents:

$$O_{feasible} = \bigcap_{a \in A} O_{feasible}^{(a)} \tag{1}$$

To be considered as feasible by all agents, however, is not sufficient for an option to be accepted. A rational agent can not be expected to agree upon an option if he fares better in not agreeing to anything. If no agreement is reached, each agent has to fall back to his best non-negotiated option which is given by:

$$\text{argmax}(\text{util}^{(a)}(o_{nn} \in O_{nn}^{(a)})) \tag{2}$$

Table 1. Option hierarchy induced by N3S

(Sub-)Space		Description
$\mathbb{O}$		includes every possible option
$O^{(a)}_{feasible}$	$\subseteq \mathbb{O}$	includes all options that agent a judges to be possible in respect to $C^{(a)}$
$O_{feasible}$	$= \bigcap_{a \in A} O^{(a)}_{feasible}$	includes all options that are perceived as possible by all agents in A
$O^{(a)}_{agreeable}$	$\subseteq O^{(a)}_{feasible}$	all options that are agreeable for agent a
$O_{agreeable}$	$= \bigcap_{a \in A} O^{(a)}_{agreeable}$	all options agreeable to all agents in A

So, every option o which (i) has a higher utility than the best of his non-negotiated options and (ii) is feasible is in the set of agreeable options:

$$\forall_o [(\text{util}(o) \geqq \max(\text{util}^{(a)}(o_{nn} \in O^{(a)}_{nn})) \wedge o \in O^{(a)}_{feasible}) => o \in O^{(a)}_{agreeable}] \quad (3)$$

The set of options agreeable to all agents in A is defined as the intersection of all agents' individual agreeable option sets:

$$O_{agreeable} = \bigcap_{a \in A} O^{(a)}_{agreeable} \quad (4)$$

From (1), (3) and (4) it follows that

$$O_{agreeable} \subseteq O_{feasible} \quad (5)$$

Table 1 summarizes the above considerations.

Example 10. Consider the care dealer situation introduced in Example (2). As we already saw in Example (4), the constraints for the customer a_b include that he cannot pay more than 60K: $\neg(price > 60000) \in C^{(a_b)}$. Thus an option $(61000, \mathsf{BMW}, \mathsf{true}, 200)$ would be in $\mathbb{O}$ but not in $O^{(a)}_{feasible}$. Against that, $(59000, \mathsf{BMW}, \mathsf{true}, 200)$ would be feasible. However, if a_b also got the offer $(58000, \mathsf{BMW}, \mathsf{true}, 200)$ from another sales agent, $(59000, \mathsf{BMW}, \mathsf{true}, 200)$ might not be in $O^{(a)}_{agreeable}$. Depending on his *ISM* and his utility function $(59000, \mathsf{BMW}, \mathsf{true}, 300)$ could again be in $O^{(a)}_{agreeable}$, if the difference in speed (300 km/h instead of 200 km/h) is worth the additional 1,000 € to a_b.

4.3   Stages of the Negotiation Process

Based on the option space hierarchy, N3S distinguishes different stages of the negotiation process and allows to assign specific negotiation strategies to these stages. These stages are described below. The stage transition diagram is shown in Figure 3.

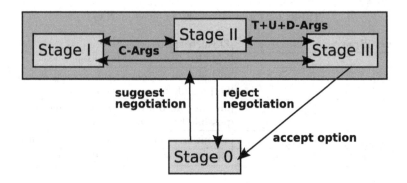

Fig. 3. The transition diagram for the N3S negotiation stages

Pre or post Negotiation (0). Before or after a negotiation process the option space of an agent does only consist of the non-negotiated options. These non-negotiated options may well include options that were made possible by preceding or still ongoing negotiations. An obvious negotiation strategy for a rational agent in this stage is to broaden his option portfolio. Thus an agent should check whether there is a prospect of making new options available through negotiation. A way to achieve this is to start a negotiation process in order get information about the options of other agents. Initiation of a negotiation process will lead to one of the tree following stages. The goal is to reach stage III which is the only stage at which each participating agent has an incentive to agree.

Negotiation without available options (I). In this stage, at least two agents are engaged in negotiation, but $O_{feasible} = \emptyset$. A strategy to identify options in this stage is to argue about the other agents' constraints $C^{(a)}$. Thus c-arguments will dominate the conversation.

Negotiation without incentive to agree (II). In this stage, at least two agents are engaged in negotiation and there are feasible options (i.e., $O_{feasible} \neq \emptyset$), but for at least one agent none of these options seems to have a higher utility than his best non-negotiated option (i.e., $O_{agreeable} = \emptyset$). There are two strategies to open up $O_{agreeable}$: to influence and change the utility functions of the other agents through u-arguments; and to attack the other agents' $O_{nn}^{(a)}$ through n-arguments.

Negotiation with incentive to agree (III). In this stage, at least two agents are engaged in negotiation and $O_{agreeable} \neq \emptyset$, and thus it is rational for both parties to successfully finish the negotiation. A negotiation protocol like the monotonic concession protocol presented in [31] to clarify specific terms of agreement and to find a fair option for all parties. Besides this, there is still the possibility to strengthen one's position with n-arguments and u-arguments. Note that it is well possible to fall back to stage II by using (e.g.) passive threads.

5 Putting the Pieces Together – A Sample Flow of Arguments

The examples provided in the preceding sections illustrate specific aspects of N3S. In this section, an example is given that illustrates the interdigitation of agent-internal and external elements of N3S. The emphasis of this example is on the overall flow of argumentation. (For the sake of clarity, the used formalism is kept as simple as possible and utilities – functions and u-arguments – are not considered. Note that there is no commonly accepted benchmark negotiation problem in the field; we think the problem introduced below may be a good benchmark candidate for both theoretical and practical analysis).

5.1 The "Polygon Negotiation Problem"

Basic Setting. Assume there are two agents (a_1, a_2) and a polygon P of order n $(y = c_0 x^0 + c_1 x^1 + \cdots + c_n x^n)$. Coefficients are in the range $[-1, 1]$ and initially set to zero: $c_0 = c_1 = \cdots = c_n = 0$. If the agents do not find an agreement, the initial values are kept. Each agent wants to maximize P at certain positions x; thus, these positions represent their interests. If a_1 is interested in maximizing P at $x = 0.5$, this is written as "$\max(P(0.5)) \in I_{a_1}$". Moreover, assume each agent has only one interest (hence, reasoning about utilities is not necessary) and possesses the capability to manipulate at least one coefficient. If a_1 can set the coefficients c_0 and c_3, this is written as "$\mathrm{canSet}(a_1, \{c_0, c_3\})$".

Instantiation. The following information determines the initial conditions:

P of order $n = 3$
$\max(P(1.0)) \in I_{a_1} \Rightarrow u_{a_1} = c_0 + c_1 + c_2 + c_3$
$\max(P(-1.0)) \in I_{a_2} \Rightarrow u_{a_2} = c_0 - c_1 + c_2 - c_3$
$\mathrm{canSet}(a_1, \{c_3\})$
$\mathrm{canSet}(a_2, \{c_0\})$

5.2 Sample Flow of Argumentation

Table 2 shows a possible sequence of arguments exchanged by the two agents involved in the Polygon Negotiation Problem. According to this sequence, agent a_2 starts right away with a proposal (1). As a_1 can not change c_1 and c_2 and thus he rejects the proposal (2a) and supplies a reason with a c-argument (2b). a_2 replies with a d-argument (3) informing a_1 about her capabilities. a_1 does the same (4). These two disclosures lead from stage I, in which no feasible options are known, to stage II. In this stage the agents can make feasible proposals (which, however, may be not acceptable for the respective other agent). a_2 brings forth a new and feasible proposal (5), but a_1 rejects for a good reason (6a), reveals his interests (6b), and tells a_2 that rejection is a non negotiated option for him (6c). At that point it is clear to a_2 that, from a_1's point of view, rejecting yields a utility of 0.0. In (7) stage III is entered, as the proposal of a_2 is acceptable

Table 2. An argumentation sequence in the Polygon Negotiation Problem

1	a_2 propose	$set(a_1, \{c_1 = -1.0, c_2 = 1.0, c_3 = -1.0\})$, $set(a_2, \{c_0 = 1.0\})$
2a a_1 reject		$set(a_1, \{c_1 = -1.0, c_2 = 1.0, c_3 = -1.0\})$, $set(a_2, \{c_0 = 1.0\})$
2b	c-argument	$\neg canSet(a_1, \{c_2, c_1\})$
3	a_2 d-argument	$canSet(a_2, \{c_0\})$
4	a_1 d-argument	$canSet(a_1, \{c_3\})$
5	a_2 propose	$set(a_1, \{c_3 = -1.0\}), set(a_2, \{c_0 = 1.0\})$
6a a_1 reject		$set(a_1, \{c_3 = -1.0\}), set(a_2, \{c_0 = 1.0\})$
6b	d-argument	$max(P(1.0)) \in I_{a_1}$
6c	t-argument	$\{c_0 = 0.0, c_1 = 0.0, c_2 = 0.0, c_3 = 0.0\} \in O_{nn}^{(a_1)}$
7	a_2 propose	$set(a_1, \{c_3 = -0.5\}), set(a_2, \{c_0 = 1.0\})$
8a a_1 reject		$set(a_1, \{c_3 = -0.5\}), set(a_2, \{c_0 = 1.0\})$
8b	propose	$set(a_2, \{c_0 = 1.0\})$
9	a_2 accept	$set(a_2, \{c_0 = 1.0\})$

(for a_1 it yields a utility greater than 0.0). Thus there is an incentive for a_1 to agree. However, instead of agreeing immediately, a_1 tries to get more: he rejects again (8) and makes a counter proposal (8b). At this point a_2 accepts (9) and thus the negotiation ended successfully with a joint agreement.

6   Discussion and Conclusions

N3S has the following main characteristics:

- it is generic negotiation framework that is problem- and domain-independent, covers both agent-internal and external elements of negotiation, and is applicable to a broad range of negotiation scenarios;
- it introduces the concept of option and option hierarchy as a key ingredient of rational negotiation;
- it provides a practical taxonomy of arguments that supports an agent in arguing in a systematic way; and
- it effectively structures the negotiation process into distinct stages.

Together these characteristics make N3S unique and distinct from other ABN frameworks (see Section 2). Based on these characteristics, a particular strength of N3S is its potential to serve both as a "technical " guideline for implementing automated negotiation processes and a "conceptual" guideline for unifying and integrating available approaches addressing selected aspects of negotiation. Examples of such aspects are society protocols for agent interaction [7], nested argumentation [12], argumentation based on strategic reasoning [2], commitment based argumentation [8], trust-based negotiation [22], and learning of argumentation strategies [4]. The integration of all these approaches, or of components of them, into a coherent whole is, in our opinion, *the* research and application challenge in the field.

The key contribution of N3S to the state of the art lies in the identification of both arguments types and negotiation stages. Specifically, to our knowledge no other currently available ABN framework, including those mentioned in Section 2, *relates* argument types and negotiation stages in a structured way as N3S does. By relating argument types and negotiation stages, N3S opens new possibilities in enabling rational agents to argue *efficiently*. We already addressed facets of this issue in our approaches to practical and social reasoning (e.g., [24,25]), and our current research aims to explore the relationships and possible synergies between N3S and these approaches as well as related ones (e.g., [2,18,23]).

An important question raised by N3S is how its argument taxonomy, which essentially is "option-centered", compares to the argument taxonomies proposed within the related frameworks. Another relevant research topic raised by N3S is how the negotiation stages of N3S are related to the types of conflicts among agents identified in other frameworks (e.g., [17]). Last but not least, it is currently an open issue how the N3S negotiation strategies (see 4.3) are related to negotiation strategies, tactics and heuristics proposed elsewhere (see, e.g., the various papers on strategic aspects of negotiation in [11]).

Acknowledgements. The authors thank Achim Rettinger, Ana Balevic and Florian Echtler for many useful and inspiring discussions automated negotiation in general and N3S in particular. Special thanks goes to the Chair of Systems Design of ETH Zurich, headed by Frank Schweitzer, for generous financial support of our joint research efforts on N3S.

References

1. Andersson, M.R., Sandholm, T.W.: Leveled commitment contracts with myopic and strategic agents. In: Proceedings of the 15th National Conference on Artificial Intelligence (AAAI-98), pp. 38–45 (1998)
2. Bentahar, J., Mbarki, M., Moulin, B.: Strategic and tactic reasoning for communicating agents. In: Maudet et al. MPR06a. Available at: http://homepages.inf.ed.ac.uk/irahwan/argmas/argmas06/
3. Durfee, E.H., Lesser, V.R.: Partial global planning: A coordination framework for distributed hypothesis formation. IEEE Transactions on Systems, Man, and Cybernetics, SMC 21(5), 1167–1183 (1991)
4. Emele, C.D., Norman, T., Edwards, P.: A framework for learning argumentation strategies. In: Maudet et al. MPR06a. Available at: http://homepages.inf.ed.ac.uk/irahwan/argmas/argmas06/
5. Fisher, R., Ury, W., Patton, B.: Getting to Yes: Negotiating Agreement Without Giving In. Random House Business Books, 2nd edition (January 2004)
6. Jennings, N.R.: Commitments and conventions: The foundation of coordination in multi-agent systems. The Knowledge Engineering Review 8(3), 223–250 (1993)
7. Kakas, A., Maudet, N., Moraitis, P.: Layered strategies and protocols for argumentation-based agent interaction. In: Rahwan., et al. (eds.) RMR05a, pp. 64–77.
8. Karunatillake, N., Jennings, N., Rahwan, I., Norman, T.: Argument-based negotiation in a social context. In: Parsons et al. PMMR06a [14]

9. Kraus, S., Sycara, K., Evenchik, A.: Reaching agreements through argumentation: a logical model and implementation. Artificial Intelligence 104(1-2), 1–69 (1998)
10. Lesser, V.R., Decker, K.R., Wagner, T., Carver, N., Garvey, A., Horling, B., Neiman, D., et al.: Evolution of the GPGP/TAEMS domain-independent coordination framework. Autonomous Agents and Multi-Agent Systems 9(1/2), 87–144 (2004)
11. Maudet, N., Parsons, S., Rahwan, I. (ed.): Argumentation in Multi-Agent Systems. Proceedings of the Third International Workshop (2006), Available at http://homepages.inf.ed.ac.uk/irahwan/argmas/argmas06/
12. Modgil, S.: Nested argumentation and its application to decision making over actions. In: Parsons., et al. (eds.) [14] PMMR06a,
13. Müller, H.J.: Negotiation principles. In: O'Hare, G.M.P., Jennings, N.R. (eds.) Foundations of Distributed Artificial Intelligence, pp. 211–230. Wiley, New York (1996)
14. Parsons, S., Maudet, N., Moraitis, P., Rahwan, I. (eds.): Argumentation in Multi-Agent Systems. Proceedings of the Second International Workshop. In: Parsons, S., Maudet, N., Moraitis, P., Rahwan, I. (eds.) ArgMAS 2005. LNCS (LNAI), vol. 4049, Springer, Heidelberg (2006)
15. Parsons, S., Sierra, C., Jennings, N.: Agents that reason and negotiate by arguing. Journal of Logic and Computation 8(3), 267–274 (1998)
16. Pasquier, P., Rahwan, I., Dignum, F., Sonenberg, L.: Argumentation and persuasion in the cognitive coherence theory: preliminary report. In: Maudet., et al.: [11] MPR06a. Available at, http://homepages.inf.ed.ac.uk/irahwan/argmas/argmas06/
17. Rahwan, I.: Interest-based Negotiation in Multi-Agent Systems. PhD thesis, Department of Information Systems, University of Melbourne, Australia (2004)
18. Rahwan, I., Amgoud, L.: An argumentation-based approach for practical reasoning. In: Proceedings of the Fifth International Joint Conference on Autonomous Agents and Multiagent Systems (AAMAS 2006), pp. 347–354 (2006)
19. Rahwan, I., Moraïtis, P., Reed, C. (eds.): ArgMAS 2004. LNCS (LNAI), vol. 3366. Springer, Heidelberg (2005)
20. Rahwan, I., Ramchurn, S., Jennings, N., McBurney, P., Parsons, S., Sonenberg, L.: Argumentation-based negotiation. Knowledge Engineering Review 18(4), 343–375 (2004)
21. Raiffa, H.: The Art and Science of Negotiation. Harvard University Press, Cambridge, Mass (1982)
22. Ramchurn, S.D.: Multiagent negotiation using trust and persuasion. PhD thesis, Faculty of Engineering and Applied Science, University of Southampton, UK (2005)
23. Roth, B., Riveret, R., Rotolo, A., Governatori, G.: Strategic argumentation: a game theoretical investigation. In: Proceedings of the 11th International Conference on Artificial Intelligence and Law (ICAIL 2007) (2007)
24. Rovatsos, M., Rahwan, I., Fischer, F., Weiß, G.: Practical strategic reasoning and adaptation in rational argument-based negotiation. In: Parsons, S., Maudet, N., Moraitis, P., Rahwan, I. (eds.) ArgMAS 2005. LNCS (LNAI), vol. 4049, pp. 122–137. Springer, Heidelberg (2006)

25. Rovatsos, M., Weiß, G., Wolf, M.: An approach to the analysis and design of multi-agent systems based on interaction frames. In: Alonso, E., Kudenko, D., Kazakov, D. (eds.) Proceedings of the First International Conference on Autonomous Agents and Multiagent Systems (AAMAS 2002). LNCS (LNAI), vol. 2636, Springer, Heidelberg (2003)
26. Sadri, F., Toni, F., Torroni, P.: Abductive logic programming architecture for negotiating agents. In: Flesca, S., Greco, S., Leone, N., Ianni, G. (eds.) JELIA 2002. LNCS (LNAI), vol. 2424, pp. 419–431. Springer, Heidelberg (2002)
27. Saha, S., Sen, S.: A Bayes net approach to argumentation based negotiation. In: Rahwan, et al. (eds.) [19] RMR05a, pp. 208–222.
28. Sierra, C., Jennings, N.R., Noriega, P., Parsons, S.: A framework for argumentation-based negotiation. In: Singh, M.P., Rao, A., Wooldridge, M.J. (eds.) ATAL 1997. LNCS, vol. 1365, pp. 177–192. Springer, Heidelberg (1998)
29. Smith, R.G.: The contract-net protocol: High-level communication and control in a distributed problem solver. IEEE Transactions on Computers 29(12), 1104–1113 (1980)
30. Weiß, G.: Multiagent Systems. A Modern Approach to Distributed Artificial Intelligence. The MIT Press, Cambridge, MA (1999)
31. Wooldridge, M.J.: An Introduction to MultiAgent Systems. John Wiley and Sons Ltd, Chichester (2002)

The Effect of Mediated Partnerships in Two-Sided Economic Search

Philip Hendrix and David Sarne

School of Engineering and Applied Sciences
Harvard University
Cambridge, MA 02138 USA
{phendrix,sarned}@eecs.hrvard.edu

Abstract. In this paper we investigate the effect of mediated partnerships over agents' equilibrium strategies in two-sided economic search. A *mediated partnership* is formed when an agent acts as a mediator, establishing a partnership between a pair of agents it encountered along its search, thereby reducing the other agents' amount of search. Surprisingly, this reduction in market friction induced by mediated partnerships does not always improve market efficiency. Use of mediated partnerships changes the equilibrium strategies used by agents in two-sided search models and introduces substantial computational complexity. This computational complexity is overcome with an innovative algorithm that facilitates equilibrium calculation.

1 Introduction

Two-sided economic search concerns distributed problems in which self-interested agents search for appropriate partners to form mutually acceptable pairwise partnerships [10,4,1,6,3]. The concept of economic search differs from search in AI. AI search typically involves an agent finding a sequence of actions that take it from an initial state to a goal state, while economic search refers to the identification of the best agent for a partnership. Since the agents are self-interested, the analysis of two-sided economic search models is inherently equilibrium-based.

The goal of each agent in two-sided economic search is to find a partner that is optimally beneficial. Each agent is associated with a specific type that captures the benefit of partnering with it[1]. During each stage of search, agents randomly interact pairwise and learn each others' type. A partnership between two interacting agents will be formed if they both commit to it. The process of initiating and maintaining an interaction with another agent is associated with a cost (i.e., search cost) incurred by both agents. Therefore, a key challenge for each agent in such an environment is to identify the set of agent types with whom it is willing to partner.

---

[1] This concept of "type" is different than the concept of "type" used in many AI domains where an agent's type is derived from utility gained through different opportunities.

M. Klusch et al. (Eds.): CIA 2007, LNAI 4676, pp. 224–240, 2007.

This paper extends the traditional two-sided economic search model to mediated partnerships. A *mediated partnership* is formed when an agent acts as a mediator, using its memory of past interactions with two other agents to form a partnership between them. Specifically, we base our model on the traditional two-sided search model where utilities are non-transferable, an agent's utility is fully correlated with its partner's type, agents incur a search cost for each round of search, and agents' true types are revealed once paired [10,6].

Multi-Agent System (MAS) environments in which agents must interact with each other to evaluate potential partners and form viable partnerships are of interest because of their potential economic and strategic importance. An example of such a system is the dual backup application [13] in which agents representing different servers seek to form partnerships for purposes of mutual offsite backup. Unlike tradition two-sided search models, when considering the process in a MAS, one must take into consideration the unique capabilities of autonomous agents for enhancing the search process. One important capability of this kind is an agent's ability to maintain a full recollection of past interactions with other agents and utilize this ability to act as a mediator for forming partnerships as described in this paper.

The contributions of the paper are threefold. First, it introduces and models a new distributed two-sided search enhanced with mediated partnerships. Second, the introduction of mediated partnership capabilities to two-sided economic search potentially creates an equilibrium which differs from that obtained in the traditional two-sided search model, and consequently can change individual and collective agent performance. The paper shows that the new model prevents a direct calculation of this new equilibrium. A new algorithm is introduced to find the new equilibrium. Finally, it is shown that although the memory and matchmaking extensions generally reduce market frictions and increase the overall efficiency of forming partnerships, in some rare cases overall system performance is decreased. This latter result is specifically important for market makers and multi-agent designers.

In Section 2 we formally introduce the two-sided search model with memory and mediated partnering. Section 3 gives an equilibrium analysis and discusses its cluster based partnering structure. We then describe a set of equations in Section 4 for the two-sided search model with mediated partnering for scenarios where agents form exactly into two-clusters. This analysis enables us to introduce examples in Section 5 showing the effect that memory can have on the traditional two-sided search model. Section 6 outlines an algorithmic based approach for determining agents' equilibrium strategies followed by experimental results that demonstrate its phenomenal efficiency. We conclude with reviewing related work (Section 7) and a discussion (Section 8).

2   The Model

We base our model on the traditional two-sided search model [10,6]. In its most basic form the two-sided search model considers an environment populated with an infinite number of self-interested fully rational agents[2]. Each agent A_i has a type, defined over the continuum $[\underline{t}, \overline{t}]$, that captures special properties that characterize it. This type determines the utility that any other agent A_j gains if partnered with agent A_i. For simplicity, we assume that the expected utility gained by partnering with an agent of type t ($t \in [\underline{t}, \overline{t}]$) is equal to t (i.e., $U(t) = t$). An agent can be of only one type, although many agents may be of the same type.

At the beginning of search, agents have no prior information concerning the type of specific agents in its environment. Agents randomly initiate pairwise interactions (i.e., search) with other agents to learn their type. The model assumes non-transferable utilities, thus no bargaining takes place as part of the interaction. Upon interaction, both agents reveal their type and the set of agent types with which it is willing to form a partnership. This set of acceptable types is called the *acceptance set*. A partnership is formed only if both agents are willing to commit to it (i.e., if each agent has the other agent's type in its acceptance set). Otherwise both agents resume their search in the same manner.

The process of initiating and maintaining an interaction is associated with a cost (i.e., search cost) incurred by both agents. This cost is equivalent to a reduction c in each agent's overall utility. Therefore, an agent chooses its acceptance set based on the expected utility gained from partnering and the cost of continued search[3].

While agents have no prior information concerning the types of a specific agent at the beginning of search, they are assumed to be acquainted with the general distribution of types in their environment, described by the probability distribution function $f(x)$ and cumulative distribution function $F(x)$. The nature of the two-sided search suggests that the agents are satisfied with a single partner, thus once a partnership is formed the two agents forming it terminate their search process and leave the environment. The two-sided search model assumes that the pair is replaced with two identical agents (i.e, having the same types) that start the search from scratch[4].

Our model extends the interaction protocol of two-sided search by utilizing information gleaned from previous interactions with potential partners. If a direct partnership is not formed, then instead of simply resuming search, as in the traditional two-sided search model, each agent records in memory the type and acceptance set of the other. Moreover, each agent seeks to act as a mediator (match maker) for the other by searching its memory, seeking a suitable partner for the other from its records of other agents already encountered. If such a mediator agent finds a partner for its newly rejected potential partner among

---

[2] The infinite number of agents assumption is common in two-sided search models (see [5,17]). In many domains (e.g., eCommerce) this is derived from high entrance and exit rates, thus the probability of running into the same agent in a random pairing is negligible.

[3] In the absence of search costs, the unique competitive equilibrium has assortative joining, i.e., joined partners are identical (in type) [2].

[4] This assumption is commonly used in two-sided search literature in order to maintain the fixed distribution function $f(x)$ [3,1].

the agents it has previously encountered, then a partnership is arranged for the newly rejected agent and the remembered agent, if neither has already entered a partnership. We call this process *mediated partnering*.

The model assumes that the information the agents exchange in their interactions is true, that it is not subject to change, and that the penalty for rejecting an agent of a type in the acceptance set is significantly greater than any utility that can be achieved. Thus, mediated partnering always takes place unless one of the agents has already partnered. We further assume that a mediated partnering is costless because it does not require any additional exploration or data exchange, except for checking that both agents are still in the market.

3 Analysis

The analysis of the classical two-sided search model suggests a "complete segregation" solution in equilibrium [4,6,10]. Agents form clusters, based on their type, in which every agent in a cluster is always willing to partner with any other agent in the cluster. Consequently, the agents' acceptance (partnering) strategy is reservation-value based. The reservation value of an agent corresponds to the lowest type that it is willing to accept. This reservation value is different from the reservation price commonly found in e-commerce models associated with a buyer or a seller that is not involved in a search and denotes the buyer's or seller's true evaluation for the opportunity. The agent accepts all agents with types greater than or equal to its reservation value. Furthermore, the agents' search strategy is stationary (i.e. an agent will not change its reservation value).

In the traditional two-sided search model agents' reservation values in equilibrium can be found using backward induction. Consider the agent $A_{\bar{t}}$ of the highest type, $\bar{t}$, that needs to set its optimal reservation value. Obviously this agent's optimal reservation value will not be affected by the reservation value used by other agents, since all other agents will always accept type $\bar{t}$ (having no better type to strive for). Therefore, the expected utility of agent $A_{\bar{t}}$, denoted $V_{A_{\bar{t}}}(x)$, as a function of its reservation value x can be expressed as

$$V_{A_{\bar{t}}}(x) = -c + \int_{y=x}^{\bar{t}} yf(y)dy + F(x)V_{A_{\bar{t}}(x)} \qquad (1)$$

Cost of meeting another agent   Expected value of partnering with paired agent   Expected value of continued search (no partnering)

We use $x_{\bar{t}}^*$ to denote the reservation value that maximizes $V_{A_{\bar{t}}}(x)$. Now consider an agent of type $\bar{t} - \epsilon$. If this agent is accepted by agents of type $\bar{t}$, then it necessarily uses the same Equation 1 to set its own reservation value (substituting $V_{\bar{t}}(x)$ by $V_{\bar{t}-\epsilon}(x)$). Thus, the optimal reservation value for this agent is the same as for the highest type agent (i.e., $x_{\bar{t}}^* = x_{\bar{t}-\epsilon}^*$). The same logic holds for all other agents of types belonging to the interval $[x_{\bar{t}}^*, \bar{t}]$. Since none of the agents of types $[x_{\bar{t}}^*, \bar{t}]$ accepts any other agents of types in the interval $[\underline{t}, x_{\bar{t}}^*)$ then

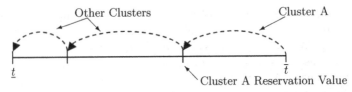

Fig. 1. Illustration of three clusters and the reservation value for the top cluster A

the process can be replicated, resulting with additional clusters. The complete segregation scenario is illustrated in Figure 1.

The integration of mediated partnerships with the traditional two-sided search model introduces several new dynamics that must be added to the above analysis. The information the agents supply to each other is binding, therefore their acceptance strategy is reservation-value based. We begin by formulating the expected utility $V_{(\bar{t},M)}(x)$ of an agent of type $\bar{t}$ having a memory M when using a reservation value x, which can be formulated as

$$V_{(\bar{t},M)}(x) = -c + \int_{y=x}^{\bar{t}} y f(y) dy + \int_{z=0}^{x} \left(\int_{y=x}^{\bar{t}} y G_{(M,z)}(y) dy + \left(1 - \int_{y=x}^{\bar{t}} G_{(M,z)}(y) dy\right) V_{(\bar{t},M\cup z)}(x) \right) f(z) dz \quad (2)$$

Cost of   Expected value of   Expected value of   Expected value of
search   partnering with   being partnered by   continued search
            paired agent   mediated partnering   with updated memory

where $G_{M,z}(y)$ is the probability of forming a mediated partnership with an agent of type y, initiated by the currently met agent or a perviously met agent. Notice that Equation 2 is an extensive modification of Equation 1 used for the traditional two-sided search.

Using similar logic as above, we obtain that all agents of types $[x_{\bar{t}}^*, \bar{t}]$ use the same optimal reservation value $x_{\bar{t}}^*$. For all agents of these types, the probability $G_{M,z}(y)$ receives a similar value. Thus clusters are formed in the same way as before, and each agent will only partner with other agents having a type in its own cluster.

The probability $G_{M,z}(y)$ is affected by the division of other agents into clusters. Agents of types that belong to small clusters are associated with longer searches and consequently are more likely to have a relevant agent in their memory upon being met. Therefore, unlike in the traditional model, each cluster in the new model depends on the number and the size of the other clusters. Hence, a backward induction solution as the one often suggested for the traditional two-sided search [10] is not applicable in our case.

4   Two Cluster Solution

Solving a set of expected utility Equations like the one given in equation 2 is impractical for the general case. In the general case, the probability of being

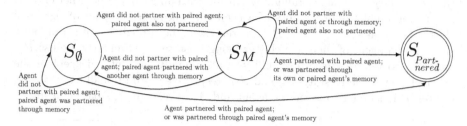

Fig. 2. Finite state machine for the two-cluster case of the two-sided search model with reputation

partnered through mediated partnerships changes from one search round to another (based on the probability that other agents are still in the market). The extraction of the probability $G_{M,z}(y)$ requires understanding all the possible states the agent can be in and the transitions between them. In a finite state machine for an agent A_i, a state encapsulates information about which agents A_i holds in memory, and to which clusters they belong. The number of states require for an agent operating in an environment with N clusters is $2^{N-1} + 1$, representing the combinations of having memory for the $N - 1$ other clusters, and the end state where the agent has partnered and left the game[5]. The probability of transitioning from one state to another changes as the length of time the agents are kept in memory increases, because agents are likely to be partnered with other agents over time.

Fortunately, solving the problem analytically for environments where only two clusters are formed in equilibrium is possible. We will illustrate the two-cluster solution to show the dynamics of the system. In the two-cluster model the agent can have a memory of one agent in the other cluster. The agent can never have more than one agent in its memory, because having two un-partnered agents belonging to the same cluster would result in a partnering. Additionally, an agent will be in another agent's memory for at most one round of the search process. This derives from the fact that unless this agent is matched directly, it will be matched with whomever the agent storing it in memory is paired with the following search round (unless the storing agent is matched with its own kind, and leaves the market).

The finite state machine in Figure 2 explains the transitions of an agent in the two cluster solution using mediated partnering. In the two-cluster model, the finite state machine for an agent has three states: $S_\emptyset$ for the state where there is no agent in memory, S_M for the state where there is an agent of the other cluster in memory, and $S_{Partnered}$ for the state where the agent has joined with another agent of its own cluster and left the market.

Throughout the paper we will use the notation P_c to represent the probability of an agent randomly pairing with an agent in its own cluster, $\overline{P_c}$ to represent

---

[5] Due to the unique cluster-base structure of the equilibrium, the agent will never have two agents from the same cluster in its memory, since these can always be matched.

the probability of an agent randomly pairing with an agent of the other cluster, $P_{S_\emptyset}$ to represent the probability of a randomly encountered agent of the same cluster being in state $S_\emptyset$, and P_{S_M} to represent the probability of a randomly encountered agent of the same cluster being in state S_M. Similarly, we use $\overline{P_{S_\emptyset}}$ and $\overline{P_{S_M}}$ to represent the probability of a randomly encountered agent of the other cluster to be in states $S_\emptyset$ and S_M, respectively. Additionally we will use the term A_s to refer to a specific agent and A_p to refer to A_s's paired agent.

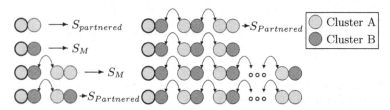

Fig. 3. Examples of how chains can form in a two cluster environment from the perspective of an agent in Cluster A with no memory. This agent with no memory is outlined with a bold circle. Arrows between agents signify a memory link. Arrows to states show that the agent in bold will transfer to that state with probability 1.

The pairing of agents with memory can lead to having chains of pairs of agents in which none of the pairs can directly join, however each agent in the sequence (except for the first and the last) has a memory link to another unjoined agent with whom its paired agent can join (see illustration in Figure 3). We use L_s to denote the number of memory links in a sequence originating from the agent A_s. Similarly, we use L_p to denote the number of memory links in a sequence originating from the paired agent A_p. The probability for having exactly k links in a sequence originating from an agent that has memory can be calculated using

$$P(L_s = k) = \begin{cases} \overline{P_c} & \text{if } k = 0 \\ P_c(P_{S_\emptyset} + P_{S_M}P(L_s = 0)) & \text{if } k = 1 \\ P_cP_{S_M}P(L_s = k - 1) & \text{if } k > 1 \end{cases} \qquad (3)$$

$$P(L_p = k) = \begin{cases} P_c & \text{if } k = 0 \\ \overline{P_c}(\overline{P_{S_\emptyset}} + \overline{P_{S_M}}P(L_p = 0)) & \text{if } k = 1 \\ \overline{P_cP_{S_M}}P(L_p = k - 1) & \text{if } k > 1 \end{cases} \qquad (4)$$

The probability of an agent being partnered, or staying in the market follows from the Padovan [18] sequence defined as

$$Padovan(i) = \begin{cases} 1 & \text{if } i \in \{0, 1, 2\} \\ Padovan(i - 2) + Padovan(i - 3) & \text{if } i \in Z, i > 2 \end{cases} \qquad (5)$$

In addition, we use $P_{partner}(L_p, L_s)$ to denote the probability of being partnered when having L_p and L_s memory links sequences originating from the agent and from his partner, respectively. The value of $P_{partner}(L_p, L_s)$ can be calculated using

$$
P_{partner}(L_p, L_s) = \begin{cases} 0 & \text{if } L_p = 0, L_s = 0 \\ 1 & \text{if } L_p = 0, L_s = 1 \\ 1 & \text{if } L_p = 1, L_s = 0 \\ 0.5 & \text{if } L_p = 1, L_s = 1 \\ 1 - \frac{Padovan(L_s-2)}{Padovan(1+L_s)} = \frac{Padovan(L_s-1)}{Padovan(1+L_s)} & \text{if } L_p = 0, L_s > 1 \\ 1 - \frac{Padovan(L_p-2)}{Padovan(1+L_p)} = \frac{Padovan(L_p-1)}{Padovan(1+L_p)} & \text{if } L_p > 1, L_s = 0 \\ 1 - \frac{Padovan(L_p-2)Padovan(L_s-2)}{Padovan(1+L_p+L_s)} & \text{if } L_p > 1, L_s > 1 \end{cases}
\tag{6}
$$

Agent A_s will transition from state $S_\emptyset$ back to state $S_\emptyset$ if it cannot partner with agent A_p or an agent from A_p's memory, and the agent in A_p's memory is able to perform a mediated partnership to partner A_p. This probability is defined as

$$
P_{S_\emptyset \to S_\emptyset} = (1 - P_c)\overline{P_{S_M}} \sum_{i=0}^{\infty} P(L_p = i)(1 - P_{partner}(i, 0))P_{partner}(0, i)
\tag{7}
$$

Agent A_s will transition from $S_\emptyset$ to the memory state S_M if it cannot partner with A_p, and neither A_s or A_p can be partnered through a mediated partnership. This transition probability is defined as

$$
P_{S_\emptyset \to S_M} = (1 - P_c)\left(\overline{P_{S_\emptyset}} + \overline{P_{S_M}} \sum_{i=0}^{\infty} P(L_p = i)(1 - P_{partner}(i, 0))(1 - P_{partner}(0, i))\right)
\tag{8}
$$

Agent A_s will transition from $S_\emptyset$ to $S_{partnered}$ if it can partner with A_p, or if it can be partnered with an agent from A_p's memory. This transition probability is defined as

$$
P_{S_\emptyset \to S_{partnered}} = P_c + (1 - P_c)\overline{P_{S_M}} \sum_{i=0}^{\infty} P(L_p = i)P_{partner}(i, 0)
\tag{9}
$$

Agent A_s will transtion from S_M to $S_\emptyset$ if it cannot partner with A_p or an agent from A_p's memory, but either A_s or the agent in A_p's memory is able to perform a mediated partnering and join A_p. This probability is defined as

$$
P_{S_M \to S_\emptyset} = (1 - P_c)(A + B)
\tag{10}
$$

where

$$
A = \overline{P_{S_\emptyset}} \sum_{i=0}^{\infty} P(L_s = i)(1 - P_{partner}(0, i))P_{partner}(i, 0)
$$

$$
B = \overline{P_{S_M}} \sum_{i=0}^{\infty} \sum_{j=0}^{\infty} P(L_s = i)P(L_p = j)(1 - P_{partner}(j, i))P_{partner}(i, j))
$$

Agent A_s will transition from S_M back to S_M if it cannot partner with A_p, and neither A_s or A_p can be partnered through a mediated partnership. This transition probability is defined as

$$P_{S_M \to S_M} = (1 - P_c)(C + D) \tag{11}$$

where

$$C = \overline{P_{S_0}} \sum_{i=0}^{\infty} P(L_s = i)(1 - P_{partner}(0, i))(1 - P_{partner}(i, 0))$$

$$D = \overline{P_{S_M}} \sum_{i=0}^{\infty} \sum_{j=0}^{\infty} P(L_s = i)P(L_p = j)(1 - P_{partner}(L_p = j, L_s = i))(1 - P_{partner}(i, j))$$

Finally, agent A_s will transition from S_M to $S_{partnered}$ if it can partner with A_p, or if it can be partnered through a mediated partnership initiated by either A_p or the agent in A_s's memory. This transition probability is defined as

$$P_{S_M \to S_{partnered}} = P_c + (1 - P_c)(E + F) \tag{12}$$

where

$$E = \overline{P_{S_0}} \sum_{i=0}^{\infty} P(L_s = i)P_{partner}(0, i)$$

$$F = \overline{P_{S_M}} \sum_{i=0}^{\infty} \sum_{j=0}^{\infty} P(L_s = i)P(L_p = j)P_{partner}(j, i)$$

The probabilities of an agent being in a specific state can be calculated using the above transition probabilities as follows:

$$P_{S_0} = P_{S_0}P_{S_0 \to S_0} + P_{S_M}P_{S_M \to S_0}$$

$$P_{S_M} = P_{S_0}P_{S_0 \to S_M} + P_{S_M}P_{S_M \to S_M}$$

Consequently, the expected utilities for an agent in the different states it may be in are defined as

$$V_{S_0} = -c + P_{S_0 \to S_M}V_{S_M} + P_{S_0 \to S_{partnered}}V_{S_{partnered}}$$

$$V_{S_M} = -c + P_{S_M \to S_0}V_{S_0} + P_{S_M \to S_{partnered}}V_{S_{partnered}}$$

$$V_{S_{partnered}} = E[cluster]$$

The above set of equations is solvable for any interval of agent types and any search cost. The expected performance of the agent is V_0, as this is its state upon entering the market. An equilibrium division into two-clusters is a clustering that results with (a) the expected utility for agents of type $\bar{t}$ (and all other agents in their cluster) using a reservation value x is equal to x; and (b) the expected utility of all the agents in the second cluster is less than $\underline{t}$.

5   The Mediated Partnering Effect

The integration of mediated partnering with the two-sided search model is meant to help market designers further reduce market friction and increase average overall agent utility. Market friction is the reason agents do not have complete information. In the two-sided search model, market friction is the search cost. If search cost is eliminated completely, then the unique competitive equilibrium has assortative joining, meaning that agents only select partners of their own type [2]. In the latter case, the overall social utility is maximized. The overall social utility can always be decomposed to the aggregate utility each individual agent gains through partnership and the accumulated cost along the agents' search. The first component is fixed, since the agents partner amongst themselves. The second component, is a parameter of the search cost and the division of agents into clusters (which affects the extent of agent search and thus the accumulated search cost). Similarly, when the search cost is significantly large, each agent is willing to accept all other agents, thus an immediate partnership is formed upon executing the first search round. By introducing the mediated partnering mechanism, we seemingly accelerate the agents' search process without adding any additional cost. Instead of having the ability to form partnerships based on single random interactions, each agent is now subject to additional mediated, targeted opportunities along its search. This reduces market frictions and according to economic theory should improve agents' performance (since less effort needs to be invested in costly search).

Intuitively, the method appears to improve performance since it reduces market frictions, thus market makers and agent designers may attempt to use it unquestionably. Nevertheless, as we demonstrate in the following paragraph, reducing market inefficiencies does not necessarily improve overall market performance. Using the analysis given in the former section we produce three contrasting examples to illustrate the effect that mediated partnering has over the system's overall performance. The first example uses a uniform distribution with agents having 128 distinct types ranging from 0 (inclusive) to 128 (exclusive), and with a cost of 32. The average performance (which is an indicator for the overall market performance) in this case when using mediated partnering is 20. This is an improvement in comparison to the average performance achieved in the traditional two-sided search model, with an average performance of 18.6. This effect of improvement is the most common result found when applying mediated partnering in a two-sided search model.

The second example is obtained using a search cost of 0.25 and a probability distribution with 0.75 of the agents having types uniformly distributed between 64 and 128, and the remaining 0.25 of the agents having types uniformly distributed between 0 and 64. The overall performance is 83.5 in the classical model, and reduced to 79.7 with the use of the new technique. Here, the reduction of market friction benefits some agents, but also hinders other agents. The agents that are hindered are those agents that initially were at the base of a cluster, but with the addition of mediated partnering fall into a lower cluster. These agents make up a small proportion of the distribution, but are significantly and negatively impacted by the use of mediated partnerships in such a manner that their reduced performance reduces the overall performance. Last, we consider the trivial case where the search cost c is zero (where all agents are accepting only agents of their own type) and the case where the search cost is prohibitively large (where only one cluster is formed). Here, the mediated partnering does not affect the system (nor any individual agent's) performance at all. The implications of the above are discussed towards the end of the paper.

6   Algorithmic Based Approach

Determining the number of clusters and cluster boundaries analytically for the general case is not feasible for two-sided search with mediated partnering. Fortunately, in most environments, the agent types are discrete rather than continuous (e.g., the number of available Gigabytes a server can offer to its partner in the dual backup application). Therefore, theoretically we could have tried all possible clustering combinations, and determined the equilibrium sets. Whether a set of clusters is in equilibrium will be based on the following rule: the expected utility of the agents in a cluster $(i...j)$ (i.e., the agents that use type i as their reservation value) should not exceed i and should not be smaller than $i-1$ (otherwise, each single agent of types $(i...j)$ would have an incentive to deviate from accepting all agents of type i and greater to a strategy of accepting all agents of types $i-1$ and greater or $i+1$ and greater, respectively). The expected utility of agents in the different clusters, in this case, can be extracted using simulation of the two-sided mediated partnership-based search in a specific environment given.

The main disadvantage of the above method, is the exponential number of simulation runs. We use $R(N)$ to denote the number of possible clusters division when having N unique agent types. $R(N)$ can be calculated as follows

$$R(N) = 1 + \sum_{i=1}^{N-1} R(i)$$

This is simply by analyzing the first (highest cluster) - it can either cover all types (i.e., 1) or can cover $i \in \{i = 1, ..., N-1\}$ subsequent types, in which case this should be multiplied by the number of clusters that can be formed with the remaining types $(R(N-i))$. Now we prove that $R(N) = 2^{N-1}$

Proof by induction. Base case: when $N = 1$, $R(1) = 1$. Inductive case: assume for $N = k$, $R(k) = 2^{k-1}$. We want to show that $R(k+1) = 2^k$.

$$R(k+1) = 1 + \sum_{i=1}^{k} R(i) = 1 + \sum_{i=1}^{k-1} R(i) + R(k) = 2R(k) = 2^k$$

By showing the base case and the inductive case we have shown that $R(N) = 2^{N-1}$ Q.E.D.

We have developed a more sophisticated algorithmic-based solution for the purpose of extracting the equilibrium set of clusters using simulation (see pseudocode below). According to our algorithm, all agents are first (step 1) initialized to accept only their own type (i.e., each type forms its own cluster). This is done using the method *generateClusters(clusters, lower, upper)*, which sets the reservation value of each cluster in the interval [lower, upper] to itself. The method *runSimulation* simulates an environment clustered according to the division given in the variable *clusters* and returns the expected utility of the agents in each cluster. Then, for the cluster with the highest type that is not satisfied[6] (step 4), decreases its lower bound to accept the next lowest type (step 5), and initializes all agents below this type to accept only their own type (using *generateClusters* in step 6). The process repeats until all clusters are satisfied or there exists a single cluster that covers the entire range of types.

Algorithm 1. CLUSTER(A, T, c) — Calculate optimal equilibrium clustering

> **Input:** The set of all agents A, a vector of all distinct types T, the search cost c.
> **Output:** A Vector of clusters ordered from the cluster with the highest reservation value to the lowest.
> (1)   generateClusters(clusters, T[0], T[T.length-1]);
> (2)   runSimulation(A, clusters, c);
> (3)   **while** (! satisfied(clusters))
> (4)   Cluster s = satisfied(clusters);
> (5)   s.lowest−−;
> (6)   generateClusters(clusters, T[0], s.lowest-1);
> (7)   runSimulation(A, clusters, c)
> (8)   **return** *clusters*

Theorem 1. *The clustering algorithm will always terminate in finite time and always result with an equilibrium (if at least one exists).*

Proof:
First we will show that given a newly updated reservation value r for a cluster, the expected performance of agents in clusters defined with reservation values greater than r is an upper bound to the equilibrium performance.

Let us denote agents with types greater than or equal to r as $A_{t \geq r}$, and agents with types less than r as $A_{t < r}$. After r is updated, all clusters with types below r are set so that their reservation value is equal to the maximum acceptance value, meaning agents of only one type are in each cluster. The probability of an agent $A_{t \geq r}$ meeting an agent $A_{t < r}$ with memory is $\int_{i=t}^{i < r} P(i) P_{M,i^*}$, where i^* is the reservation value of an agent with type i and P_{M,i^*} is the probability that an agent in a cluster with a reservation value of i^* will have memory. If any cluster with reservation value i, $i < r$, is expanded to accept agents in $[i-1, i]$ then the probability of an agent with type i having memory is reduced such that $P_{M,i-1} < P_{M,i}$. This is because the probability of an agent with type i directly pairing with an agent of its own cluster is increased by $P(i-1)$, thereby increasing the probability that an agent with type i will meet an agent in its own cluster, leave the market, and be replaced with a new agent with no memory. Therefore,

---

[6] A cluster is *satisfied* if the expected value of the agents in the cluster falls below the reservation value but above the next lowest type of the reservation value.

Table 1. The three tables depict the number of iterations for a given number of distinct types and a cost for the uniform distribution. Sketches of each distribution have been placed above their corresponding table.

$P = 1.0$; $P = 0.75$ / $P = 0.25$; $P = 0.75$ / $P = 0.25$

	Num Iterations - Unif Distribution Number of Types							Num Iterations - 25-75 Skew Number of Types							Num Iterations - 75-25 Skew Number of Types					
Cost	16	32	48	64	80	96		16	32	48	64	80	96		16	32	48	64	80	96
0.25	0	0	0	32	40	370		0	0	12	16	46	100		0	0	12	16	46	669
0.5	0	0	24	42	366	72		0	8	28	24	57	3248		0	8	28	352	367	1106
1	0	16	32	48	196	603		0	18	58	52	188	349		0	18	65	271	340	456
2	8	21	38	84	272	242		4	24	38	62	352	98		4	66	91	173	332	339
4	10	25	42	85	158	274		10	28	53	194	92	104		19	26	81	229	239	398
8	12	47	49	114	128	166		12	31	44	69	105	238		18	32	93	125	199	322
16	13	48	45	61	108	123		13	29	60	61	77	93		15	34	49	78	95	141
32	14	30	46	62	78	94		14	30	46	62	78	94		14	30	46	62	78	94

Table 2. Comparison of the number of iterations to reach an equilibrium in three different two-sided search environments to the theoretical number of possible clusterings

	Iteration Comparison			
# of Types	Uniform	25-75 Skew	75-25 Skew	Theoretical
16	0-14	0-14	0-19	3.28×10^4
32	0-48	0-31	0-66	2.15×10^9
48	0-49	12-60	12-93	1.41×10^{14}
64	32-114	16-194	16-352	9.22×10^{18}
80	40-272	46-352	46-367	6.04×10^{23}
96	72-370	93-3248	94-1106	3.96×10^{28}

the probability of an agent $A_{t \geq r}$ meeting an agent $A_{t < r}$ with a memory with whom it can partner is reduced.

Therefore, if a cluster is not satisfied (the expected value of agents in the cluster falls below the next lowest type of the reservation value) the clustering of agents in $A_{t < r}$ in an attempt to increase the average agent performance in this cluster will not succeed. Similarly, combinations of clusters with reservation values greater than r cannot be modified and continue to be satisfied since these combinations were eliminated in former algorithm rounds. This leaves only the reservation value to be reduced to satisfy the cluster.   Q.E.D.

We executed the algorithm varying the number of agent types and the distribution of types in the environment, measuring the number of simulations required before reaching a clustering in equilibrium. The results are shown in Table 1. Here, the search cost was varied between 0.25 and 32, and the number of distinct types varied between 16 and 96. The smallest type used was 0, and the largest 128.

We used three arbitrary distributions for illustration purposes, and found no significant differences as the number of iterations, search cost and number of distinct types were varied.

The number of iterations required for all scenarios tested was well below the theoretical 2^{N-1} maximum combinations, where N is the number of distinct types. As reflected in Table 2, the number of simulation runs required for extracting the equilibrium using the proposed algorithm is significantly smaller in magnitude.

7   Related Work

The two-sided search process is practically pairwise partnership formation application and can be related to the broader coalition formation domain [16,15,20,21]. Nevertheless, coalition formation literature commonly assumes that agents can scan as many agents as needed (with no associated cost) or can make use of a central matcher or middle agents [7].

The analysis of two-sided search models is mostly found in economic search theory. These models can be distinguished according to several assumptions made. The first, is the payoff for each agent associated with a given partnership. While some of these models assume that the utility is a function of the other agent's type exclusively [10,4], others assume a function defined over both types [1]. The second is the way according to which the search friction (cost) is modeled. This can be either the discounting of future flow of gains [4] or additive explicit search costs [10,6,1]. Lastly, the models are distinguished by the nature of the utility earned by each of the agents (transferable [1] and non-transferable [4,6]). Our model was developed for autonomous software agents, operating in dynamic fast-paced environments. Therefore it assumes non-transferable utilities, explicit search costs and a payoff that depends on the other agent's type exclusively. For this model, it has been shown that in equilibrium, the agents perform a complete segregation[7] [10].

None of the above two-sided search models have made use of information agents collect throughout their search. The concept of matchmakers does arise occasionally in this literature, however always in the form of centralized rather than a distributed mechanism [3]. Mechanisms that made use of formerly collected "reputation" information can be found in MAS literature.

Reputation literature in AI focuses on reputation models in which agents classify the potential worth of agents through many rounds of direct and indirect interaction [19,9]. Agents in these models usually form pairs, complete a task, and then at a later time period have the option of partnering with the same agent again. An agent generally judges the expected utility of joining with another agent using statistical models and information gathered from many agents [9,22,11,19]. However, none of these works suggested a two-sided search model in a costly environment or presented results based on equilibrium agent strategies. Our model's use of mediated partnerships is similar to a form of reputation known as witness reputation. Witness reputation is information reported by an agent about its direct interactions with another agent [12]. In general, reputation

---

[7] Complete segregation can also be found in many other model variants [1,3,6].

information is noisy (potentially deceitful) and is generated from a history of interactions. The form of witness reputation we used assumes that agents gather exact information from only one interaction and that agents do not deceive when revealing their type or attempting to form mediated partnerships.

Very little work has been done in the AI reputation literature with measuring the cost of reputation data or finding beneficial agents [8]. Instead, most work only associates a reward for completing tasks in marketplace environments [19,14,22] or social networks [11]. Our work does not associate a cost directly with reputation data, but rather the cost of search. Other research [14] looks into the truthfulness of reputation data and provides models for judging the validity of reputation information.

8   Discussion, Conclusions and Future Work

The introduction of mediated partnering to two-sided search is an important extension whenever applying two-sided search in MAS. In contrast to the traditional two-sided search model, where formerly collected data has no value, mediated partnering allows a non-direct partnering based on agents' former experiences in the market. This approach fits well with autonomous agents that can easily support extensive memory storage and immediate costless interaction with formerly encountered agents. The proposed extension to the agents' search has not appeared in either economic search theory or MAS research, despite the fact that similar mechanisms in which agents use formerly collected data to improve system performance have been suggested (e.g., reputation systems and witness reputation).

As illustrate throughout our analysis, the integration of mediated partnering to two-sided search is not straightforward, and should be done with great caution. Despite the many advantages of witness reputation, in some settings it worsens overall performance (measured as aggregated social welfare). This is both surprising and important. It is surprising because we would have expected that reducing friction would improve efficiency and important because system designers should be careful not to automatically allow this option, but rather first calculate how reducing market friction can affect the performance of the system.

One important issue that ought to be discussed is the agents' incentive to use the proposed model. Here we distinguish between two aspects of the mediated partnering. The first is the agent's willingness to reveal its acceptance rule to the other agents it meets along its search. The agent will always benefit from revealing its acceptance rule since it enables the agent to be exposed to additional targeted partnering opportunities with no cost. The second is the incentive of agents to initiate the mediated partnering. Indeed, the agent is indifference between initiating the mediated partnering and simply resuming its search (since there is no direct benefit from doing this). However, an incentive can be easily created externally by enforcing the mediated partnering protocol by the market maker / agent designer, paying some of the market surplus to the agent in order to initiate mediated partnering or simply fining those agents that deviate from this protocol.

While the new model is commonly favorable, it adds a significant computational complexity to the equilibrium analysis. This prevents an analytic solution and requires approximation through appropriate algorithms and heuristics that embed simulation. The algorithm we present for this purpose in Section 6 suggests exceptional performance improvement within any range of agent types.

Finally, the attempt to integrate "search theory" techniques with day-to-day applications brings up the question of applicability. Justification and legitimacy considerations for this integration were discussed in the literature we referred to throughout the paper. This paper is not focused on re-arguing applicability, but rather on the improvement of the core two-sided search model.

This research is a first and important step towards developing far more advanced information sharing mechanisms for enhancing two-sided search applications in MAS environments. Extensions are numerous: from gossip (sharing all information available to the agent) to information bargaining (trading with the collected information with other agents).

Acknowledgments

We thank Barbara J. Grosz for her helpful comments and discussions. The work reported in this paper was supported in party by NSF grant CNS-0453923.

References

1. Atakan, A.E.: Assortative matching with explicit search costs. Econometrica 74(3), 667–680 (2006)
2. Becker, G.: A theory of marriage. Journal of Political Economy 81, 813–846 (1973)
3. Bloch, F., Ryder, H.: Two-sided search, marriages, and matchmakers. International Economic Review 41, 93–115 (2000)
4. Burdett, K., Coles, M.G.: Marriage and class. The Quarterly Journal of Economics 112(1), 141–168 (1997)
5. Burdett, K., Wright, R.: Two-sided search with nontransferable utility. Review of Economic Dynamics 1(1), 220–245 (1998)
6. Chade, H.: Two-sided search and perfect segregation with fixed search costs. Mathematical Social Sciences 42(1), 31–51 (2001)
7. Decker, K., Sycara, K., Williamson, M.: Middle-agents for the internet. In: Proc. of IJCAI (1997)
8. Hendrix, P., Grosz, B.: Reputation in the venture games. In: National Conference on Artificial Intelligence, AAAI, Vancouver, Canada (2007)
9. Huynh, T.D., Jennings, N., Shadbolt, N.: Fire: An integrated trust and reputation model for open multi-agent systems. In: Proceeding of 16th European Conference on Artificial Intelligence, pp. 18–22 (2004)
10. Morgan, P.: A model of search, coordination, and market segmentation. SUNY Buffalo, mimeo (1995)

11. Pujol, M., Sanguesa, R., Delgado, J.: Extracting reputation in multi agent systems by means of social network topology. In: Proceedings of the First International Joint Conference on Autonomous Agents and Multiagent sytems: Part 1, pp. 467–474 (2002)

12. Sabater, J., Sierra, C.: Reputation and social network analysis in multi-agent systems. In: Proceedings of the First International Joint Conference on Autonomous Agents and Multiagent Systems: Part 1, pp. 475–482 (2002)

13. Sarne, D., Kraus, S.: Time-variant distributed agent matching applications. In: Proceedings of AAMAS-2004, pp. 168–175. NY (2004)

14. Sen, S., Biswas, A., Debnath, S.: Believing others: Pros and cons. In: Proceedings of the Fourth International Conference on MultiAgent Systems (ICMAS-2000), p. 279 (2000)

15. Sen, S., Dutta, P.: Searching for optimal coalition structures. In: Proceedings of the Fourth International Conference on Multiagent Systems, July 2000, pp. 286–292 (2000)

16. Shehory, O., Kraus, S.: Methods for task allocation via agent coalition formation. Artificial Intelligence Journal 101(1-2), 165–200 (1998)

17. Shimer, R., Smith, L.: Assortative matching and search. Econometrica 68(2), 343–370 (2000)

18. Stewart, I.: Tales of a neglected number. Scientific American 274(6), 102–103 (1996)

19. Teacy, W., Patel, J., Jennings, N., Luck, M.: Coping with inaccurate reputation sources: Experimental analysis of a probabilistic trust model. In: Proceedings of Fourth International Joint Conference of Autonomous Agents and Multiagent Systems, pp. 997–1004 (2005)

20. Tsvetovat, N., Sycara, K., Chen, Y., Ying, J.: Customer coalitions in electronic markets. In: Proceedings of AMEC 2000, pp. 121–138. Barcelona (2000)

21. Yamamoto, J., Sycara, K.: A stable and efficient buyer coalition formation scheme for e-marketplaces. In: Proceedings of the Fifth International Conference on Autonomous Agents, Montreal, Canada, pp. 576–583 (2001)

22. Zacharia, G., Maes, P.: Trust management through reputation mechanisms. Applied Artificial Intelligence 14, 881–907 (2000)

Who Works Together in Agent Coalition Formation?

Vicki H. Allan and Kevin Westwood

Utah State University, Logan, Utah 84332-4205
vicki.Allan@usu.edu, kevwestwood@cc.usu.edu

Abstract. Coalitions are often required for multi-agent collaboration. In this research, we consider tasks that can only be completed with the combined efforts of multiple agents using approaches which are both cooperative and competitive. Often agents forming coalitions determine optimal coalitions by looking at all possibilities. This requires an exponential algorithm and is not feasible when the number of agents and tasks is large. We propose agents use a two step process of first determining the task, and secondly, the agents that will be solicited to help complete the task. We describe polynomial time heuristics for each decision. We measure four different agent types using the described heuristics. We explore diminishing choices and performance under various parameters.

Keywords: Coalition Formation, Multiagent Systems, Coordination and Collaboration.

1 Introduction

Recently there has been considerable research in the area of multi-agent coalitions [1, 2, 5, 7, 9, 10]. A *coalition* is a set of agents that work together to achieve a mutually beneficial goal [4].

In the request for proposal (RFP) domain [6], a collection of agents is challenged to complete specific tasks, each of which can be divided into subtasks. In this format, the task requirements and utility are specified and requests for proposals from agents are sought. In the model considered here, no single agent possesses all the skills necessary to complete a task, and so coalitions must be formed. The agents attempt to form coalitions that will be able to complete tasks in a way which maximizes the payoff received. In addition to considering cooperative and competitive strategies, a blend of the two is also considered. Coopetitive is a phrase coined by business professors [3] to emphasize the need to consider both competitive and cooperative strategies as agents which are solely competitive not be able to secure the partners required for coalition formation. Agents use heuristics to facilitate accepting less than optimal personal benefit if the global utility is increased – hence the term, coopetitive.

M. Klusch et al. (Eds.): CIA 2007, LNAI 4676, pp. 241–254, 2007.

1.1 Skilled Request for Proposal Domain

We use the term *Skilled Request For Proposal* (SRFP) to denote a domain in which both skills and levels of skill are important. Initially, a requester provides a set of tasks to be completed along with a fixed payoff. Task i is divided into a list of subtasks $T_i = (t_{i1}...t_{im})$. In order to simplify the analysis, each task in the system has the same number of subtasks. A subtask t_{ij} is associated with a skill-level pair (ω_{ij}, ℓ_{ij}) where a task skill, ω, represents a skill taken from a skill set S, and ℓ represents a specific level of skill, $\ell \in \{1...\textbf{LevelMax}\}$. A set of service agents A exists such that each agent k has an associated fee, f_k, and agent skill-level pair $(\omega_k, a\ell_k)$.

A coalition $C_i = <T_i, C_i, P_i>$ consists of a task vector for task i, consisting of each of the subtasks required to perform the task, a vector of agents comprising the coalition for the task, and a payment vector indicating the payoff for each agent of the coalition. Since the system is pareto optimal, all task payoff is distributed. The actual cost f_{ij} is associated with agent i performing the jth subtask. A rational agent will not consent to performing a subtask for less than its cost.

To facilitate the discussion, we introduce some terminology. The task $T_i = (t_{i1}...t_{im})$ is *satisfied* by the coalition consisting of agents $C_i = (C_{i1} ... C_{id} ...C_{im})$ ($C_{id} \in A$) if each agent d of the coalition (C_{id}) has associated skill-level pair $(\omega_{id}, a\ell_{id})$ such that $\omega_{id} = \omega_{id}$ and $a\ell_{id} \geq \ell_{id}$. The difference $a\ell_{id} - \ell_{id}$ is termed the *underutilization* of the dth agent in performing task i. We term a task *viable* if there are available agents to satisfy the requirements of the task. We say an agent *supports* a task if the agent can perform an (as of yet) unassigned subtask. A level equal to or greater than the required level is termed an *acceptable level*. The level of skill which is actually required (due to possibly missing skill-level pairs) to perform a subtask is called the *effective level*.

For a particular skill, the base fees are a non-decreasing function of level and are publicly known. In our domain, the set of fees for a given skill-level pair is taken from a normal distribution with known mean and standard deviation.

2 Previous Work

Lau and Zhang [8] explore the complexity of exhaustive methods for coalition formation relying on comparisons to the multi-dimensional knapsack problem. In their model, each agent is group rational - the goal is to maximize system profit. The auction protocol proposed by [6] is conducted by a central manager. Agents are randomly ordered at the beginning of each round $r = \{0,1,2...\}$ so that no agent is disadvantaged due to its position in the proposing order. In each round, the agents are asked (in turn) to respond to previous proposals or initiate a proposal of their own. For each satisfied proposal, the task and the associated agents are removed from the auction. At the end of a round, any unsatisfied proposals are discarded. Thus, an agent is only allowed to propose to agents that follow it in the current agent ordering.

In the domain proposed by [6], each subtask is completely unique from all other subtasks encountered. A Boolean function $\Phi(A_k, t_{ij})$ determines whether agent A_k can perform subtask t_{ij}. An agent has a 40% chance of having the skills required to complete a "normal task" and a 15% chance of having the skills to complete a "specialized task". This skill structure is limited in its ability to model realistic skill requirements as there is only a rough concept of supply and demand on an individual skill basis. We improve the model by paying more for a highly skilled agent doing a less demanding task. Kraus, et.al., [6] consider incomplete information as agent costs are unknown, but because averages are known, incomplete information provides merely random changes in the overall behavior, increasing the variance but not changing the cumulative behavior. Compromise is introduced in [5] to increase the speed at which coalitions can be formed.

Scully, Madden, and Lyons [11] use a skill vector that uses a variety of skill metrics. We propose a combined skill-level pair. Blankenburg, Klusch, and Shehory [2] introduce the idea forming coalitions when there is fuzzy knowledge. An agent's evaluation of the coalition is uncertain, and agents can be (fuzzy) members of multiple coalitions to various degrees.

3   Auctioning Protocol

We employ a protocol similar to [6] which is used as a simplification of a parallel model. For simplicity of analysis, tasks and agents are not allowed to leave the auction. We use a reverse English auction in which the bidders compete to provide a service. Each agent has knowledge of a set of tasks $T = \{T_1,...,T_n\}$. Each task has a value $V(T_i)$ that will be paid when the task is complete. The auctioneer posts all tasks and associated values. The auction consists of rounds until no remaining task is viable or the time expires.

One by one, each agent can either propose (termed the *proposer*) a coalition or accept a proposal (termed a *joiner*). A proposal specifies the task to be done and the set of agents to complete the task. After agent k has decided whether to propose a coalition or accept a previous proposal, the turn passes to the next agent. All agents accept or reject the proposal only when it is their turn in the round.

An important feature of this mechanism is that an agent is motivated to make reasonable proposals. An agent who makes a proposal cannot be involved in other proposals until every agent has had a turn, during which time successful agents and tasks are removed from the system. Over time, the possible tasks that can be completed diminish even if the original choice for tasks is large. Thus, agents are motivated to accept less that optimal personal benefit. In Figure 1, the test was run with 75 agents and 50 tasks, each task requiring three agents. Thus, there were twice as many tasks as could be performed. The figure shows the number of eligible tasks for a *single* representative agent with high skill level. Originally, the agent can support 28 of the 50 tasks (56%). However, as the agent pool shrinks, the agent can support only 11 of the 27 remaining tasks (41%).

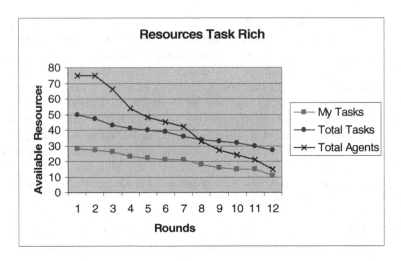

Fig. 1. Data for sample agent showing how resources diminish over time

When an agent i decides between accepting one of the current offers or making an offer of its own, it has no knowledge of proposed coalitions which do not involve i nor does it know how many agents have accepted or rejected the current proposals. This lack of current information reflects the state of information in a parallel proposal environment in which acceptances happen simultaneously. Since agents and tasks are removed from the auction when the initiating agent has a turn in the following round, the information is updated with predictable delay. For the current offers, the task and the net gain (given the proposed coalition) is known.

Division of the profit is an important issue. When a task is completed, each agent k is paid its fee f_k plus a portion of the *net gain* of the coalition. The net gain is the amount remaining after the agents are paid their advertised fees. Net gain is divided among the agents. Historically, the portion paid to each agent is divided equally, proportionally according to the agent's individual cost, or differentially, based on demand for a particular skill. The Shapley value [12] is a popular method for distributing cost, but is inappropriate in this setting as it requires that the value of every possible combination of agents be known. In our model, we divide the net gain proportionally.

4 Agent Algorithms

In many coalition formation algorithms, the best coalition structure is calculated by considering all possible coalitions to complete each task, an exponential algorithm. Because this quickly becomes unreasonable for even moderately sized problems, we divide the problem into two components: task selection and coalition selection, and propose the following polynomial time heuristics for each component. In the first component, the agent selects a task to complete. In the second component, the agent selects the set of agents for the coalition to complete the task.

4.1 Task Selection Heuristics

Local Profit Task Selection: An agent selects a task that will maximize its own profit. For each task, the agent first computes the expected net gain by subtracting the average fee of the effective level of each required skill. Tasks are ranked by comparing the agent's individual utility in each possible task.

Global Profit Task Selection: An agent selects a task that will maximize global profit. For each task, the agent computes the expected net gain by subtracting the average fee of the effective level of required skill. The task with the maximum net gain is selected.

Best Fit Task Selection: An agent picks the task for which the sum of the agent underutilization is minimized. The motivation is that when the resources are used wisely, future coalitions will be less likely to face resource constraints. Ties are broken using maximum local profit.

Coopetitive Selection: Select the best fit task as long as it is within a given percent of the maximum local profit.

4.2 Coalition Selection Heuristics

For each subtask, select the supporting agent with the smallest underutilization or fees – to agree with the motivation used in task selection.

4.3 Acceptance Policy

To decide between (a) accepting one of the various proposals and (b) initiating a new proposal, our agents compare existing proposals with the proposal they would make and select the proposal most closely fitting their criteria.

Accepting the best proposal may give the agent a better chance of joining a coalition, particularly when the number of subtasks per task is small, as an agent knows that at least two agents (itself and the initiator) accept the proposal. Since agent i has no information about proposals not involving i and no knowledge of how many agents have accepted the current proposal, the advantage of accepting a current proposal over initiating a proposal is not great as one might think.

To maximize expected utility, the *compromising heuristic* assumes that the increased chance of a contending proposal being satisfied is worth some decreased utility. [Note, compromising applies to acceptance while coopetition applies to task selection.] If the contending proposal is within C% of the ideal proposal (based on the criteria the agent values), the contending proposal is accepted. Otherwise, the proposal is rejected. Several different values for the compromising ratio, C, are evaluated in our experiments.

5 Results

Results represent the average of tests run 5000 times. Each subtask has a randomly generated agent skill-level pair. For each skill, the base fee is level*5. Actual fees are assigned using a normal distribution with a median value of the base fee and a standard deviation of 2.5 (half the fee difference from one skill level to the next). Payoffs of the tasks are assigned as a random percentage (100-200%) of the expected cost of completing the task.

5.1 Compromising Ratio

To determine the compromising ratio, we conduct the following experiment. The number of tasks is fixed at 30 with 3 subtasks each. The number of agents in the various tests ranges from 45 to 270. Using an equal distribution of each agent type, we track the individual profit and the number of tasks completed by each agent type.

Figure 2 shows the profit ratio for the Local Profit agent under a variety of compromising ratios and loads. The other agent types use a compromising ratio of .7. The profit ratio is the profit achieved divided by the ideal profit. However, the ideal profit is not achievable for all agents due to high demand for inexpensive resources and the fact there are more agents than subtasks. A profit ratio of one is only achievable for infinite agents and infinite tasks. Thus, the individual profit ratio depends on the scarcity of tasks. As the compromising ratio is increased, profits actually decrease. This same pattern is seen in other agent types as well. One would expect that, when the numbers of agents and tasks are compatible, profits would rise as the agent becomes more demanding, but the number of tasks completed by all Local Profit agents decreases. See Figure 4. As the compromising ratio increases, agents with differing goals are unable to find the common ground necessary for an agreement. The profit ratio (profit/ideal profit) for *scheduled* agents does increase significantly as shown in Figure 3. The affect is most pronounced for the Local Profit agents, but even agents who do not change their compromising ratio slightly increase the profit ratio. It is interesting that even in an agent rich environment (where demanding agents could just be ignored) the number of tasks completed decreases when only the Local Profit agents become more demanding. Agent performance is affected by the mixture of agents. Demanding Local Profit agents reject the proposals of others, and the others have no mechanism to learn from their failure. These results are also colored by the fact that the majority of the proposals an agent receives have a high profit ratio. The agent types are *all* prone to pick high profit coalitions - as even Best Fit generally increases profits. While Local Profit agents are "blind" about whether they belong in a coalition, they are not blind in selecting other agents for the coalition.

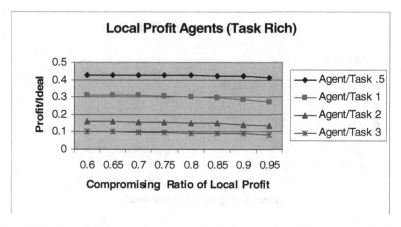

Fig. 2. Local Profit Agent performance under various compromising ratios and loads

We fix the compromising ratio at .90 for all agent types in the following tests as it represents a middle ground between achieving individual performance and system performance.

5.2 Context

We conduct tests to see which agents choose to work together. In Figure 5, the results represent the averages of 5000 tests in a *balanced* environment (the number of agents available is equal to the number of agents required by the tasks). There are an equal number of agents of each of the four types. Each coalition consists of three agents – one proposer and two joiners. For each coalition, we tally the number of times each pair of types of proposer/joiner occurs. The y axis is the number of pairings normalized by dividing by the number that would occur in an unbiased pairing. Note that Local Profit and Global Profit are less likely to accept proposals by Coopetitive or Best Fit agents. Coopetitive and Best Fit agents are more likely to accept regardless of the proposer.

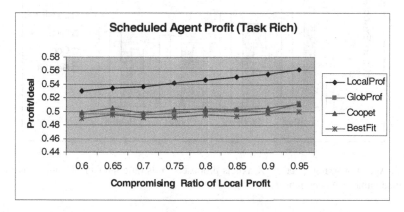

Fig. 3. Average profit ratio of the scheduled Local Profit agents

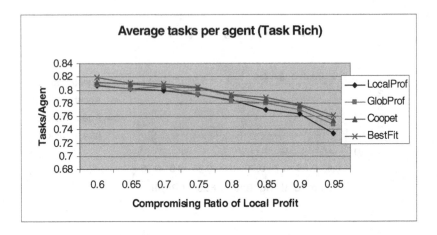

Fig. 4. Average tasks completed per agent in task rich environment

We see that Local and Global Profit Agents function primarily as proposers (Figure 5). It is interesting, that there is no propensity for Local Profit or Global Profit agents to propose to themselves, but that both Coopetitive and Best Fit agents have success in proposing to agents of their same type. Notice that Local Profit Agents not only excel at proposing, but for every agent type, the team joined is most likely proposed by a Local Profit Agent. A likely joiner for a coalition proposed by a Local Profit agent is a Coopetitive Agent, yet a Local Profit Agent is less likely to accept a proposal by a Coopetitive Agent.

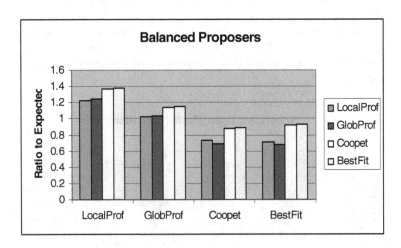

Fig. 5. Agents working together. X axis is proposers. Y is ratio number of joiners divided by expected number of occurrences.

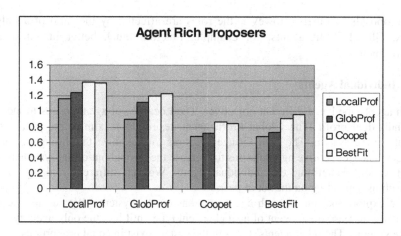

Fig. 6. Agents working together in an agent rich environment (number of agents is three times that required by tasks). X axis is proposers. Y is ratio number of joiners divided by expected number of occurrences.

We repeat the tests changing the ratio of tasks to agents. *Task rich* environments have three times the number of doable tasks. *Agent rich* environments have three times the number of needed agents. There are significant load-based differences. Recall that Coopetitive agents are sometimes looking at fit and sometimes looking at profit. Coopetitive agents function better as proposers to Local Profit agents in balanced or task rich environments. This occurs for two reasons. First, since there are more choices of tasks, they tend to pick tasks having better profit (as profit is used to discriminate between tasks having equal underutilization). Second, more tasks give a Coopetitive agent a better sense of profit-potential (which is used to determine whether a best fit solution is acceptable). Both of these facts mean that Coopetitive agents are more sensitive to profit – which makes them more acceptable to Local Profit and Global Profit agents. Notice how much better Global Profit agents do as proposers in a task rich environment.

In Figure 7, we examine the likelihood that a Local Profit agent serves as a proposer or joiner as the proportion of each type of agent is varied separately in a task rich environment. Proposing success appears to be negatively correlated to increasing percent of agents like itself. While space does not permit inclusion of other charts, it is interesting to know that Global profit agents are similarly negatively impacted by increasing percent of Global Profit agents. However, both Coopetitive and Best Fit agents are stable with increasing percentages of their own type and negatively correlated to increasing percentages of agents of other types. Note that both Local Profit and Global Profit agents primarily function as proposers.

The role of Global Profit agents as proposers is more pronounced in a balanced environment. Agents with a wider range of acceptable coalitions function better as joiners as they are easier to satisfy. Similarly, those with a restricted definition of acceptability function better at proposing as they are less likely to be satisfied by the proposals of others. However, the roles are affected by the ratio of agents to work. Global Profit agents, for example, function much better in a task rich environment.

5.3 Individual Agents

Each agent type has a different objective. The Local Profit agent desires to increase its individual profit. The Global Profit agent desires to maximize the profit of all agents in the system. The Best Fit agent desires to maximize the throughput of the system. The Coopetitive agent desires to increase the throughput of the system as long as it is still benefiting the individual agent. We see different rankings of the algorithms depending on the set of peers.

We explore the impact each agent type has on the system. We set up tests that gradually increase the percent of a single agent type until it is the only agent type left in the system. The other agents' types in the system exist in equal proportions.

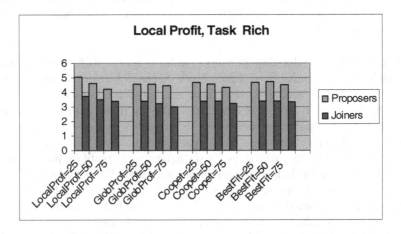

Fig. 7. Likelihood that a Local Profit agent proposes (or joins) given varying proportion of one agent type (while other agent types are equal)

Figure 8 shows the individual profit ratio for the specific agent type as its percent in the system increases. It can be seen that the Local Profit agent does very well when less than 70% of the agents are Local Profit agents. As its percentage in the system increases, the profit for the agent type decreases. The primary cause of this decrease is that there are fewer tasks completed. When there are few Local Profit agents in the system, Local Profit agents do well. They make proposals that the other agents (with different goals) find acceptable – at least within 90% of the desired goal. The greedy, Local Profit agent is able to capitalize on the generosity of the other agents in not requiring 100% of the utility to be achieved. When the percent of Local Profit agents increases, the tasks that are proposed maximize what the majority of the agents are looking for – high individual profits. Yet, fewer proposals are accepted. Why? The answer lies in the fact that the Local Profit agents are objective in picking which task to select and which agents to ask to participate. The only blindness is in deciding that the agent itself belongs in the coalition. In other words, the agent says, "Given that I am going to perform a task, which task is best?" Once the task has been selected, the agent asks, "Given that I am going to participate in the task, which agents should work with me?" Thus, the logic always presupposes that the agent should participate. The agent never considers that *it* may not be a good match for the task. Thus, when the majority of the agents are greedy (each seeking the ideal in the task and **other** members of the coalition), they reject each other's proposals as not meeting their expectations. They are saying, "I may not be perfect but I expect you to be." The only coalitions that can be accepted are the ones proposed by agents ideally matched to the task (ideal skill level and/or cost).

Individual Profit Ratio

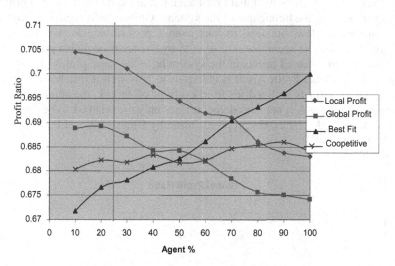

Fig. 8. Individual Profit ratio. The vertical line indicates equal numbers of each agent type.

Agents that deliberately go after individual profit are most successful at achieving it *until the agents compete with too many other agents with the same selfish goal.* The Best Fit agents have the opposite effect. As the percentage of Best Fit agents increase, the number of tasks completed and the individual profit for the Best Fit agents increase as well. The number of tasks that are completed is maximized by Best Fit agents having a high compromising ratio. Unlike Local Profit agents, Best Fit agents are not oblivious about their own fit. While Local Profit agents subtract their fixed fee from every possible profit (making their fee irrelevant in their decision as it represents a linear transformation of the rewards), Best Fit agents are attracted to tasks for which they are a good match (in terms of level) and every Best Fit agent sees the match the same way. More proposals by Best Fit agents results in better conservation of resources and, hence, more tasks completed. In addition, profit goes up as cheaper agents within the same level are preferred. (Cheaper agents maximize local profit in proposals having the same under-utilization.) Thus, a proposal always includes the cheapest agent within a needed skill level.

The profit of Coopetitive agents increases slightly with percent of Coopetitive agents, but is fairly steady. With less than 50% of the same agent type, the Coopetitive agent has a higher individual profit than the Best Fit agent. The throughput of the Coopetitive agent is fairly steady as well. As the percentage of Coopetitive agents increases, the average number of tasks completed stays about the same.

Consider the vertical line of Figure 8. At 25% of the primary agent, all agents are equally represented in the auction. Thus, when agents are equally represented, we see that Local Profit agents do the best in individual profit followed by Global Profit, Coopetitive, and Best Fit

Figure 9 shows the global profit ratio of the system, calculated as the total profit of all agents divided by maximum possible profit of all agents (using a hill-climbing algorithm). The objective of Global Profit agents is to maximize the global profit and in doing so increase the throughput of the system. Global Profit Agents do well when their numbers are few, but tend to hurt overall performance as they become the majority. Why does an agent whose goal is to try to maximize global profit do worse than others, even when all agents in the system have the same goal? Like the Local Profit agent, a Global Profit Agent is oblivious to how well it matches the proposed task. Since it subtracts its fixed fee from every possible task, the appropriateness of the agent to do the task is not considered; it finds only the best task to be done *not* the best task for *it* to do. Thus, the proposals made by Global or Local Profit agents are often less than ideal.

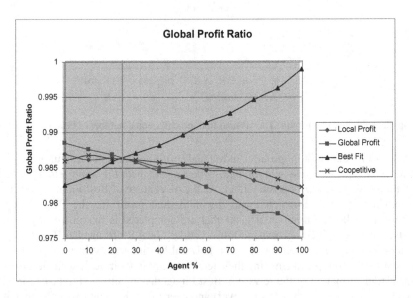

Fig. 9. Global Profit Ratio

The achieved global profit is dependent on the number of tasks that are completed. The Best Fit agent maximizes the global profit as its percentage increases past 25%. Both Local and Global profit agents are similar in performance. As their percentage increases, the global profit decreases. The Coopetitive agent does a little better than the Local and Global profit agents, but not nearly as well as the Best Fit agent. This is a bit surprising as Coopetitive Agents act just like Best Fit agents as long as profits are not reduced too much.

6 Conclusions and Future Work

Agent types differ in their measure of utility, making it difficult to find common ground without the ability to compromise. Some agents appear to other agents to be non-cooperative simply because their utility function is so different. Because agents do not know the types of other agents, a "non-cooperative class" of agents hurts global profit as repeated attempts are made to form coalitions that fail.

As expected, agents perform differently depending on the environment: task rich, balanced agents and tasks, and agent rich. Some agent types are very good in selecting between many tasks, but are not as impressive when there are only a few choices. In any environment, the choices diminish rapidly over time so those that do not join coalitions early are negatively impacted. This has the surprising effect that being more selective actually hurts performance.

Agents interested in Global profit or Local Profit function well as proposers. Best Fit agents or Coopetitive agents both do much better as joiners than proposers. It appears that a coalition that maximizes profit is more likely to achieve good skill utilization than the converse. Coopetitive agents and Best Fit agents have the most success in proposing to themselves, but Local and Global Profit agents have better success in proposing to other types of agents.

While agent types are somewhat affected by changing the percentage of one other agent type in the system, the effects are most pronounced when it is the agent whose percentage in the system is being changed. In the case of Local Profit agents, they actually are negatively impacted by increasing numbers of Local Profit agents.

Because agents are fairly similar in what they value, agents can succeed just by accepting other offers. In future work, we will introduce side payments to motivate an agent to select a task when other agents receive no special motivation.

References

[1] Belmonte, M.-V., Conejo, R., Pérez-de-la-Cruz, J.L., Triguero, F.: A Robust Deception-Free Coalition Formation Model. ACM symposium on Applied computing, Nicosia, Cyprus , 469–473 (2004)

[2] Blankenburg, B., Klusch, M., Shehory, O.: Fuzzy Kernel-Stable Coalitions between Rational Agents. In: Autonomous agents and multiagent systems, Melbourne, Australia, pp. 9–16 (2003)

[3] Brandenburger, A.M., Nalebuff, B.J.: The Game Theory That's Changing the Game of Business. Doubleday Publishing, New York, NY, U.S.A (1996)

[4] Klusch, M., Shehory, O.A.: Polynomial Kernel-Oriented Coalition Formation Algorithm for Rational Information Agents. In: International Conference on Multi-Agent Systems (ICMAS), Kyoto Japan, pp. 157–164 (1996)

[5] Kraus, S., Shehory, O., Taase, G.: The Advantages of Compromising in Coalition Formation with Incomplete Information. Autonomous Agents and Multiagent Systems (AAMAS) , 588–595 (2004)

[6] Kraus, S., Shehory, O., Taase, G.: Coalition Formation with Uncertain Heterogeneous Information. In: Autonomous Agents and Multi-Agent Systems (AAMAS), Melbourne, Austrailia (2003)

[7] Lau, H., Lim, W.: Multi-Agent Coalition via Autonomous Price Negotiation in a Real-Time Web Environment. In: Intelligent Agent Technology (IAT), pp. 580–583 (2003)

[8] Lau, H., Zhang, L.: Task Allocation via Multi-Agent Coalition Formation: Taxonomy, Algorithms and Complexity. In: International Conference on Tools with Artificial Intelligence (ICTAI'03) (2003)

[9] Lau, H.C., Zhang, L.: A Two-Level Framework for Coalition Formation Via Optimization and Agent Negotiation. In: International Conference on Intelligent Agent Technology (IAT), pp. 441–445 (2004)

[10] Li, X., Soh, L.: Learning How to Plan and Instantiate a Plan in Multi-Agent Coalition Formation. In: Li, X., Soh, L. (eds.) Intelligent Agent Technology, (IAT), Beijing, China, pp. 479–482 (2004)

[11] Scully, T., Madden, M.G., Lyons, G.: Coalition Calculation in a Dynamic Agent Environment. In: International Conference on Machine Learning, Banff, Canada (2004)

[12] Shehory, O., Kraus, S.: Methods for Task Allocation Via Agent Coalition Formation. Artificial Intelligence 101, 165–200 (1998)

Using Ant's Brood Sorting to Increase Fault Tolerance in Linda's Tuple Distribution Mechanism

Matteo Casadei[1], Ronaldo Menezes[2], Mirko Viroli[1], and Robert Tolksdorf[3]

[1] Università di Bologna, DEIS
Cesena (FC), Italy
{m.casadei,mirko.viroli}@unibo.it
[2] Florida Tech, Computer Sciences
Melbourne, Florida, USA
rmenezes@cs.it.edu
[3] Freie Universität Berlin, Institut für Informatik
Berlin, Germany
tolk@inf.fu-berlin.de

Abstract. Coordination systems have been used in a variety of different applications but have never performed well in large scale, faulty settings. The sheer scale and level of complexity of today's applications is enough to make the current ways of thinking about distributed systems (e.g. deterministic decisions about data organization) obsolete. All the same, computer scientists are searching for new approaches and are paying more attention to stochastic approaches that provide good solutions "most of the time". The trade-off here is that by loosening certain requirements the system ends up performing better in other fronts such as adaptiveness to failures. Adaptation is a key component to fault-tolerance and tuple distribution is the center of the fault-tolerance problem in tuple-space systems. Hence, this paper shows how the tuple distribution in LINDA-like systems can be solved by using an adaptive self-organized approach à la Swarm Intelligence. The results discussed in this paper demonstrate that efficient and adaptive solutions to this problem can be achieved using simple and inexpensive approaches.

1 Introduction

We could start this paper discussing the *era of uncertainty* in computer systems. But by now, most computer science researchers have realized that the sheer scale of current systems and applications is "forcing" them to look away from standard approaches to deal with these systems and concentrate their efforts in the search for solutions inspired from other areas such as biology and chemistry. Why have not they done this before? It is quite simple: *(i)* computer scientists are used to having total control over their systems' workings, and *(ii)* the scale of the problems they face does not require unconventional solutions. The consequence is that antiquate ways of thinking flood today's applications leading to solutions that are complex and brittle. It is outside the scope of this paper to discuss in length who is to blame for this scenario. In fact, no one may be to blame since it would have been hard to foresee such an increase in computer usage let alone people's dependency on computer systems.

M. Klusch et al. (Eds.): CIA 2007, LNAI 4676, pp. 255–269, 2007.

Nevertheless, the aforementioned issues could be improved if we better understand that uncertainty is not a synonym for "incorrectness". We all live in a world where the causality of events makes non-determinism/uncertainty a norm. Although far from the level of complexity of the real world, computer applications exhibit complexity that is hard to deal with even by today's most powerful computers. For a sample of challenging problems we are facing, see the grand challenges listed by the Computing Research Association (CRA) [16] and by the UK Computing Research Committee (UKCRC) [19].

Coordination systems are constantly being pointed as a mechanism to deal with some of the complex issues of large-scale systems. The so-called separation between computation and coordination [8] enables a better understanding of the complexity of distributed applications. Yet, even this abstraction has had difficulties in overcoming hurdles such as fault-tolerance present in large-scale applications. A new path in dealing with complexity in coordination systems has been labeled *emergent coordination* [13]. Examples of this approach include mechanisms proposed in models such as Swarm-Linda [12] and TOTA [11]. This paper explores one mechanism proposed in Swarm-Linda that refers to the organization of data (tuples) in distributed environments using solutions borrowed from natural forming multi-agent swarms, more specifically based on ant's brood sorting behavior [7].

In tuple-space coordination systems, the coordination itself takes place via generative communication using tuples. Tuples are stored and retrieved from distributed tuple spaces. As such, the location of these tuples is very important to performance—if tuples are maintained near the processes requiring them, the coordination can take place more efficiently. Standard systems base this organization on hash functions that are quite efficient but inappropriate to large distributed applications because they are rather difficult to adapt to dynamic scenarios; particularly to scenarios where distributed nodes/locations can fail. In this paper we devised a mechanism to implement the algorithm suggested in the original SwarmLinda. We evaluated the performance of our approach using a metric based on the spatial entropy of the system. We argued that this approach is more appropriate to faulty systems because it does not statically depend on the existence of any particular node (as in hashing). In fact, the organization pattern of tuples emerges from the configuration of nodes, tuple templates, and connectivity of nodes. This emergent pattern adapts to variations in environments including node failure.

2   SwarmLinda

The LINDA coordination model is based on the associative memory communication paradigm. LINDA provides processes with primitives enabling them to store and retrieve tuples from tuple spaces. Processes use the primitive out to store tuples. They retrieve tuples using the primitives in and rd; these primitives take a template (a definition of a tuple) and use associative matching to retrieve the desired tuple—while in removes a matching tuple, rd takes a copy of the tuple. Both in and rd are blocking primitives, that is, if a matching tuple is not found in the tuple space, the process executing the primitive blocks until a matching tuple can be retrieved.

SwarmLinda uses several adaptations of algorithms taken from the abstraction of natural multi-agent systems [1,14]. Over the past few years, new models originating from biology have been studied in the field of computer science [1,9]. In these models, actors sacrifice individual goals (if any) for the benefit of the collective. They act extremely decentralized, carrying out work by making purely local decisions and by taking actions that require few computations, thus improving scalability. These models have self-organization as one of their main characteristic. Self-organization can be defined as a process where the entropy of a system (normally an open system) decreases without the system being guided or managed by external forces. It is a phenomenon quite ubiquitous in nature, in particular in natural forming swarms. These systems exhibit a behavior that seems to surpass the sum of all the individuals' abilities [1,14].

To bring the aforementioned ideas into SwarmLinda, we interpret the "world" of nodes as a graph in which ants search for food (similar tuples), leaving trails to successful searches. Adopting this approach, one can produce a self-organized pattern in which tuples are non-deterministically stored in specific nodes according to their characteristics (e.g. size, template).

The above is just an illustration. A more realistic SwarmLinda should consider a few principles that can be observed in most swarm systems [14]:

Simplicity: Swarm individuals are simple creatures, doing no deep reasoning and implementing a small set of simple rules. These rules lead to the emergence of complex behaviors.

Dynamism: Natural swarms adapt to dynamically changing environments. In open distributed LINDA systems, the configuration of running applications and services changes over time.

Locality: Swarm individuals observe their direct neighborhood and make decisions based on their local view.

LINDA systems do not define the idea of ants or food. The description of Swarm-Linda is based on the following abstractions: the *individuals* are active entities that are able to observe their neighborhood, move in the environment, and change the state of the environment in which they are located; the *environment* is the context in which individuals work and observe; the *state* is an aspect of the environment that can be observed and changed by individuals.

We aim at optimizing the distribution and retrieval of tuples by dynamically determining storage locations for them based on the template of that particular tuple. It should be noted that we do *not* want to program the clustering but rather make it emerge from the algorithms implemented through the mechanism of self-organization. Note that according to Camazine [2], self-organization refers to the various mechanisms by which patterns, structures and order emerge spontaneously within the system. In the case of this paper, the clustering of tuples is the pattern that emerges in the system.

SwarmLinda describes four algorithms [18]: tuple distribution, tuple retrieval, tuple movement, and balancing template. In this paper we work on issues related to tuple distribution.

2.1 Tuple Distribution

The process of tuple distribution or tuple sorting is one of the most important tasks to be performed in the optimization of coordination systems. It relates primarily to the primitive out as this is LINDA's way of allowing a process to store information. Tuple distribution in SwarmLinda stores tuples based on their type (template), so that similar tuples stay close to each other. To achieve this abstraction we see the network of SwarmLinda nodes as the terrain in which out-ants roam. These ants have to decide at each hop in the network, whether or not the storage of the tuple should take place. The decision is made stochastically but biased by the amount of similar tuples around the ant's current location.

For the above to work, there should be a guarantee that the tuple will eventually be stored. This is achieved by having an aging mechanism associated with the out-ant. For every unsuccessful step the ant takes, the probability of storing the tuple in the next step increases is guaranteed to eventually reach 1 (one).

The similarity function is another important mechanism. Note that it may be to restrictive to have a monotonic scheme for the similarity of two tuples. Ideally we would like to have a function that says *how similar the two tuples are* and not only if they are exactly of the same template. Later in Section 4, we describe the similarity mechanism used in our experiments.

Now compare the above with a standard approach based on hashing. As a tuple needs to be stored, a hash function is used to determine the node where to place the tuple. This approach, although efficient in closed systems, is inappropriate to dynamic open cases because:

1. The modification of the hash function to include new nodes is not trivial.
2. The clustering of tuples is quite excessive. All the tuples of the same kind will (deterministically) be placed on the same node.
3. Due to Item 2, the system is less fault tolerant. The failure of a node can be disastrous for applications requiring the tuple kind(s) stored in that node.

Our approach is based on the brood sorting algorithm. We demonstrate that it is possible to achieve good entropy for the system (level of organization of tuples) without resorting to static solutions such as hashing. The approach proposed here adapts to changes in the network, including failures.

3 A Maude Library for Simulation

We developed in [4] a framework using the MAUDE term rewriting language to speed up the process of modeling and simulating stochastic systems specially the ones with a high degree of distribution and interaction, such as complex systems. We here briefly describe some features of this framework.

MAUDE is a high-performance reflective language supporting both equational and rewriting logic, for specifying a wide range of applications [5]. Other than specifying algorithmic aspects through algebraic data types, MAUDE can be used to provide *rewriting laws*—i.e. transition rules—that are typically used to implement a concurrent *rewriting semantics*, and are then able to deal with aspects related to interaction and system evolution. In the course of finding a general simulation tool for stochastic systems, we found MAUDE a particularly appealing framework for it allows the direct modeling of a system structure and dynamics, as well as the prototyping of new domain-dependent languages to have more expressiveness and compact specifications.

Inspired by the work of Priami on the stochastic π-Calculus [15], we realized in MAUDE a general simulation framework for stochastic systems which does not mandate a specification language as e.g. π-Calculus, but is rather open to any language equipped with a stochastic transition system semantics. In this tool a system is modeled as a LTS (labeled transition system) where transitions are of the kind $S \xrightarrow{r:a} S'$, meaning that the system in state S can move to state S' by action a, where r is the *(global) rate* of action a in state S. The rate of an action in a given state can be understood as the number of times action a could occur in a time-unit (if the system would rest in state S), namely, its occurrence frequency. This idea generalizes the activity mechanism of stochastic π-Calculus, where each channel is given a fixed local rate, and the global rate of an interaction is computed as the channel rate multiplied by the number of processes willing to send or receive a message. Our model is hence a generalization, in that the way the global rate is computed can be customized, and ultimately depends on the application at hand—e.g. the global rate can be fixed, or can depend on the number of system sub-processes willing to execute the action.

Given a transition system of this kind and an initial state, a simulation is simply executed by: *(i)* checking the available actions and their rate for each cycle of the simulation framework; *(ii)* picking one of the actions probabilistically (the higher the rate, the more likely the action should occur); *(iii)* accordingly changing the system state; and finally *(iv)* advancing the time counter according to an exponential distribution, so that the average frequency is the sum of the action rates.

4 A Solution for Tuples Distribution

Tuple distribution that can work well when failures are commonplace is still an open issue in the implementation of large scale distributed tuple spaces [12]. Several approaches for distributing tuple spaces have been proposed [17,6,20], but none of them has proven useful (without modifications) in the implementation of failure-tolerant distributed tuple spaces [12].

In order to find a solution to this problem, we took inspiration from self-organization, in particular from *swarm intelligence*. In the past years, many models deriving from biology and natural multi-agent systems—such as ant colonies and swarms—have been studied and applied in the field of computer science [10]. The solution to the distribution problem adopted in SwarmLinda is dynamic and based on the concept of *brood sorting* as proposed by Deneubourg *et al.* [7].

We consider a network of distributed tuple spaces, in which new tuples can be inserted using the LINDA primitive out. Tuples are of the form $N(X_1, X_2, ..., X_n)$, where N represents the tuple name and $X_1, X_2, ..., X_n$ represent the tuple arguments. A good solution to the distribution problem must guarantee the formation of clusters of tuples in the network. More precisely, tuples belonging to the same template should be stored in the same tuple space or in neighboring tuple spaces. Furthermore, tuples with a similar template should be stored in near tuple spaces. Note that this should all be achieved in a dynamically .

In order to decide how similar a tuple and a template are, we need a *similarity function* defined as `similarity(tu,te)`, where `tu` and `te` are input arguments: `tu` represents a tuple and `te` represents a tuple template. The values returned by this function are floating point values between 0 and 1:

- 0 means that tuple `tu` does not match template `te` at all; complete dissimilarity.
- 1 means that tuple `tu` matches template `te`; complete similarity.
- Other values mean that `tu` does not match perfectly `te` but, nonetheless, they are not completely dissimilar.

In the following, we provide a description of the process involved in the distribution mechanism, when an `out` operation occurs in a network of distributed tuple space. For the sake of simplicity, we assume only one tuple space for each node of a network.

Upon the execution of an `out(tu)` primitive, the network, starting from the tuple space in which the operation is requested, is visited in order to decide where to store tuple `tu`. According to *brood sorting*, the `out` primitive can be viewed as an ant, wandering in the network searching the right tuple space to drop tuple `tu`, that is, the carried food. The SwarmLinda solution to the distribution problem is composed of the following phases:

1. *Decision Phase.* Decide whether to store `tu` in the current tuple space or not.
2. *Movement Phase.* If the decision taken in the previous phase is not to store `tu` in the current tuple space, choose the next tuple space and repeat the process starting from 1.

4.1 Decision Phase

During the decision phase, the `out`-ant primitive has to decide whether to store the carried tuple `tu`-food in the current tuple space. This phase involves the following steps:

1. Using the *similarity function*, calculate the concentration F of tuples having a template similar to `tu`.
2. Calculate the probability P_D to drop `tu` in the current tuple space.

The concentration F is calculated considering all the templates for which tuples are stored in the tuple space. F is given by:

$$F = \sum_{i=1}^{m} (q_i \times similarity(tu, T_i)) \tag{1}$$

where m is the number of templates in the current tuple space, q_i is the total number of tuples matching template T_i. Note that $0 \leq F \leq Q$, where Q is the total number of tuples within the current tuple space (for all templates).

According to the original brood sorting algorithm used in SwarmLinda [7], the probability P_D to drop `tu` in the current tuple space is given by:

$$P_D = \left(\frac{F}{F + K} \right)^2 \tag{2}$$

Differing from the original idea in brood sorting, here the value of K is not a constant; K represents the number of steps remaining for the out(tu) primitive, namely, the number of tuple spaces that an out-ant can visit. When an out operation is initially requested on a tuple space, the value of K is set to *Step*, that is, the maximum number of tuple spaces each out-ant can visit. In the experiments in this paper, we assume *Step* as a parameter specific for the network topology we are using.

Each time a new tuple space is visited by an out-ant without storing the carried tuple tu, K is decreased by 1. When K reaches 0, P_D becomes 1 and tu is automatically stored in the current tuple space independently of the value of F. K is adopted to implement an *aging mechanism* to avoid having an out-ant wandering forever without being able to store the carried tuple tu. If $K > 0$ and the tuple tu is not stored in the current tuple space, a new tuple space is chosen among the neighbors of the current one. The following section provides a description of the process involved in the choice of the next tuple space.

4.2 Movement Phase

The movement phase occurs when the tuple carried by an out is not stored in the current tuple space. This phase has the goal of choosing, from the neighborhood of the current tuple space, the best neighbor for the next hop of the out-ant. The best neighbor is the tuple space with the highest concentration F for the carried tuple tu. According to self-organization principles, the choice of the next tuple space is local and based only on the *neighborhood* of the current tuple space.

If we denote by n the total number of neighbors in the *neighborhood* of the current tuple space, and F_j the concentration of tuples similar to tu in neighbor j (obtained using Equation 1), we can then say that the total number of tuples similar to tu in the neighborhood is given by:

$$\text{Tot}_{tu} = \sum_{j=1}^{n} F_j \tag{3}$$

The probability P_j of having an out-ant move to neighbor j, is calculated by:

$$P_j = \frac{F_j}{\text{Tot}_{tu}} \tag{4}$$

Adopting this equation for each neighbor, we obtain:

$$\sum_{j=1}^{n} P_j = 1 \tag{5}$$

The higher the value of P_j for a neighbor j, the higher the probability for that neighbor to be chosen as the next hop of the out-ant. After a new tuple space is chosen, the whole process is repeated starting from the decision phase (as described in Section 4.1).

4.3 A Case Study

In order to provide an example of the SwarmLinda distribution mechanism, we adopt the case study presented in Figure 1. The figure shows a network of four tuple spaces TS_1, TS_2, TS_3, TS_4: in TS_1 the insertion of a new tuple $a(1)$ is being executed. As reported in the figure, the tuple spaces contain tuples of four different templates: $a(X)$, $b(X)$, $c(X)$ and $d(X)$.

For didactical purposes, we assume $Step = 1$, and we use a very simple *similarity function*, given by:

$$similarity(tu, T) = \begin{cases} 1 & \text{if tuple } tu \text{ matches template } T, \\ 0 & \text{otherwise.} \end{cases}$$

Using the SwarmLinda distribution mechanism, the first step is the execution of the decision phase on tuple space TS_1. Tuple space TS_1 is characterized by the following concentration values:

$$q_{a(X)} = 6, \ q_{b(X)} = 3, \ q_{c(X)} = 2, \ q_{d(X)} = 0$$

where $q_{a(X)}$ is the concentration of tuples matching template $a(X)$ and so on. Given these values, the concentration F_1 (of tuples similar to $a(1)$) is calculated using Equation 1:

$$\begin{aligned} F_1 = \ & similarity(a(1), a(X)) \times q_{a(X)} \\ & + similarity(a(1), b(X)) \times q_{b(X)} \\ & + similarity(a(1), c(X)) \times q_{c(X)} \\ & + similarity(a(1), d(X)) \times q_{d(X)} \end{aligned}$$

and, replacing the symbols with the values we get:

$$F_1 = 1 \times 6 + 0 \times 3 + 0 \times 2 + 0 \times 0 = 6$$

Then, according to Equation 2, the probability P_D to drop $a(1)$ in TS_1 is:

$$P_D = \left(\frac{F_1}{F_1 + K} \right)^2 = \left(\frac{6}{6 + 1} \right)^2 \approx 0.734$$

Supposing that $a(1)$ is not stored in TS_1, we have to choose the next tuple space among the neighbors of TS_1. The neighborhood of TS_1 is composed of two tuple

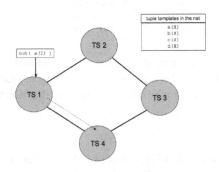

Fig. 1. Case study: a network composed of four tuple spaces

spaces: TS_2 and TS_4. If we assume $F_2 = 2$ and $F_4 = 8$, the probabilities to move in TS_2 or TS_4 are:

$$P_2 = \frac{F_2}{F_2 + F_4} = \frac{2}{10} = 0.2$$

$$P_4 = \frac{F_4}{F_2 + F_4} = \frac{8}{10} = 0.8$$

Before moving in the next tuple space, the value of K is decreased by 1: in this case the new value of K is 0. Since $K = 0$, $a(1)$ will be stored in the next tuple space, independently of the chosen tuple space.

5   Experimental Results

The results described in this section are based on an executable specification of Swarm-Linda developed in the stochastic simulation framework presented in Section 3. As previously explained, our stochastic simulation framework enables a rapid prototyping and modeling of many complex systems, like SwarmLinda and the Collective Sort problem presented in [4,3]. The next section briefly describes the executable specification of SwarmLinda. Then, before showing the results, we report a description of the methodology adopted for running the simulations. In particular, these simulations were performed on two different instances characterized by different network topologies.

5.1   Methodology

We want our distribution mechanism to achieve a reasonable distribution of tuples. Tuples having the same template should be clustered together in a group of nearby tuple spaces. The concept of *spatial entropy* is an appropriate metric to describe the degree of order in a network. Denoting by q_{ij} the amount of tuples matching template i within tuple space j, n_j the total number of tuples within tuple space j, and k the number of templates, the entropy associated with tuple template i within tuple space j is

$$H_{ij} = \frac{q_{ij}}{n_j} \log_2 \frac{n_j}{q_{ij}} \tag{6}$$

and it is easy to notice that $0 \leq H_{ij} \leq \frac{1}{k} \log_2 k$. We want to express now the entropy associated with a single tuple space

$$H_j = \frac{\sum_{i=1}^{k} H_{ij}}{\log_2 k} \tag{7}$$

where the division by $\log_2 k$ is introduced in order to obtain $0 \leq H_j \leq 1$. If we have t tuple spaces, then the *spatial entropy* of a network is

$$H = \frac{1}{t} \sum_{j=1}^{t} H_j \tag{8}$$

where the division by t is used to normalize H, so that $0 \leq H \leq 1$. The lower the value of H, the higher the degree of order in the network considered. For each simulated network, we performed series of 20 simulation, using each time different values for the *Step* parameter (representing the maximum number of steps a tuple can take). One run of the simulator consists of the insertion of tuples in the network of tuple spaces—via out primitive—until there are no pending out to be executed in the entire network.

After the execution of a series of 20 simulation for a given network, the value of the *spatial entropy* H of the network is calculated as the average of the single values of H resulting from each simulation; we call this value *average spatial entropy* (H_{avg}). For each network topology presented in the next section, we considered only tuples of four different templates: $a(X)$, $b(X)$, $c(X)$, and $d(X)$. Therefore, the possible values returned by the *similarity functions* are: *(i)* 0.6, if a tuple matches the considered template, *(ii)* 0, otherwise.

The following section presents the results obtained by simulating two different instances of network.

- *Collective sort instance*: a network of 4 tuple spaces completely connected, that is equal to the *collective sort case* discussed in [4] and [3].
- *Star instance*: a network of 6 tuple spaces characterized by a star-like topology. Note that this is used because it contrasts with the collective sort instance. In the collective sort we have nodes forming a clique. The star contrasts the results with networks that are not fully connected.

Despite the simplicity of the networks, they let us see the behavior of the proposed scheme in quite distinct scenarios. Note that the parameters (such as the *Step* parameter) are set to reflect the size of the network.

5.2   Star Instance

The network topology for the *star instance* is reported in Figure 2. As depicted in the figure, we verified the performance of our SwarmLinda distribution mechanism simulating the insertion of 15 tuples of each template. The insertion of the tuples takes place in each of the tuple spaces in the network. Consequently, at the end of a simulation, we have 90 tuple for each template stored the network.

A first set of simulations were run considering the network initially empty. We considered different values of the *Step* parameter in the range $[0 - 40]$, performing for each value a series of 20 simulation. In particular, the situation with $Step = 0$ corresponds to the case in which our distribution mechanism is turned off. Indeed, since no steps are allowed, the tuple carried by each out is stored in the tuple space in which the out operation is initially requested.

The results related to this first set of simulations are reported in Figure 3 (a). The figure shows the trend of H_{avg} considering different values for the *Step* parameter. As expected, using $Step = 0$, we obtained a value of H_{avg} equals to 1, since our distribution mechanism is not used. In other words, we are in a deterministic situation, since each tuple is stored in the tuple space in which the corresponding out occurs. Increasing the value of $Step$ causes H_{avg} to decrease. In particular, with $Step = 16$,

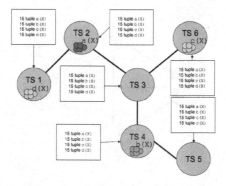

Fig. 2. Star instance. This network was used in two different experiment: with seeding and without seeding. The picture depicts the seeds in 4 of the nodes but the network is assumed empty when we perform the no-seeding experiments.

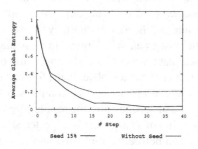

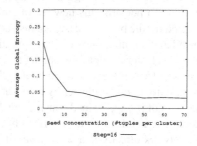

(a) Trend of *average spatial entropy* H_{avg} with and without clusters (seeds) formed.

(b) Trend of *average spatial entropy* H_{avg} considering clusters of different size.

Fig. 3. Charts showing the results for the Star instance

H_{avg} reaches the value 0.2, that is the lowest value for this set of simulations. Further increasing the value of $Step$ has minimum effect to H_{avg}.

The SwarmLinda distribution mechanism works dynamically—each out has to decide the location to store the carried tuple tu in a tuple space with high concentration of tuples similar to tu without any previous knowledge about the status of the network of tuple spaces. For this reason, even though $H_{avg} = 0.2$ does not correspond to complete clustering, nonetheless it represents a good result. Moreover, it is important to find a good balance between the value of the $Step$ parameter and the value of H_{avg} we want to achieve. Indeed, a too high value of $Step$ can eventually leads to a high network traffic. Hence, when we choose a $Step$ value, it is important considering not only the performance in terms of H_{avg}, but also the cost in terms of network traffic eventually generated. Another point to be made is that complete clustering may actually not be desirable given that it makes the system less tolerant to failures. Excessive dependency to a given node lead to the infamous single-point-of-failure problem.

The other experiment we performed measures the behavior of the tuple distribution mechanism when faced with a network where clustering is already present. In Figure 2

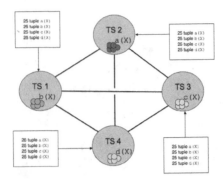

Fig. 4. Collective sort instance. Again here the network shows nodes with seeds but these are only considered in the experiments that consider the existence of seeds.

this consists of the configuration where the clusters depicted are considered in the execution. To perform this new set of simulations, we used the same number of tuples adopted for the previous case, but instead of considering the network initially empty, we considered an initial network configuration with clusters already formed. In particular, for each cluster of tuples belonging to a given template, we considered a concentration equals to 15% of the total number of tuples expected for that template. In the following, we refer to an initial cluster by the term *seed*. Figure 3 (a) also shows a comparison of the trend of H_{avg} obtained considering the presence of seeds against the results without seeds. The results are characterized by a better value of H_{avg}. This is expected since the network already had some level of organization before the tuple distribution was executed.

Finally, we performed an analysis of the sensitivity of the tuple distribution in relation to the seed concentration when $Step = 16$. A series of 20 simulations for different initial seed concentrations in the range $[0\% - 80\%]$ were performed. The results related to this case are shown in Figure 3 (b). As expected, the higher the value of seed concentration, the lower the value of H_{avg}.

5.3 Collective Sort Instance

The network topology for the *collective sort instance* is shown in Figure 4. This instance corresponds to the same case already studied in [4] and [3], even though the approach adopted here to address the distribution problem is different.

It is worth comparing the two solutions that, although characterized by the same goal, are based on different approaches. Indeed, in the case discussed here the clustering process in realized *dynamically* as soon as an out operation is executed on a tuple space. Conversely, the approach described in [4] and [3] is executed in background, after the execution of an out operation is completed. This means that the tuple carried by an out is initially stored in the tuple space in which the operation occurs. At the same time, a set of software agents work in background with the task of moving tuples from a tuple space to another: the global goal is to achieve the complete clustering of the different tuple templates in the network.

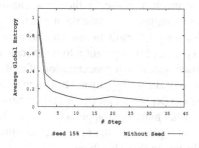 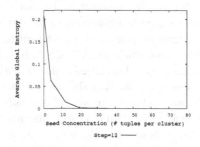

(a) Trend of *average spatial entropy* H_{avg} with and without clusters (seeds) already formed.

(b) Trend of *average spatial entropy* H_{avg} considering clusters of different size.

Fig. 5. Charts showing the results for the Collective Sort instance

We performed on the collective sort instance the same simulations executed previously on the star instance, using 100 tuples for each tuple template (Figure 4). Figure 5 (a) reports the trend of H_{avg} considering the network initially empty (without seed). Figure 5 (a), also shows the trend of H_{avg} with seeds compared with the one resulting from the simulation without seeds already formed in the network. Finally, Figure 5 (b) shows the trend of H_{avg} considering different seed concentrations, using the same criterion described in Section 5.2. The results show the same trend of H_{avg}, already described in Section 5.2 for the star instance.

Now, we can compare the approach used here with the one presented in [4] and [3]. First of all, considering the two solutions from the viewpoint of the performance, we can notice that, while the approach in [4] and [3] can achieve a perfect clustering of the tuples ($H_{avg} = 0$), the SwarmLinda solution can reach 0.2 as minimum value of H_{avg}. However, the SwarmLinda solution tries to store a tuple in the right tuple space *upon* the execution of an out operation, whereas in the solution developed in [4] and [3] realizes the clustering using software agents that work in background. Consequently, this solution seems to have a higher computational price than the SwarmLinda solution presented in this paper.

The previous discussion has only the goal of highlighting the differences between the two solutions. For this reason, the two approaches do not have to be viewed as conflicting solutions, but as a different alternative, or even as complementary approaches. Therefore, the choice of the solution to adopt depends on the particular application domain.

6 Conclusion and Future Work

In this work, we focused on the tuple distribution problem in distributed tuple spaces, designing a solution based on *self organization*. Then, the solution was tested using an executable specification, developed on the stochastic simulation framework discussed in [4]. We demonstrated the emergence of organization in two different network topologies: *(i)* a star-like network and *(ii)* a completely connected network (collective sort instance).

The results demonstrate that our distribution mechanism—even though it cannot achieve a perfect clustering of the different tuple templates—is capable of reaching a low value of *spatial entropy*. This can be considered a good result, because the proposed mechanism works dynamically, that is, for every out(tu) operation to be executed, it tries to store the carried tuple tu in a tuple space with a high concentration of tuples similar to tu. We have also pointed out that complete clustering may not be desirable in dynamic networks given that processes become too dependent on certain nodes being available all the time. In fact, this avoidance is what makes our approach more fault-tolerant. Note that if the node with the majority of tuples fails, the few tuples in other nodes will act as "seeds" for the formation of clusters in those locations.

With regard to the collective sort instance, we compared this results with the ones resulting from the application of a different solution described in [4] and [3]. In particular, the solution proposed here seems to be more general and more suitable for a wider range of network topologies. All the same, there are still issues to be dealt with:

- We want to perform further simulations using networks with an higher number of tuple spaces, in order to see if the trend of the *spatial entropy* is maintained.
- We plan to test the proposed solution on other network topologies, in particular on scale-free topologies. *Mutatis mutandis*, our solution applies to more general networks including scale-free.
- In order to see how tuples aggregate around clusters, we are going to run simulations on network instances with more tuple templates: in other words, we plan to use not only templates totally different from each other, but also templates featuring diverse degrees of *similarity*.
- Finally, an important issue is the *over-clustering*: since we want to avoid tuple spaces containing clusters of tuples that are too large. We need to devise a *dynamic similarity function* that takes into account the current number of tuples stored in a tuple space.

The last point is particularly interesting to allow the application of SwarmLinda in the domain of *open* networks. We believe self-organization will also play a role in the control of over-clustering. We are starting to experiment with solutions inspired by bacteria molding as they appear to have the characteristics we desired. We aim at achieving a balance in the system where clustering occurs but it is not excessive.

References

1. Bonabeau, E., Dorigo, M., Theraulaz, G.: Swarm Intelligence: From Natural to Artificial Systems. In: Santa Fe Institute Studies in the Sciences of Complexity, Oxford University Press, Inc, Oxford (1999)
2. Camazine, S., Deneubourg, J.-L., Franks, N., Sneyd, J., Theraula, G., Bonabeau, E.: Self-Organization in Biological Systems. Princeton Univ Press, Princeton (2003)
3. Casadei, M., Gardelli, L., Viroli, M.: A case of self-organising environment for MAS: the collective sort problem. In: 4th European Workshop on Multi-Agent Systems (EUMAS'06), EUMAS, Lisbon, Portugal (15 December 2006) Warsaw University. Proceedings (2006)

4. Matteo, C., Luca, G., Mirko, V.: Simulating emergent properties of coordination in Maude: the collective sorting case (Proceedings). In: Carlos, C., Mirko, V. (eds.) 5th International Workshop on Foundations of Coordination Languages and Software Architectures (FOCLASA'06). CONCUR 2006, Bonn, Germany, pp. 57–75. University of Málaga, Spain (2006)
5. Clavel, M., Durán, F., Eker, S., Lincoln, P., Martí-Oliet, N., Meseguer, J., Talcott, C.: Maude Manual. Department of Computer Science University of Illinois at Urbana-Champaign, 2.2 edition, Version 2.2 is available online (December 2005), at http://maude.cs.uiuc.edu
6. Corradi, A., Leonardi, L., Zambonelli, F.: Strategies and protocols for highly parallel Linda servers. Software Practice and Experience 28(14), 1493–1517 (1998)
7. Deneubourg, J.-L., Goss, S., Franks, N., Sendova-Franks, A., Detrain, C., Chretien, L.: The dynamic of collective sorting robot-like ants and ant-like robots. In: Proceedings of the First International Conference on Simulation of Adaptive Behavior: From Animals to Animats 3, pp. 356–365. MIT Press, Cambridge, MA (1991)
8. Gelernter, D., Carriero, N.: Coordination languages and their significance. Communications of the ACM 35(2), 96–107 (1992)
9. Kennedy, J., Eberhart, R.C.: Swarm Intelligence. Morgan Kaufmann, San Francisco (2001)
10. Mamei, M., Menezes, R., Tolksdorf, R., Zambonelli, F.: Case studies for self-organization in computer science. Journal of Systems Architecture 52(8-9), 443–460 (2006)
11. Mamei, M., Zambonelli, F., Leonardi, L.: Tuples on the air: A middleware for context-aware computing in dynamic networks. In: Proceedings of the 23rd International Conference on Distributed Computing Systems, p. 342. IEEE Computer Society Press, Los Alamitos (2003)
12. Menezes, R., Tolksdorf, R.: A new approach to scalable linda-systems based on swarms. In: Proceedings of the ACM Symposium on Applied Computing, ACM Press, New York (2003)
13. Ossowski, S., Menezes, R.: On coordination and its significance to distributed and multi-agent systems. Concurrency and Computation: Practice and Experience 18(4), 359–370 (2006)
14. Parunak, H.: Go to the ant: Engineering principles from natural multi-agent systems. Annals of Operations Research 75, 69–101 (1997)
15. Priami, C.: Stochastic pi-calculus. The Computer Journal 38(7), 578–589 (1995)
16. Spafford, E., DeMillo, R.: Grand research challenges in information systems. Technical report, Computing Research Association (2003), http://www.cra.org/reports/gc.systems.pdf
17. Tolksdorf, R.: Laura — A service-based coordination language. Science of Computer Programming 31(2–3), 359–381 (1998)
18. Tolksdorf, R., Menezes, R.: Using swarm intelligence in linda systems. In: Omicini, A., Petta, P., Pitt, J. (eds.) ESAW 2003. LNCS (LNAI), vol. 3071, Springer, Heidelberg (2004)
19. UK Computing Research Committee. Grand challenges in computer research. Technical report, British Computing Society (2004), http://www.bcs.org/upload/pdf/gcresearch.pdf
20. Wyckoff, P., McLaughry, S.W., Lehman, T.J., Ford, D.A.: T Spaces. IBM Systems Journal (Special Issue on Java Technology), 37(3) (1998)

Agent Behavior Alignment: A Mechanism to Overcome Problems in Agent Interactions During Runtime

Gerben G. Meyer and Nick B. Szirbik

Department of Business & ICT, Faculty of Economics, Management and
Organization, University of Groningen, Landleven 5, P.O. Box 800,
9700 AV Groningen, The Netherlands
Tel.: +31 50 363 {7194 / 8125}
{g.g.meyer,n.b.szirbik}@rug.nl

Abstract. When two or more agents interacting, their behaviors are
not necessarily matching. Automated ways to overcome conflicts in the
behavior of agents can make the execution of interactions more reliable.
Such an alignment mechanism will reduce the necessary human intervention. This paper shows how to describe a policy for alignment, which
an agent can apply when its behavior is in conflict with other agents.
An extension of Petri Nets is used to capture the intended interaction of
an agent in a formal way. Furthermore, a mechanism based on machine
learning is implemented, to enable an agent to choose an appropriate
alignment policy with collected problem information. Human intervention can reinforce certain successful policies in a given context, and can
also contribute by adding completely new policies. Experiments have
been conducted to test the applicability of the alignment mechanism
and the main results are presented here.

1 Introduction

This paper presents a method for agents to mutually adjust their behaviors in an
automatic way. Behavioral alignment is implemented as a mechanism that allows
the agents to successfully complete an interaction. The method was implemented
and experiments have been conducted. These experiments have shown that automated procedures to overcome problems in agent-to-agent interactions can make
the whole multi-agent system execution more reliable, and have resulted in less
human intervention.

When two or more agents (human or software) interact as part of a business
process instance (BPI) [17], they can either follow an established protocol, or
use their own experience and judgment. In the first case, the agents should learn
or be instructed about the protocol beforehand. In the second case, there either is no protocol, or it does not cover special circumstances that occur in this
particular instance. Furthermore, the agents have to use their own experience,
acquired in previous similar interaction. Before the interaction, the agents will

M. Klusch et al. (Eds.): CIA 2007, LNAI 4676, pp. 270–284, 2007.
© Springer-Verlag Berlin Heidelberg 2007

build an *intended behavior* (their own course of action) and they will also assume what the other agents are doing (*expected behavior*). Together these beliefs form a description of an *intended interaction*, which is also called an *interaction belief* [18].

When two agents are going to perform an interaction, and they have inconsistent behaviors, the resulting interaction will not achieve its goal, especially when the agents do not have mechanisms to change their behaviors. If both agents are able to realize, after the interaction has started, that there is a conflict in their beliefs about the interaction, they need to be able to change their behavior in a collaborative manner that promises success in ending the whole interaction. In successive steps of trial and error, they will be able to manage to finish the interaction. If exactly the same agents are interacting again in the future, they can use the final behavior (which they memorize) that led to success.

The process of collaboratively changing behaviors will be called *alignment*. Two or more behaviors that have been aligned successfully are named *matching behaviors*. If the agents do not manage to align their behavior on the fly, their complete behaviors have to be compared externally. Because agents do not know each other's behaviors, a superior agent who can access both behaviors is the most suitable to align these behaviors. This agent can be a software agent, but can also be human. This procedure, if one of the agents is calling for a higher level agent to align the behaviors, is called *escape* mode. A higher level agent can also *intervene* in the interaction process a priori or during the interaction, to align the agents' intended behaviors. In both cases, the higher level agent adapts the behavior of the agents [14]. A third possibility, which can often be found in real life, is to align their behavior a priori of the interaction, by exchanging each intended behavior and analyze and discuss beforehand. Analysis can reveal potential deadlock and conflicts. The interacting agents can align their behaviors before the interaction starts.

This paper only focuses on how agents can individually align their behavior *on-the-fly*. A discussion about the other types of alignment can be found in [11]. The paper is structured as follows. In the next section related work is discussed. Section 3 explains how the behavior of agents are modeled and executed. Section 4 discusses how primitive and advanced operations allow for behavior modifications. In section 5 a new mechanism for automatic behavior alignment is presented. Section 6 unveils some of the conducted experiments, with their result. Section 7 ends the paper with discussion and conclusions.

2   Related Work

There are two long-established agent-research areas that contributed to the architecture of the communication and coordination mechanisms in MAS. These are the agent communication language (ACL) research, and the interaction protocol (IP) research. However, there are scarce results in terms of how IPs a and ACLs specifications could be coherently linked together. Most approaches concentrate on one of these two sides, either limiting via IPs the agents' autonomy (agents obey central protocols), either leaving open the consequences of communication, leading to loss of coherence in the overall interactional process executed by the agents. One of the promising approaches that tries to bridge the communication specifications to the way agents adhere to previously created committed commitments is proposed in [15], by using *Conversation Patterns*. These are symbolic constructs that regulate agent-to-agent interaction, but they are modified via machine learning. The question remains if these frames-based models are appropriate for complex business processes, where the execution semantics derived from the distributed structure of the frames that are derived from the conversation patterns can become combinatorial complex.

Dignum [5] emphasizes the importance of the link between Petri net based analysis techniques in the context of protocolized agent communication, and their link to the description of dynamic properties of agent conversations. A non-agent approach that tackles similar problems is the modeling and execution of inter-organizational workflows [2]. However, this approach is based on a central perspective, similar to the IP approach. The main advantage of the approach is that it is possible to determine when a workflow is sound [3]. Another related problem is how to dynamically change workflows, and it has been addressed by Ellis et al. [6]. Ellis and Keddara [7] describe how a workflow can be dynamically changed using another workflow, which defines the change. Recently, distributed workflows expressed as agent behaviors have been investigated by Meyer and Szirbik [11] who have developed a formal way to model agent behavior as executed in agent to agent interaction as Behavior Nets. Because these Behavior Nets are an extension of Petri Nets, workflow techniques as proposed by Van der Aalst and Ellis can also be applied to these Behavior Nets. Behavior Nets are part of the agent modeling language TALL [17], which is currently under development. As for the overlapping research themes in workflows and agents, a thorough review of is presented in [19].

3   Modeling and Executing Agent Behavior

As mentioned before, agents' behaviors as exhibited in agent to agent interactions are modeled by Behavior Nets, as formally defined by Meyer and Szirbik in [11], which is an extension of Petri Nets. Petri Nets are a class of modeling tools, which originate from the work of Petri [13]. The advantage of Petri Nets is that they have a well defined mathematical foundation, but also a clear graphical notation [16]. Because of the graphical notation, Petri Nets are powerful design tools, which can be used for communication between the people who are engaged in the design process. On the other hand, because of the mathematical foundation, mathematical models of the behavior of the system can be set up. The mathematical formalism also allows for various analysis techniques of the Petri Net. Moreover, as Petri Nets are commonly used for modeling business processes (see for example [1]), applying Petri Nets for modeling agent behavior makes the approach suitable for simulating and supporting business processes with agents.

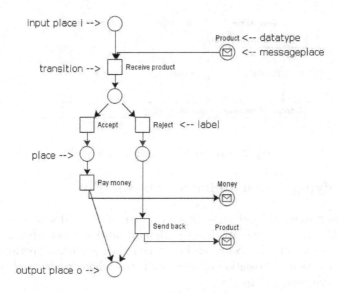

Fig. 1. Example of a Behavior Net

An example of a Behavior Net can be seen in figure 1. The definition of Behavior Nets is partially based on (and an extension of) Workflow nets [3], Self-Adaptive Recovery Nets [8] and Colored Petri Nets [10]. The Behavior Net as shown in figure 1 illustrates most of the Behavior Net constructs. The figure depicts the behavior of a buyer, in a buyer-seller interaction, who will receive a product, and either accepts the product and pays the money, or reject the product and sends it back.

The modeled behavior of the agent can be executed directly, however, to enable on-the-fly problem detection, change, and alignment of behaviors, an extended version of the behavior is used. For this, a two-layered approach is used, based on [9]. These two layers, and the relations between them, are shown in figure 2. The top layer is the original agent behavior, where the bottom layer is the extended behavior, for enabling the alignment mechanism. On both layers, the behavioral descriptions are based on the Petri Net extension as described above. In this way, when the behavior is executed in the compiled layer, it is still possible to visualize the progress of this execution in the original modeled behavior.

During run time, the state of the compiled layer will change, as tokens will flow though the behavior. The state of the other layer will be updated according to the state of the compiled layer, to make it possible to visualize the progress of the execution in the original modeled behaviors. During align time (when the agent is changing its own behavior on-the-fly due to a problem with executing it in this particular interaction) the agent changes the behavior on the behavior layer. The compiled layer has to be recreated by recompiling the behavior layer. The compilation step will be further discussed in section 5.1.

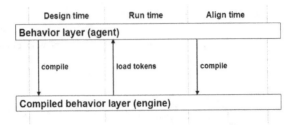

Fig. 2. Two layers to describe behavior

4   Modifying Agent Behavior On-the-Fly

Since we want to modify behaviors in a consistent and sound way, the manner in which modifying operations are defined is important. For this purpose, two types of operations on Behavior Nets can be defined. The *primitive* operations allow for simple addition and deletion of nodes and arcs, whereas *advanced* operations allow for more complex changes.

4.1   Primitive Operations

Based on [8], some primitive operations are identified for modifying Behavior Net structures. In total, 16 primitive operations have been implemented in the system, which are shown in table 1.

Table 1. The primitive operations

createPlace(x)	deletePlace(x)
createMessagePlace(x)	deleteMessagePlace(x)
createTransition(x)	deleteTransition(x)
createArc(x_1,x_2)	deleteArc(x_1,x_2)
setLabel(x,x')	setDatatype(x,d)
createGuard(x,g)	deleteGuard(x,g)
createBinding(x,b)	deleteBinding(x,b)
createToken(x,t)	deleteToken(x,t)

These primitive operations can be used and are sufficient to build Behavior Nets from scratch, and populate them with tokens. By using the `deleteToken` operation, it is even possible to simulate the flow.

Guards are an addition to the basic Petri Nets. In Behavior Nets, they can be added to and deleted from arcs, and their role is to provide certain conditions that need to be fulfilled before a token can travel along that arc. A guard can put constraints on values which are kept within the token. Guards are used to detect exceptions, that potentially lead to behavior net modifications. Furthermore, guards enable a straightforward implementation of the typical deterministic OR-split, representing a choice.

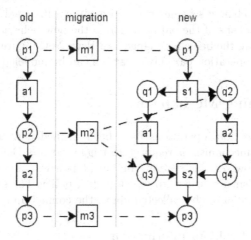

Fig. 3. Migration of old to new behavior

A second addition to the classical Petri Nets are bindings, which can be added to transitions. The role of bindings is to modify the values kept by tokens. These values can later on be used by the guards.

Applying a primitive operation does not guarantee preservation of soundness of the behavior. The discussion of soundness is outside the scope of the paper, but interested readers can find the definition of soundness in [3].

4.2   Advanced Sound Operations

In addition to the primitive operations, a set of advanced operations is defined, based on the principles presented in [3,4]. These advanced operation can also be used to modify the structure of a Behavior Net, but contrary to primitive operations, they have the advantage of preserving the soundness of the initial net. All these operations can be constructed by combining a set of primitive operations. The defined and implemented advanced operations are:

- **division** and **aggregation**, which divides one transition into two sequential transitions, or vice versa,
- **parallelization** and **sequentialization**, which puts two sequential transitions in parallel, or vice versa,
- **specialization** and **generalization**, which divides a transition into two mutually exclusive specializations (an OR-split), or vice versa,
- **iteration** and **noIteration**, which replaces a transition with an iteration over a transition, or vice versa,
- **receiveMessage** and **notReceiveMessage**, which adds or deletes an incoming message place,
- **sendMessage** and **notSendMessage**, which adds or deletes an outgoing message place.

For modeling the change schemes the approach of Ellis et al. [7] is used. The migration of the tokens of the old behavior to the new behavior can be exactly defined by modeling the behavior change as a Petri Net. Figure 3 shows how the migration for the operation `parallelization` can be modeled.

5   The Aligning Mechanism

In the previous section, operations have been defined how to alter Behavior Nets. However, a mechanism is required to trigger when a Behavior Net should be altered, and how. In this section, alignment policies will be introduced, as well as a mechanism for agents to select such a policy. For selecting a policy, certain information has to be collected about the conflict and its context.

5.1   Collecting Problem Information

As discussed previously, the modeled behavior itself is not executed, but the compiled version of it. By compiling, extra nodes are added to the original behavior, for collecting information when there is a problem in the execution of the behavior. This information can be used for selecting a proper alignment policy (this will be discussed later in this section), which is the best for dealing with the problem. Figure 4 shows what information can be collected by the extra nodes added when the behavior is compiled.

The highlighted nodes are the nodes of the original behavior. The **align** transition is the transition where all collected information is gathered, and (if

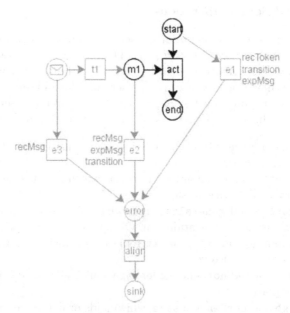

Fig. 4. Collecting problem information

necessary) an alignment policy is chosen and executed. In this example, the original behavior has only one transition act that expects one incoming message m1. This transition act failed to execute, therefore all the gathered information is about transition act, to make the align transition able to choose an alignment policy which will successfully align the behavior around the transition act. Despite the simplicity of this example, it shows all the ways of how information used for choosing an alignment policy is collected, as for more complex Behavior Nets, the same kind of nodes are added. Problem information is gathered in three different types of occurrences, as discussed next. In all three cases, a time-out mechanism is used to identify the occurrence of non-moving tokens. The information is collected separately for every transition, where problems arise.

I. When a token from a place is not moving (Place start in this example.) Here, transition e1 will send information about this occurrence to the align transition. It will send the following information:

- recToken, short for received token, to inform the align transition that there is a non-moving token on that place in the behavior
- transition, the name of the transition for which the recToken is required to enable it (act in the example)
- expMsg, the data type of the message expected by the transition (in this example the data type the place m1 may contain)

II. When a known received message is not moving (Place m1 in this example.) In this situation, transition e2 will send information about this occurrence to the align transition. It will send the following information:

- recMsg, the data type of the received message (in this example the data type which place m1 may contain, which of course equals the data type of the message, because otherwise the message would not be in this place)
- expMsg, the data type of the expected message (also the data type place m1 may contain)
- transition, the name of the transition which is connected to the message place (act in the example)

III. When an unknown message is received (The message place in this example.) When this happens, transition e3 will send information about this occurrence to the align transition. It will send the following information:

- recMsg, the data type of the received message

All the collected data is combined and used for the selection of an appropriate policy. Furthermore, this data will be used as parameters for the operations within the policy, as will be discussed next.

5.2   Alignment Policies

An alignment policy is an ordered set of primitive or advanced operations. In this approach, an agent has a set of policies in its knowledge-base from which it can choose when problems arise with the execution of the behavior within an interaction. Table 2 shows two example policies.

Table 2. Examples of alignment policies

Policy #	Operations
1	parallelization($transition, seqTransition$)
2	notReceiveMessage($transition, expMsg$)
	specialization($transition$,Receive $expMsg$, Receive $recMsg$)
	receiveMessage(Receive $expMsg, expMsg$)
	receiveMessage(Receive $recMsg, recMsg$)

Policy 1 puts two transitions in parallel. Policy 2 makes a specialization of a transition, which expects a certain message, into two transitions, one that expects the original expected message, and the other that expects the actual received message. Instead of giving fixed names as parameters for the operations, variables can be used. Also a combination of fixed names and variables is possible. When the policy is actually executed, these variables will automatically be replaced by the appropriate values for that situation. The advantage is that in this way a more general policy can be defined, that is not only restricted to a given name of the transition or message types. Table 3 shows the list of variables what can be used when defining the parameters, and with what they will be replaced.

In the current state of research, policies are manually identified by the agent-system designer. However, the agents can learn by repetitive experience when to apply a certain policy, and select the appropriate policy automatically. One of

Table 3. Variables for defining parameters

Variable	Will be replaced with
transition	the name of the transition
seqTransition	the name of the transition which is sequential placed after *transition*
parTransition	the name of the transition which is parallel with *transitions*
specTransition	the name of the transition which is mutual exclusive to *transition*, and originates from the same specialization
expMsg	the data type of the expected message
recMsg	the data type of the received message

the selection methods that was implemented is presented in the next section. An early observation in the discovery of potentially usable policies has shown that an agent does not need a large number of policies to solve most of the occurred problems in behavior executions. An intuition is that there is rather small and manageable number of alignment patterns.

5.3   Implementation

This subsection describes the implemention of the policy selection and alignment mechanism used for experiments. In this system, an agent selects a policy using a base of training examples which contain both the problem information and the policy which was chosen at that time. A machine learning technique is used to generalize over this data, and also to enable the agent to choose a policy based on the previous experience. How an agent will choose an alignment policy (or if it will choose one at all) depends on multiple factors. The factors used in the implementation are: problem information, beliefs about the agents interacting with, and the willingness to change its own behavior.

Problem information. Most of the time, a problem will occur when the agent is not receiving the message it is expecting. It can also be the case that the agent did not receive a message at all. If it did receive a message, but the wrong one, the type of received message and other information about the problem are used as attributes for selecting the proper alignment policy.

Beliefs about the agents interacting with. Beliefs about the other agents can be of importance when choosing an alignment policy. For example, when one agent completely trusts the other agents, it might be willing to make more changes in its behavior than when it distrusts the other agents.

Willingness to change behavior. When an agent has very advanced and fine-tuned behaviors, it is not considered good practice to radically change the behaviors because of a single exceptional interaction. On the other hand, when the behavior of the agent is still very primitive, changing it more often could be a good way to acquire new experience. In this way, when an agent gets "older", and the behaviors are based on more experience, the willingness to change its behavior will decrease. This approach can be compared with the way humans learn, or with the decrease of the learning rate over time when training a neural network.

The process of alignment is shown in figure 5 as a finite state automaton. The policy selection method (in this particular implementation the decision algorithm based on machine learning) uses the collected problem information and values from the belief base (like trust and willingness) to choose an alignment policy. However, it can be the case that the agent cannot find an appropriate alignment policy for this specific problem. In this case, the agent can trigger escape mode, and ask a human to indicate a policy. After this policy has been chosen by the person, the agent needs to memorize the number of the chosen policy as well as the context. In this way, the agent has gained new experience when to apply a

Fig. 5. The process of alignment

certain policy. More information about the concept of escape mode can be found in [14]. After the alignment policy is chosen, either by the decision algorithm, or by a human, this policy will be executed to adapt the Behavior Net.

The machine learning technique used in this implementation is based on artificial neural networks [12]. This is an obvious choice, given the simplicity and robustness of these classification systems, and also given the availability of well proved open-source implementations (in this system, the WEKA3 package is used [20]). However, other learning or classification methods can also be used. With neural networks not only the selected policy is returned, but also the activation levels of all policies. Therefore, the certainty of the choice can be known. The activation level of the selected policy is compared with the activation level of the runner-up (the second-best policy). The activation level of the best policy should be factor x (with $x > 1$) higher than the runner-up, otherwise, the agent will go into escape mode. In the implementation $x = 2$ has been used. This ensures that the agent will not execute an alignment policy when suitability for the current problem is questionable. When unsure, it will ask the human for advice. An example of the training data used (two learning patterns) to train the neural network is shown in table 4. The `policy #` row in the table represents the output of the network, which has one output neuron for each known policy, and the rest of the rows are representing the inputs. The neural network has to be rebuilt and retrained when a new policy is added to the known pool.

Table 4. Example of training data

Training pattern #	1	2	...
transition	Receive money	Receive money	...
seqTransition	Send product		...
parTransition		Send product	...
specTransition			...
recToken	yes	yes	...
recMsg		product	...
expMsg	money	money	...
trust	1	1	...
willingness	1	1	...
policy #	1	2	...

6 Experiments and Results

Experiments have been conducted to test the applicability of the alignment and policy choosing mechanism. This section provides an illustrative example of such an experiment. In this experiment, as shown in figure 6, a buyer and a seller already agreed on the product the buyer wants to buy, but as seen in the figure, they have different views (formalized as behaviors) in what order the delivery and the payment should occur. In this experiment, it is presumed that the behavior of the buyer is very advanced, and thus he is not willing to easily change his behavior. On the other hand, the seller is inexperienced and his behavior is still primitive, hence we are looking at the problem how the seller can align its behavior with the buyer, presuming that the seller trusts the buyer.

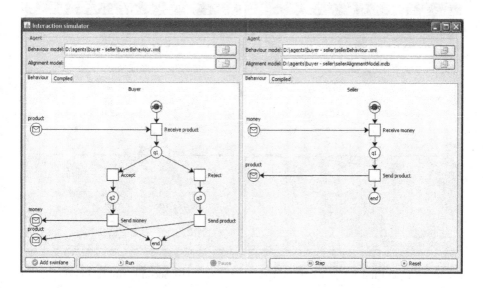

Fig. 6. Behaviors of buyer and seller

Within the experiment, when the interaction started, it immediately dead-locked; the buyer was waiting for the product, and the seller was waiting for the money. The seller collected the information in the way explained in section 5.1. Besides the name of the transition, also seqTransition, parTransition and specTransition (as explained in section 5.2) were looked up. The collected problem information is shown in table 5 (a).

This information was used by the neural network of the seller to select the appropriate alignment policy. Due to the previously trained neural network, via repeated escapes to humans in previous experiments, this led to the execution of policy 1 from table 2. The seller transformed its behavior by placing its two transitions in parallel. The result is shown in figure 7 (a).

Still, the two behaviors were not aligned. The buyer rejected the product, and sent it back. Consequently, the seller did not have the appropriate behavior to

Table 5. Collected problem information

Information	Value	Information	Value
transition	Receive money	transition	Receive money
seqTransition	Send product	seqTransition	
parTransition		parTransition	Send product
specTransition		specTransition	
recToken	yes	recToken	yes
recMsg		recMsg	Product
expMsg	Money	expMsg	Money

(a) First problem   (b) Second problem

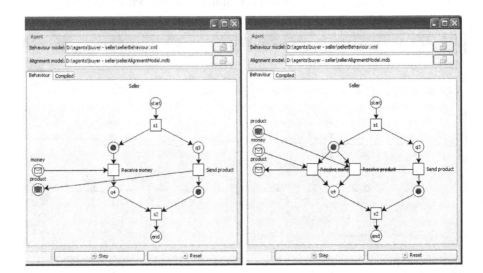

(a) First adaptation   (b) Second adaptation

Fig. 7. Adapted behaviors of seller

handle this, as the seller was expecting the payment. After the buyer sent the product back, the seller detected the problem and collected the data as shown in table 5 (b).

The neural network for selecting an alignment policy was used again, and due to previous training policy 2 of table 2 was chosen. The seller divided the transition *receive money* in two mutual exclusive transitions: *receive money* and *receive product*. This changed the behavior as shown in figure 7 (a) into the behavior as shown in figure 7 (b). The behaviors of the buyer (figure 6 (a)) and the behavior of the seller (figure 7 (b)) were now matching. Because the seller already knew two suitable alignment policies, and also when to select them, the interaction completed successfully, despite their initial conflicting behaviors.

The conducted experiments, as the one discussed above, have shown promising results concerning the applicability of the proposed alignment mechanism. The necessary human assistance by the use of the alignment mechanism has significantly decreased as the agents are better able to solve conflicting behaviors during the interaction. Human intervention reinforced certain successful policies in a given context, and also contributed with completely new policies. The net result, revealed during the experiments, is that the simulated business processes become increasingly automated.

7 Discussion and Conclusions

Currently, there are several limitations with alignment policies, which need to be studied in future research. Firstly, an alignment policy is limited to a certain scope, only the transitions defined by the problem information can be used as variables for the operations of the alignment policy when defining an alignment policy. Ways to overcome this limitation have to be investigated. Secondly, the problem data used for parameters of alignment policies can only contain one of the transitions which are in parallel with the transition where the problem is. It could however be the case that there are three transitions in parallel. Using the variable `parTransition` in defining the alignment policy would give a non-deterministic execution of the policy. This issue does not only affect the problem data, but also the used neural network, as it is not suitable for having multiple parallel transitions as input. For this reason, other machine learning techniques like decision trees have to been investigated.

In conclusion, it is possible to describe a policy for alignment that can be applied when the interaction beliefs of two or more interacting agents are not matching. An extension of Petri Nets has been used to capture the intended interaction of an agent in a formal way. Furthermore, a mechanism based on a neural network that chooses an appropriate alignment policy with the collected problem information was implemented. When this mechanism fails to choose an alignment policy, a method based on human intervention was described, which can teach the agent new ways of alignment. As the experiments have shown, this approach enables agent behavior execution to be more reliable and necessitates less human intervention in terms of alignment.

References

1. van der Aalst, W.M.P.: The application of petri nets to workflow management. The Journal of Circuits, Systems and Computers 8(1), 21–66 (1998)
2. van der Aalst, W.M.P.: Interorganizational workflows: An approach based on message sequence charts and petri nets. Systems Analysis - Modelling - Simulation 34(3), 335–367 (1999)
3. van der Aalst, W.M.P.: Workflow verification: Finding control-flow errors using petri-net-based techniques. BPMt: Models, Techniques, and Empirical Studies 1806, 161–183 (2000)

4. Chrzastowski-Wachtel, P., Benatallah, B., Hamadi, R., O'Dell, M., Susanto, A.: A top-down petri net-based approach for dynamic workflow modeling. In: van der Aalst, W.M.P., ter Hofstede, A.H.M., Weske, M. (eds.) BPM 2003. LNCS, vol. 2678, pp. 335–353. Springer, Heidelberg (2003)
5. Dignum, F.: Agent communication and cooperative information agents. In: Klusch, M., Kerschberg, L. (eds.) CIA 2000. LNCS (LNAI), vol. 1860, pp. 191–207. Springer, Heidelberg (2000)
6. Ellis, C., Keddara, K., Rozenberg, G.: Dynamic change within workflow systems. In: Proc. of COCS '95, pp. 10–21. ACM Press, New York (1995)
7. Ellis, C.A., Keddara, K.: A workflow change is a workflow. In: van der Aalst, W.M.P., Desel, J., Oberweis, A. (eds.) Business Process Management. LNCS, vol. 1806, pp. 201–217. Springer, Heidelberg (2000)
8. Hamadi, R., Benatallah, B.: Recovery nets: Towards self-adaptive workflow systems. In: Zhou, X., Su, S., Papazoglou, M.M.P., Orlowska, M.E., Jeffery, K.G. (eds.) WISE 2004. LNCS, vol. 3306, pp. 439–453. Springer, Heidelberg (2004)
9. Harel, D., Marelly, R.: Specifying and executing behavioral requirements: The play-in/play-out approach. Technical Report MSC01-15, The Weizmann Institute of Science (2001)
10. Jensen, K.: An introduction to the theoretical aspects of coloured petri nets. In: A Decade of Concurrency, Reflections and Perspectives, REX School/Symposium, pp. 230–272. Springer, London, UK (1994)
11. Meyer, G.G., Szirbik, N.B.: Anticipatory alignment mechanisms for behavioural learning in multi agent systems. LNCS (LNAI), vol. 4520 (2007)
12. Mitchell, T.M.: Machine Learning. McGraw-Hill Higher Education, New York (1997)
13. Petri, C.A.: Kommunikation mit Automaten. PhD thesis, Institut fur instrumentelle Mathematik Bonn (1962)
14. Roest, G.B., Szirbik, N.B.: Intervention and escape mode. In: Padgham, L., Zambonelli, F. (eds.) AOSE VII / AOSE 2006. LNCS, vol. 4405, pp. 109–120. Springer, Heidelberg (2007)
15. Rovatsos, M., Fischer, F., Weiss, G.: An integrated framework for adaptive reasoning about conversation patterns. In: Proc. of AAMAS '05, pp. 1123–1124 (2005)
16. Salimifard, K., Wright, M.: Petri net-based modelling of workflow systems: An overview. European Journal of Operational Research 134(3), 664–676 (2001)
17. Stuit, M., Szirbik, N.: Modelling and executing complex and dynamic business processes by reification of agent interactions. LNCS (LNAI), vol. 4457. Springer, Heidelberg (2007)
18. Stuit, M., Szirbik, N., de Snoo, C.: Interaction beliefs: a way to understand emergent organisational behaviour. In: Proc. of ICEIS'07 (2007)
19. Thimm, R., Will, T.: A state of the art analysis of workflow modelling and agent-based information systems. Technical report, University of Trier (2004)
20. Witten, I., Frank, E.: Data Mining: Practical machine learning tools and techniques, 2nd edn. Morgan Kaufmann, San Francisco (2005)

Methods for Coalition Formation in Adaptation-Based Social Networks

Levi Barton and Vicki H. Allan

Utah State University, Logan, UT 84322-4205
LeviB@cc.usu.edu, Vicki.Allan@usu.edu

Abstract. Coalition formation in social networks consisting of a graph of interdependent agents allows many choices of which task to select and with whom to partner in the social network. Nodes represent agents and arcs represent communication paths for requesting team formation. Teams are formed in which each agent must be connected to another agent in the team by an arc. Agents discover effective network structures by adaptation. Agents also use several strategies for selecting the task and determining when to abandon an incomplete coalition. Coalitions are finalized in one-on-one negotiation, building a working coalition incrementally.

Keywords: Agents, Coordination and collaboration, Swarm Intelligence and Emergent Behavior.

1 Introduction

Much of the work in the area of coalition formation has focused on forming optimal, stable coalitions. These methods require complete information and require substantial computational time [2, 7]. In a dynamic environment, it may not be possible for an agent to maintain this complete view due to inaccurate information provided by faulty sensors or unreliable communications [10]. Our environment is different from many others in that teams must form a connected component in the social network. Most coalition formation problems are specified to highlight the problems in deciding which agents should work together to form teams. Our version of the problem adds the additional constraint of the structure of the agents that form a coalition.

Coalition formation among agents in dynamic environments presents additional challenges compared to coalition formation in static environments. The use of traditional coalition formation methods that compute a kernel to determine a coalition's division of utility is NP-Hard [6] and centralized. Both the changing nature of the tasks in a dynamic environment and the restrictions of teammates imposed by the social network render traditional coalition formation methods unusable.

Dynamic events in the environment change the nature of coalition formation. Tasks enter the system over time. The utility of a task may decrease over time. We study both the self-interested agent algorithm and how the structure of the graph affects performance. Because the correct structure cannot be known a priori, we

M. Klusch et al. (Eds.): CIA 2007, LNAI 4676, pp. 285–297, 2007.

allow agents to adapt the network structure (sometimes termed *rewiring*) in order to determine the desirability of certain structures. Agent capabilities include: joining an existing team, initiating a new team, waiting, or rewiring the connections. If team formation is successful, participating agents become active in the task, complete the task, and then rejoin the set of available agents.

In previous work, sub-optimal coalitions have been studied in coalition formation with incomplete information and with limited computational resources [9, 10]. In an effort to reduce computational expense, agents can prune the list of possible coalition partners to those considered most trustworthy or beneficial. These groups of agents who come together based on trust developed over past interactions are called congregations or clans [5]. By considering fewer potential coalition partners, as in our model, agents using clans can more easily compute possible coalitions within the time constraints imposed by the environment with the value of coalitions formed being good enough, or satisficing [5, 10].

This research implements a multi-agent system that uses an Agent Organized Network to facilitate coalition formation. The agents communicate tasks to neighboring agents to initiate the formation of a coalition to perform the task. When deciding which agents to include in the coalition, the agent forming the coalition considers the agent's skills.

2 Agent-Organized Networks

An *Agent Organized Network (AON)* is a set of inter-connected agents who collectively manage the structure of this network of agents by making individual decisions about which agent to connect to based on local information [3]. In our environment, nodes represent agents and links represent social structure. The *neighbors* of an agent are all agents connected to it via a network arc. The arcs may represent a variety of physical constraints such as physical distance, limited communication, trust, or organizational hierarchy. Teams must be connected components in the social network.

The initial set of network arcs are established by connecting each agent to any agent within distance d of its location. For our tests we used d=100 in order to compare with other author's work[3].

Tasks are introduced to all agents within a region of interest, and are executed by agents who are connected in the network. A task is composed of a set of subtasks requiring specific skills. In our system, the total number of possible skills is denoted *SkillMax*. A task is specified by an array skillCt[1.. *SkillMax*] in which skillCt[i] indicates the number of agents possessing skill i that are required by the task. Thus, in an environment with five possible skills, a task with skillCt = [1,2,0,0,3] requires six agents: one with skill 1, two with skill 2, and three with skill 5. In our system, skills are generated with a uniform distribution. We define the number of skills per task as *TaskSkills*, and the number of possible agent skills as *AgentSkills* (with *AgentSkills*=1 for our tests).

In order to compare our results directly with those of [3], we perform identical tests. System parameters are summarized in Table 1. The length of the interval between task generations is *GenInterval*. If *GenInterval* = 1, one task is generated every iteration. If *GenInterval* = 5, one task is generated every five iterations. For simplicity, the number of skills required per task is constant, and all tasks have the same utility. Tasks are announced for a time interval *AnnouncePeriod*, which depends on the number of skills required. If the team for a task has not formed within *AnnouncePeriod* iterations, the task is removed from the system and counted as a failed task. If a team fails to form during a time interval of length *CommitTime*, an agent will abandon the team. If all members abandon the team, the task returns to the available list (if its *AnnouncePeriod* has not expired). The duration of each task is specified as its *TaskDuration*.

Table 1. System Parameters

System Parameter	Meaning	Default Value
Skill Max	Number of different skills in system	10
TaskSkills	Number of skills per task	10
AgentSkills	Number of skills per agent	1
GenInterval	Iterations between task generation	10
AnnouncePeriod	Number of iterations task is schedulable	10
CommitTime	Maximum time an agent will stay in a partially formed team.	10
TaskDuration	Iterations required to complete task.	10
DegreeLimit	Maximum degree of any node in system	20
RewireFrequency	Probability agent rewires in a given turn.	1/N (N=# agents)

In forming teams, an agent may only communicate with its neighbors. Once a neighbor joins the team for a task, its neighbors are allowed to join the team. We will term the *neighbors of a neighbor*, **FOAF** (friend of a friend)[1]. Each agent has a single skill, a position on the grid, and a neighbor set. Agents are in one of three states: committed, active, or uncommitted, as shown in Figure 1. A *committed* agent has joined a partial team. An *active* agent has joined a team that is now fully formed and is currently involved in completing the associated task for *TaskDuration*. An *uncommitted* agent is not associated with a team.

Tasks require a team consisting of agents with a specific set of skills. A task consists of a taskID, an associated team (if any), an *AnnouncePeriod*, and a *TaskDuration*. A team consists of a taskID and a set of agents. A *complete team* has a match between required skills for the task and agents of the team. A *partial team* requires additional agents as some skill requirements are not met. Tasks are announced to agents within a fixed radius of the event. To eliminate variability in the results, in these tests, all tasks are known by all agents.

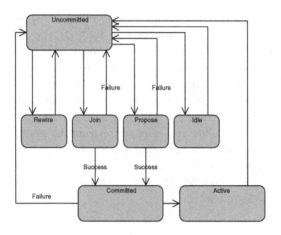

Fig. 1. AON Agent State Diagram

Arcs are initially assigned between nodes that are closer than a fixed parameter, but are changed during the process of rewiring. The degree of a node is limited to control communication. An existing partial team can only be joined by an agent if one of its neighbors has already joined the team.

The set of all uncommitted agents that are within d links of agent i and are connected to i via a chain of uncommitted agents is called a *circle of radius d.* For our tests, d is 1. Agents know the skills of the agents within their circle, termed the *circle skills.* A skill match is a measure of how closely the remaining skills for a task can be satisfied by the circle skills. *Skill match ratio* is computed as the ratio of the needed skills that can be provided by the circle divided by the total number of remaining skills to be acquired. The view of agents in the circle is limited by its radius; therefore, a skill match of less than one does not necessarily mean the team cannot be formed. Since the availability of an agent changes over time, the skill match is a rough estimate of the ability of the team to successfully form. As we increase the circle radius, we would hope to achieve a skill match of 1. *Needed agents* is the total number of agents that are missing from the partial team.

At each iteration, an uncommitted agent can remain idle, join an existing partial team, rewire, or initiate a team for a task which is a member of its *EligibleTasks* set. The definition of *best* varies depending on agent type.

Selecting the best team to join: An agent selects an eligible task (if any) from tasks joined by its set of committed neighbors.

Selecting the best task to initiate: Rather than joining an existing team, an agent may choose to start a new team. A task is *viable* for an agent if its *AnnouncePeriod* has not expired, it requires a skill the agent possesses, and the task is not completed. A task is termed *unclaimed* if no partial or full team exists which is associated with the viable task. Only unclaimed tasks are considered for initiation.

Remaining idle: An agent may decide to remain idle as a strategy to remain available for tasks which do not yet involve one of the agent's neighbors or future proposed tasks.

Rewiring: The agent type controls adaptation method, which determines the conditions for rewiring, the frequency of rewiring, and the target of rewiring.

3 Agent Types

An agent type consists of three features: tasks selection method, task patience, and adaptation approach.

Task Selection: There are two methods of picking activities: *basic* and *strategic*. Basic agents first decide if they want to rewire. If they do not rewire, they then consider the tasks, in order of task generation. This has the effect of giving preference to proposing/joining older tasks. If a neighbor has joined the team for a task, the agent joins the team (if it has a necessary skill). If a viable task is unclaimed, an agent will propose the task with probability proportional to the number of uncommitted neighbors. Note, there is no clear choice between initiating a team and joining an existing team, but since older tasks are given preference and there are more joiners than proposers needed for a coalition, joining is more likely.

Strategic agents first consider the set of tasks claimed by neighbors. The task which has the greatest ratio of committed agents (assigned skills/total skills is highest) is selected. This ratio is called the *committed ratio*. If the committed ratio is greater than some agent-specified threshold, the team for the task is joined. Thus, strategic agents give preference to joining an existing task over rewiring, proposing, or waiting. The agent will rewire based on the *rewiring frequency*. Strategic agents will choose to wait based on the desirable waiting probability (which is normally computed from the total number of skills required for a task). Agents that do not choose to rewire, wait, or join will propose the task for which the skill match ratio is the best as long as it exceeds some agent-specified threshold (.34 in our tests).

Task Patience: *CommitTime* is equal to *AnnouncePeriod* in patient agents. In impatient agents, an agent may abandon a task if certain success indicators are not present. For our tests, the task is abandoned when the count of uncommitted neighbors is less than the number of skills needed.

Adaptation Approach: An agent has the ability to remove a link to a neighbor and create a link to a new target. This rewiring is intended to increase the performance of the system. We define system performance as the fraction of tasks completed. We define *node performance* as the fraction of successful teams joined over the number of teams attempted by a node. We define *system efficiency* as the fraction of time agents spend actively executing tasks. Agents executing tasks are termed *working*.

Rewiring involves an agent selecting one of its adjacent edges to disconnect from its current target and reconnect to another target. The edge is jointly owned by source and target, so either agent can initiate the rewiring, and no agent can refuse to be connected to a particular node (unless its maximum degree has been exceeded). We study three types of rewiring approaches: *structural, performance,* and *diversity.* Structural rewiring is motivated by findings in network topology [11]. Performance rewiring, defined in [3], is based on the desire to be connected to agents which have higher performance. Diversity rewiring is based on the desire to be connected to an agent with a different skill than you possess.

In performance adaptation, the adaptation trigger is based on local performance. Local performance is only valid when the number of teams joined is larger than a set number (five in our case). Local performance is computed as the number of successful teams joined divided by the number of teams joined. When an agent's local performance is less than the average of its neighbors' performance, rewiring is triggered: the agent removes the link to the worst performing neighbor and replaces it with a link to the best performing neighbor of the best performing neighbor.

In structural adaptation, rewiring is independent of performance and occurs with probability 1/N where N is the number of agents. An agent randomly picks a link and replaces it with a link to a FOAF based on *preferential attachment.* Preferential attachment means that the probability of a particular target is its degree divided by the total degree of all the neighbor's neighbors.

In diversity adaptation, rewiring is independent of performance and occurs with probability 1/N where N is the number of agents as in structural rewiring, but the only requirement for selecting a target is that the target must have a different skill than the agent.

If we limit the degree of a node in the graph, we say the adaptation is *bounded.* The idea behind bounded rewiring is to eliminate the communication bottleneck. On network creation, arcs to a node of degree *DegreeLimit* are ignored. During adaptation, nodes with degree *DegreeLimit* are invisible to nodes seeking a new target.

4   Results

We conduct a variety of tests to evaluate both the activity selection and the adaptation approaches. To compare the methods, we repeat each test 50 times with 100 nodes and compute 95% confidence intervals. Each test executes for 2000 cycles, with adaptation beginning at time 1000. Other parameters are varied as specified. In this way, we see the affects of adaptation.

Table 2. Agent Types

Agent Type	Task Selection	Task Patience	Adaptation Approaches
Performance	Basic	Patient	Performance
Structural	Basic	Patient	Structural
Structural Strategic	Strategic	Patient	Structural
Structural Strategic Impatient	Strategic	Impatient	Structural
Diversity Impatient	Basic	Impatient	Diversity
Diversity Strategic Impatient	Strategic	Impatient	Diversity

Since the total number of arcs in all methods is unchanged by rewiring, structural rewiring creates some nodes of high degree (hubs) connected to others of small degree. Thus, equal degree nodes are sacrificed to achieve high degree nodes.

In Figure 2, adaptation is disabled showing the differences in performance resulting from agent decisions in task selection and task patience. Performance and Structural agents use Basic task selection and are Patient, resulting in the lowest performance at 0.24. The StructuralStrategic agent uses Strategic task selection and is Patient, resulting in performance of 0.29, with a 5% improvement over the Basic Patient agents. The StructuralStrategicImpatient agent uses Strategic task selection and is Impatient, resulting in performance of 0.36, with a 7% improvement over the StructuralStrategic agent. Thus, task Impatience and Strategic task selection each individually provide improvements in performance. They also complement each other, offering even better performance when used together.

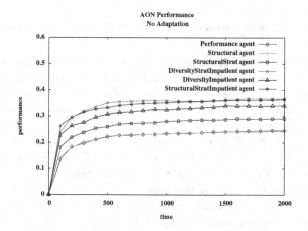

Fig. 2. AON performance with no adaptation

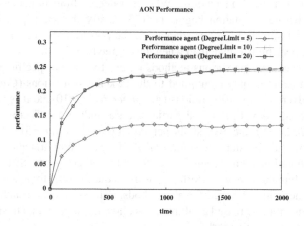

Fig. 3. The effect of limiting maximum agent degree on AON performance of Performance agents

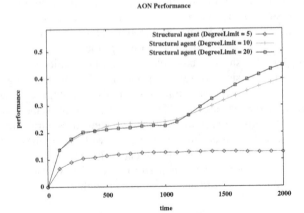

Fig. 4. The effect of limiting maximum agent degree on AON performance of Structural agents

Figure 3 shows the effect on performance when a degree limit is imposed on Performance agents in an AON network. When the degree limit is 5, the performance is much lower, even before adaptation, as agents fail to have a network structure that allows for adequate social interaction to form teams as quickly, resulting in more failed teams. By increasing the degree limit to 10, the performance increases from 0.13 to 0.24, nearly doubling the performance. This reflects the benefits of having better connectivity, resulting in a higher probability of a successful team. Diversity agents exhibit a similar pattern of response to limited degree.

Figure 4 displays the effects of imposing a degree limit on Structural agents in an AON network. Once again, when the degree limit is 5, the performance is relatively low and does not improve with rewiring. When the degree limit is increased to 10, the performance before adaptation begins (at time step 1000) also increases from 0.13 to 0.24. After adaptation, the performance increases due to the hub network structure resulting from Structural adaptation. When the degree limit is increased to 20, the performance before adaptation begins is 0.23, nearly the same performance level using a degree limit of 10. After adaptation, the performance increases at a greater rate and achieves a higher level than when the degree limit is 10.

The performance of the various methods is sensitive to the specific parameters. In one test, each agent is only committed to the task for 4 iterations (*CommitTime*=4). The task is advertised for 10 iterations (*AnnouncePeriod*=10), so it is possible that a task will succeed even if the initial partial team abandons it. There are eight skills possible (*SkillMax*=8) and five skills per task (*TaskSkillCount*=5). Two tasks are introduced per time step (*GenInterval*=0.5). StructuralStrategicImpatient and DiversityStrategicImpatient agents out-perform the other agents. StructuralStrategic and DiversityImpatient agents perform nearly equally due to a small maximum committed time being enforced for all methods. The advantage usually enjoyed by the Strategic agent is decreased as all methods allow a task less time to succeed before attempting to move to other tasks.

When we consider only the improvement occurring because of adaptation, the agent types rank differently with respect to relative improvement than with respect to performance. This indicates that agents with poor performance have more to gain from adaptation while those with better performance are unable to achieve the same level of improvement.

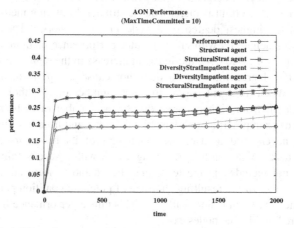

Fig. 5. AON performance with maximum time committed of 10 iterations

Once a team for a task is joined, the agents that are task Patient (Performance, Structural, and StructuralStrat agents) stay with the task until it has expired (as *CommitTime* equals *AnnouncePeriod*). An Impatient agent has a significant advantage (Figure 5) because it has the choice of abandoning an unpromising task before the task expires, which allows other agents to complete the task and frees the agent to pursue more profitable activities. Structural Strategic is still competitive as it is more selective in the initial choice of a task.

Consider the set of test cases (Figure 6 and Figure 7) which differ only in the degree limit. Agents are committed for a maximum of 10 iterations to a task (*CommitTime*=10). The tasks are advertised for 10 iterations (*AnnouncePeriod*=10).

In Figure 6, the node degree is limited to 10, allowing an increase of performance after adaptation is allowed at time step 1000. Strategic and Impatient methods have a higher initial level of performance before adaptation is allowed, but methods using Structural adaptation show more improvement in performance than other adaptation methods. The StructuralStratImpatient agent has the benefit of both a high initial performance level, a result of the combination of Strategic task selection and task Impatience. The StructuralStratImpatient agent also shows significant improvement in performance with adaptation, resulting from the structural rewiring strategy, making it the best performing method of all. We have made significant improvements over the structural results of [3] by changing the task selection algorithm and the team commitment. While strategic task selection requires increased knowledge, the team commitment limitation requires no special intelligence. Note that while StructuralStratImpatient is smarter, it increases in performance at the same rate as the less effective Structural due to their shared use of Structural rewiring.

As expected, the performance of some agent types does not increase when adaptation is begun at time step 1000 as shown by the flat curves. Increasing the degree limit to 20, the performance levels (after adaptation) increase, especially for the structural rewiring methods. Structural rewiring methods are able to achieve nodes of high degree, increasing the probability of being able to form a successful team, given the number of neighbors who possess the skills required.

Each time the degree limit is increased, the structural rewiring methods experience an increase in the highest degree of a node in the network. The other rewiring methods seek for neighbors with either greater performance or more a more diverse skill set. Therefore, they do not experience an increase in the highest node degree.

Increasing the degree limit to 30 does not cause the performance to increase significantly. When an agent has a sufficient number of neighbors, they will not continue to benefit from higher degree due to the number of skills per task (10 in these tests). For example, if a node has 20 neighbors, and needs to fill a task with 10 skills, it will never need to use more than half of its neighbors to successfully complete this task. Also, while completing a task with a successfully formed team, the neighbors not included in the team may be unable to form an additional team successfully because of the resulting shortage of potential coalition partners when the hub node connecting them is not available. This puts a performance limit on forming networks where high degree nodes exist.

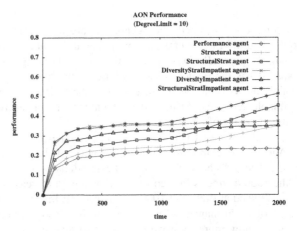

Fig. 6. Performance with degree limit of 10

We also experiment with the proportion of tasks to agents. While holding the agent count constant, we increase the number of tasks introduced at each time step. In these tests, there are 100 agents. Tasks need 10 skills and there are 10 possible skills. Since tasks have duration 10 and task formation takes an average of three iterations, the system of agents cannot possibly complete more than 1.3 tasks per iteration. Thus, introducing two tasks per iteration gives the agents many tasks for selection (Figure 8).

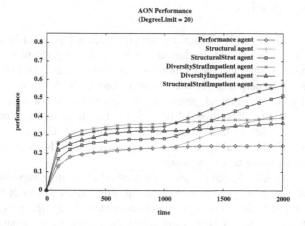

Fig. 7. Performance with Degree Limit of 20

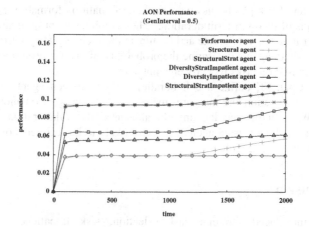

Fig. 8. Performance with two tasks per iteration

Agents who use Strategic task selection do very well as they can pick the best tasks. As before, agents using Structural adaptation also do well, as the resulting structure means that the skills are available to complete many tasks. Performance agents perform poorly, likely because clusters of under-performing agents are not motivated to rewire because they *are* doing as well as their neighbors.

Notice, however, that the performance is less than 0.12 (compared to earlier tests showing performance levels of nearly 0.6). The performance necessarily decreases, as it is not possible to complete all the tasks and performance measures percent of tasks completed. However, the performance is even lower than one might guess. This can be attributed to the fact that many competing tasks seduce more agents to join a team for a task, resulting in many partially formed teams with no waiting agents to complete them.

5 Previous Work

Gaston and des Jardins [3] implement multiagent team formation in which tasks are globally known. Gaston and des Jardins rewire based on performance and structural metrics. Under performance adaptation, an agent whose performance is less than the performance of its neighbors is allowed to reconnect to the best of the neighbor's neighbors. Under structural adaptation, agents rewire to select a target from the set of neighbor's neighbors. The probability of connecting to an agent is proportional to a node's degree. Our results show the performance of Gaston and des Jardins' performance and structural agents can be improved upon by making strategic task selections and by using task impatience.

Soh and Tsatsoulis [9, 10] present their method of creating sub-optimal coalitions in a dynamic environment to deal with partial information and meet time constraints imposed by the real-time nature of the environment. They apply their method in a distributed target-tracking environment where the agents are stationary. Our method is based on their work, with an added challenge of forming coalitions among agents who navigate within a dynamic environment.

Griffiths and Luck [4] discuss their method of coalition formation for clans whose cooperation is of medium-term duration. They combine agent motivations and agent trust to determine when to create and terminate these clans. They introduce a kinship motivation, where agents determine the probability of encountering tasks that require cooperation based on tasks recently encountered.

Soh and Chen [8] define a collaboration utility between agent-based history of successful collaboration and measures the reliance of one agent on another in forming coalitions. While no discrete labeling of clan is used, the collaboration utility specifies a loose clan in which trusted agents are more likely to be recipients of proposals and are more likely to be accepted.

6 Conclusion

We have introduced Strategic task selection, task Impatience, and Diversity adaptation. We have shown the benefits of using these agent strategies in an AON network. Agent performance is improved when using Strategic task selection versus Basic task selection. Agent performance is also improved when using task Impatience versus task patience. The best agent performance has been shown to result from using Strategic task selection, task Impatience, and Strategic rewiring, the rewiring method that provides the greatest increase in performance in AON networks. Future research should examine the possibility of using a variety of agent strategies within the same network to improve performance.

Strategic agents have been shown to perform well under high task loads due to their ability to choose the task most likely to succeed. Task Impatient agents have been shown to improve agent performance, but are affected more by high task loads.

AON networks containing nodes of high degree show increased levels of performance. However, high degree nodes also incur higher communication costs. We have shown that moderate maximum degree limitations do not significantly affect AON performance. This allows implementations of real-world AON applications to impose maximum degree limitations to decrease communication costs without experiencing degraded performance.

References

[1] Brickley, D., Miller, L.: FOAF: the 'friend of a friend' vocabulary (2007), http://xmlns.com/foaf/0.1/

[2] Dang, V., Jennings, N.: Generating Coalition Structures With Finite Bound From The Optimal Guarantees. In: Kudenko, D., Kazakov, D., Alonso, E. (eds.) Adaptive Agents and Multi-Agent Systems II. LNCS (LNAI), vol. 3394, Springer, Heidelberg (2005)

[3] Gaston, M.E., Jardins, M.d.: Agent-Organized Networks for Dynamic Team Formation. In: Autonomous Agents and Multi-Agent Systems (AAMAS), pp. 230–237. Utrecht, Netherlands (2005)

[4] Griffiths, N., Luck, M.: Coalition Formation through Motivation and Trust. In: Autonomous Agents and Multiagent Systems (AAMAS), Melbourne, Australia, pp. 17–24 (2003)

[5] Horling, B., Lesser, V.: A Survey of Multi-Agent Organizational Paradigms. The Knowledge Engineering Review 19, 281–316 (2005)

[6] Lau, H.C., Zhang, L.: Task Allocation via Multi-Agent Coalition Formation: Taxonomy, Algorithms and Complexity. In: 15th IEEE International Conference on Tools with Artificial Intelligence (ICTAI), pp. 346–350 (2003)

[7] Sandholm, T., Lesser, V.: Coalitions Among Computationally Bounded Agents. Artificial Intelligence, Special Issue on Economic Principles of Multi-Agent Systems 94, 99–137 (1997)

[8] Soh, L.-K., Chen, C.: Balancing ontological and operational factors in refining multiagent neighborhoods. In: Autonomous agents and multiagent systems (AAMAS), The Netherlands, pp. 745–752 (2005)

[9] Soh, L.-K., Tsatsoulis, C.: Reflective Negotiating Agents for Real-Time Multisensor Target Tracking. In: International Joint Conference on Artificial Intelligence (IJCAI), Seattle, WA (2001)

[10] Soh, L.-K., Tsatsoulis, C.: Utility-Based Multiagent Coalition Formation with Incomplete Information and Time Constraints. In: IEEE International Conference on Systems, Man, and Cybernetics (SMC) (2003)

[11] Thadakamalla, H.P., U. N. K. Raghaven, S., Albert, R.: Survivability of Multiagent-based Supply Networks: A Topological Perspective. IEEE Intelligent Systems 19, 24–31 (2004)

Trust Modeling with Context Representation and Generalized Identities

Martin Rehák and Michal Pěchouček

Dept. of Cybernetics and Center for Applied Cybernetics, Faculty of Electrical Engineering,
Czech Technical University in Prague
Technická 2, Prague 6, 166 27 Czech Republic
{mrehak,pechouc}@labe.felk.cvut.cz

Abstract. We present a trust model extension that attempts to relax the assumptions that are currently taken by the majority of existing trust models: (i) proven identity of agents, (ii) repetitive interactions and (iii) similar trusting situations. The proposed approach formalizes the situation (context) and/or trusted agent identity in a multi-dimensional Identity-Context feature space, and attaches the trustworthiness evaluations to individual elements from this metric space, rather than to fixed identity tags (e.g. AIDs, addresses). Trustworthiness of the individual elements of the I-C space can be evaluated using any trust model that supports weighted aggregations and updates, allowing the integration of the mechanism with most existing work. Trust models with the proposed extension are appropriate for deployment in dynamic, ad-hoc and mobile environments, where the agent platform can't guarantee the identity of the agents and where the cryptography-based identity management techniques may be too costly due to the unreliable and costly communication.

1 Introduction

The purpose of our work is to extend the applicability of the agent trust modeling towards open and unmanaged systems common in the ubiquitous computing environments. Instead of devising a novel trust model, we propose an algorithm that can be combined with existing trust models [1,2,3,4] to extend their applicability.

Therefore, we need to address the underlying assumptions of the existing trust models, that assume that *well-identified agents repetitively engage in similar interactions*. An example of a domain where the current trust models are adequate is a supply-chain management, where the agents can be associated to companies or their entities that maintain long-term relationships with repetitive deliveries. A counter-example is a wireless sensor network or a group of Unmanned Aerial Vehicles (UAVs) that are (i) hard to identify and where (ii) the barriers of entry are quite low thanks to the inherent openness of the system. Agents labeled as untrustworthy can therefore easily change their identity and enter the system with a new, uncompromised one. Furthermore, possible interactions are diverse and depend heavily on the context (e.g. sensor performance that varies between day and night).

M. Klusch et al. (Eds.): CIA 2007, LNAI 4676, pp. 298–312, 2007.
© Springer-Verlag Berlin Heidelberg 2007

1.1   Assumptions of the Existing Trust Models

Most existing trust model are based on an assumption of the **proven identity**: it is assumed that the agents can't change their identity during their lifetime and that they can't have multiple identities at once. This assumption is acceptable in purely multi-agent systems, where the agent platform manages and oversees the agents, but fails in case of open systems (featuring mobile agents and ad-hoc networks), where the agents are distributed across independent platforms and can join or leave the system at their own will. Cryptography-based identification methods [5] can address the problem rather efficiently, but the need to verify the chain of identity to the mutually recognized authority may pose severe problems in a ubiquitous computing context, where the resources and network connectivity are scarce. In such environments, the delegation of the identity management to a dedicated infrastructure is costly, possibly leaving the system designers with a choice between secure inflexible system, or an unsecured, but more robust open system. This contribution suggests how to complement the existing methods and addresses this gap.

Identity problem is closely related with a **"first time offender"** problem. Current trust models concentrate on managing ongoing relationships with repetitive interactions. While the reputation (i.e. witness reputation, trust communication) and social dimension [6] of certain models alleviate the situation, they can't by any means protect the open system against the agents that connect, build-up their trustworthiness and use it to defect when the stakes are higher.

Last, but not least, existing trust models often neglect the problem of **context** (e.g. situation) – they are mostly based on the assumption that the interactions with a given provider are similar and that the previously acquired experience is relevant for future interactions. We argue that the inability to take the context into account limits the practical use of current trust models in all domains where the agents perform diverse tasks in a highly dynamic environment – this being a precise definition of the ideal environment for deployment of intelligent social agents[7].

In Section 2, we will propose an extension of general trust models that will allow them to consider the situational trust by means of context modeling and to efficiently exploit the similarities between the various situations. In Section 3, we will extend the context representation framework to include an identity of the trusted agent as well and we provide an outline of the algorithm. Section 4 provides an evaluation of our approach in a simulated logistics scenario, before the discussion of the related work and conclusion in Sections 5 and 6.

2   Context Representation

In most existing trust models[3,8,9,10], the trustfulness of an agent X as evaluated by agent A is a weighted aggregation of past observations of X's performance, either direct or indirect, that are available to A. In most models, recent experience, direct observations and coherent reputations provided by trusted agents [11,12] are emphasized, but as these issues are orthogonal to the topic of this paper, we will not explicit them in our notation.

The goal of the formalism presented in this section is to provide an efficient mechanism for situation representation and to show how can be such mechanism integrated with existing trust models. Therefore, we need to capture relevant aspects of the situation and to represent them as a context, a point c_i in the context space $\mathbb{C}$. The context space is a Q-dimensional metric space with one dimension per each represented situation feature, and the metrics $d(c_1, c_2)$ defined on $\mathbb{C}$ describes similarity[1] between the contexts c_1 and c_2 (or rather lack of similarity).

Please note that while the definition of the $\mathbb{C}$ dimensions and distance function are necessarily domain dependent, the model that we propose only imposes minimal requirements on their definition, most notably the properties of the metric space. When these conditions are fulfilled, the algorithm we propose below can be applied.

Example: A trivial example of a trusting decision that is influenced by the context is a territory surveillance by autonomous UAVs. We can consider the territory size as one dimension of the situation, and visibility (day/night/cloudy/clear) as another one (or even as two features). We shall then find good individual metrics - here, a radius of the area to cover (or its logarithm, to emphasize the ratios rather than absolute differences).To represent the visibility, we may consider any standard metrics from the aviation domain or any metrics that appropriate for intended goal.

The distance metrics allows us to integrate the context representation mechanism with the existing trust models. Therefore, we define a set $\mathcal{R}$ of *reference contexts* r_i, the elements of the $\mathbb{C}$ to which we will associate the trustfulness values, e.g. trust model instances. We shall note that while it is highly desirable to share a common $\mathbb{C}$ definition between all the trusting agents in the system, the definition of the set $\mathcal{R}$ is up to each trusting agent and will be probably even different for each partner evaluated by the agent.

The price that we have to pay for context representation is that instead of maintaining a single instance of the trust model structure (representing the general trust) per partner, we shall maintain one instance per partner for each relevant reference context. On the other hand, when we further generalize this approach in Section 3, we will apply the same approach to identity representation as well, potentially *decreasing* the amount of data to hold.

The following relations will present the formulas for trustfulness update and trustfulness aggregation before taking the trusting decision. We will denote as $\Theta_A(X|r_i)$ the

---

[1] Any distance function $d : \mathbb{C} \times \mathbb{C} \to R$ must respect following properties: **non-negativity**:

$$d(c_1, c_2) \geq 0, \tag{1}$$

symmetry:
$$d(c_1, c_2) = d(c_2, c_1), \tag{2}$$

zero distance $\Leftrightarrow$ **identity**:
$$d(c_1, c_2) = 0 \Leftrightarrow c_1 = c_2, \tag{3}$$

triangle inequality:
$$d(c_1, c_3) \leq d(c_1, c_2) + d(c_2, c_3). \tag{4}$$

trustfulness of agent X in the situation represented by reference context r_i, assessed by agent A. To obtain a weight decreasing with the distance, we introduce a domain-dependent weight function $w_i = f(d(c_d, r_i))$, where f is a non-increasing function on $[0, +\infty)$. This function represents the decay of the observation usefulness with increasing distance d of the particular reference context r_i – obviously, it is most useful when its distance $d(c_d, r_i)$ from the reference context is zero. A convenient form of such function can be for example $w_i = e^{-d(c_d, r_i)}$.

In the following, we will use a shorthand definition for the aggregate weight of the p past observations associated with each reference context r_i, denoted W_i^p.

$$W_i^p = \sum_{j <= p} w_i^j \tag{5}$$

Each new **observation** $\tau_A(X|c_o)$ is integrated into the apriori trustfulness evaluation $\Theta_A^p(X|r_i)$ associated with each r_i (if $w_i^p + 1 > 0$) using the following formula:

$$\Theta_A^{p+1}(X|r_i) = WeAg((\Theta_A^p(X|r_i), W_i^p, (\tau_A(X|c_o), w_i^{p+1})) \tag{6}$$

The implementation of the $WeAg()$ operator depends entirely on the trust model used to represent $\Theta_A^{p+1}(X|r_i)$. In a simple example, when the $\Theta_A^p(X|r_i)$ is just a w_i weighted average of all p previous observations, we obtain:

$$\Theta_A^{p+1}(X|r_i) = \frac{W_i^p \cdot \Theta_A^p(X|r_i) + w_i^{p+1} \cdot \tau_A(X|c_o)}{W_i^p + w_i^{p+1}} \tag{7}$$

In order to take a trusting decision in a situation represented by context c_d, we must **query** the model. The result of the query, a shown in Eq. 8, is obtained as a weighted combination of trustfulness associated with relevant (i.e. close) reference contexts from $\mathcal{R}$.

$$\Theta_A(X|c_d) = WeAg_{r_i \in \mathcal{R}}(\Theta_A(X|r_i), w_i) \tag{8}$$

In the simple weighted average case, we obtain:

$$\Theta_A(X|c_d) = \frac{\sum_{r_i \in \mathcal{R}} w_i \cdot \Theta_A(X|r_i)}{\sum_{r_i \in \mathcal{R}} w_i} \tag{9}$$

The query process is illustrated in Fig. 1, where we aggregate the result from four reference contexts in the vicinity, showing the role of the weight w_i as obtained from $d_i = d(c, r_i)$.

2.1 Reference Set

While the above relations are sufficient to explain the update and aggregation process of the trustfulness information associated with the individual reference contexts, they don't address a critical issue of the model: positioning of the reference contexts in the context space.

The method we propose above doesn't require any particular shape of the reference set. The requirements on the reference set maintenance method are relatively severe, as

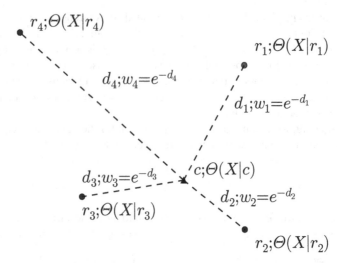

Fig. 1. Aggregation of the trustfulness of agent X in the situation represented by context c from four reference contexts in the vicinity

we need to handle: (i) real-time reference context placement, without apriori knowledge of the future data distribution, (ii) specificity of the reference contexts positions for each modeled agent and general uncertainty of the distribution of interactions in the space $\mathbb{C}$.

Our current solution is based on Leader-Follower clustering algorithm [13], presented in Fig. 2. The advantages of this particular clustering method are obvious - it is a clustering method that allows on-line approach, without pre-specifying the number of expected clusters beforehand, as is the rule for most other clustering methods. Furthermore, it requires only a single parameter as its input: the cutoff distance of the new context from the closest centroid, when the new context will become a part of the existing cluster. If the new point is farther away, it becomes a centroid of the new cluster.

The biggest disadvantage of this method is that it may easily under or over estimate the number of clusters. In our case, this is not a big problem, as the overestimation results only in maintenance of two or more reference contexts instead of one. Therefore, it is always better to slightly underestimate the cutoff distance, avoiding the data blurring.

3   Including Generalized Identity Management

To represent the uncertain identities, we will extend our context representation formalism to cover the identity as well. Instead of representing the identity as a unique tag, we will decompose the identity into relevant, measurable features, that will form an **identity space** $\mathbb{I}$. The identity space is also a metric space, where the metrics describes lack of similarity between two identities. Please note that the properties of the metrics apply to the formal identities only – the fact that two elements of the space $\mathbb{I}$ are identical (i.e. have 0 distance following 3) doesn't necessarily imply that the respective agents are identical as well – this only means that the features accessible to the model doesn't allow us to differentiate between these identities.

```
class LFClustering:
  def __init__(self, threshold):
    self.centers = []
    self.thresh = 0.5
  def newSample(self, sample):
    closest,dist = self.findClosestCenter(sample)
    if closest != '0' and dist <= self.thresh:
      closest.aggregate(sample, 1, 0.01)
    else:
      self.centers.append(sample)
  def findClosestCenter(self, sample): ...
```

Fig. 2. L-F Clustering algorithm outlined in Python

The fact that two agents with presumably distinct identities can be considered identical by the model provides the protection against changes of identity. When the properties that define the $\mathbb{I}$ are set correctly, the new identity of the agent will fall close to the original one, allowing the evaluating agent to re-use the previous trustfulness data. For example, in sensor networks, we may use the power of the signal or other radio properties of the partner node, combined with the type of services it provides. In the UAV use-case, we may use visual properties of the craft (possibly type identification), radar properties and typical flight patterns to distinguish between evaluated entities.

This approach is (to some extent) also effective against first time offenders. We assume that to take down an open, distributed system working on p2p basis, the attack will be based on similar principles and will consist of relatively large number of similar nodes[2]. Due to the similarity of attacking nodes, the mechanism will treat them as a single identity, or a small number of identities. This means that it will be able to react to the novel threat in a same manner as a biological immune system, where the knowledge gained on a single hostile entity influences the interactions with the other entities of the same type, provided that an efficient reputation-sharing component is a part of the underlying trust model.

Formally, it is preferable to consider the spaces $\mathbb{I}$ and $\mathbb{C}$ as two subspaces of the identity-context space $\mathbb{IC}$, where each vector ix belonging to this space contains all the information associated with an observation or a decision.

For consistency, we will maintain the notation of the reference set $\mathcal{R}$ and reference identity-context r_i, even if their dimension is increased to incorporate the identity subspace dimensions as well.

$$\Theta_A^{p+1}(r_i) = WeAg((\Theta_A^p(r_i), W_i^p, (\tau_A(ix_o), w_i^{p+1}))) \tag{10}$$

$$\Theta_A^{p+1}(r_i) = \frac{W_i^p \cdot \Theta_A^p(r_i) + w_i^{p+1} \cdot \tau_A(ix_o)}{W_i^p + w_i^{p+1}} \tag{11}$$

---

[2] In this model, we deliberatively ignore the attacks based on mobile code and environment saturation, that we address in [14].

When we **query** the model to obtain a situation-relevant trustfulness before taking a trusting decision, the current context c_d is determined and the trustfulness is obtained as a weighted combination of trustfulness associated with reference contexts from $\mathcal{R}$.

$$\Theta_A(ix_d) = WeAg_{r_i \in \mathcal{R}}(\Theta_A(X|r_i), w_i) \tag{12}$$

In the weighted average case, we obtain:

$$\Theta_A(ix_d) = \frac{\sum_{r_i \in \mathcal{R}} w_i \cdot \Theta_A(X|r_i)}{\sum_{r_i \in \mathcal{R}} w_i} \tag{13}$$

Before the discussion regarding the implementation of the above-defined relationships, we shall list the domain dependent functions that are used by the model in Table 1.

Table 1. Domain dependent parameters of the model

Code	Description
identityCx(ag,sit)	Creates formal identity and context
distance(x,y)	$d(x,y)$, Distance between $x, y \in \mathbb{IC}$
weight(d(rc,x))	w_i, Weight function
clustsize	Max distance for cluster attachment
treshold	Min weight for trust update

The implementation of the system is straightforward, outlined in Fig. 3 and Fig. 4. In Fig. 3, we can notice that the update of the trustfulness of individual reference contexts r_i (denoted rc) is performed in the same time as the update of the set $\mathcal{R}$ (rclist) (either by appending the new reference context id or by updating the position of the centroid of the cluster to which the new observation was assigned. Therefore, the complexity of the algorithm was increased by a linear factor $|\mathcal{R}|$(i.e. the size of the rclist). We can also note that the context representation is integrated seamlessly, only by means of domain dependent functions identityCx(ag,sit) and distance(rc,id). Therefore, removing the context representation altogether or altering its resolution (by lowering/increasing the weight of the context-relative parts of the distance function) will directly reduce the parameter $|\mathcal{R}|$. Such operation can be performed at runtime and can help the agents to manage the computational requirements of their trust model. On the other hand, we can easily use the model with crisp identities, by imposing a trivial distance function in the identity subspace, using the formula $d(x,y) = 0$ iff $x = y$, else $d(x,y) = \inf$. Such definition will effectively split the model into n independent models, where n is the number of evaluated agents[3]. These "submodels" will then function exactly as the context-only mechanism variant presented in Section 2.

---

[3] When applying this approach in the environments with significant number of agents, it is advantageous to replace the linear list rclist in Fig. 3 by appropriate partially indexed indexed structure to iterate only over the contexts relative to a particular modeled agent.

```
def newObservation(agent,situat,trustObs):
    # get the identity
    id = identityCx(ag,situat)
    mindist = Infinity
    closest = '0'
    for rc in rclist:
        dist = distance(rc,id)
        # clustering
        if dist < mindist:
            closest = rc
        # update the trustfulness
        wei = weight(dist)
        if wei > treshold:
            rc.trust.update(trustObs,wei)
    # create new reference context if necessary
    if mindist > clustsize:
        rclist.append(id)
        id.trust.update(trustObs,1)
    else:
        closest.updatePosition(id)
```

Fig. 3. Processing of the new observation outlined in Python (simplified)

```
def query(agent,situat):
    id = identityCx(agent,situat)
    result = Trustfulness()
    for rc in rclist:
        wei = weight(distance(rc,id))
        if wei > treshold:
            result.aggregate(rc.trust,wei)
    return result
```

Fig. 4. Processing of the query by the model, (Python, simplified)

While the complexity of the trust model increases by incorporation of the proposed mechanism, the Proposition 1 shows that the memory requirements will in most cases actually *decrease*.

Proposition 1. *Provided that (i) we use the identity representation mechanism as specified by Eq. 10 to Eq. 13, (ii) with the Leader-Follower clustering algorithm as outlined in [13] and Fig. 2, the trust model with uncertain identities will be smaller (in terms of required memory) than the corresponding model with crisp identities. We also assume* *that (iii) both models use the same (possibly none) mechanism for context management and (iv) that the amount of memory used to represent the typical agent identifier in a crisp model (e.g. address, AID,local name) the same or bigger than the size of a single element of the* $\mathbb{I}$.

Proof. Using the fact that the size of the structure representing the Θ_A is the same in both cases (iii), as we use the same trust model, and that the size of the associated identities is comparable as well (from (iv)), we only need to compare the number of elements Θ_A in the model. From (ii) and Fig. 2, we can see that we create at most a single centroid per each new identity. Therefore, the size of the model with uncertain identities must be smaller or equal to the crisp model.

The proposition is intuitive from information-theoretic perspective – the trust model based on agent identities is the most detailed model possible, and will therefore constitute an upper bound on the size of any generalizing model. With increasing size of the clusters (and therefore their non-increasing number), the amount of data in the model decreases.

4   Experimental Evaluation: Context Representation

In this section, we will present results of the benchmark performed in a simulated logistics scenario. To evaluate our approach experimentally, we model the trust reasoning of a humanitarian aid organization agent that acquires transportation services from several local transporters after major disaster, and we will enhance it by use of context. We will then compare the performance of the agents who use the baseline model with the performance of the agents using the model enhanced with context representation part as specified in Section 2.

4.1   Context Space Definition

To illustrate the abstract notions of metric space $\mathbb{C}$, we introduce an example of such space for our logistics scenario, where we model each trusting situation (observation or decision) by three parameters: cargo type, cargo size and road quality. Cargo *type* defines the product we transport: medical supplies, food or durable goods. Each cargo type has specific handling requirements – medical supplies are the most sensitive to carry, while the durables require less care. *Size* of the transport is simply a quantity to carry, while the *road quality* represents the quality of the roads to use for transport. It is interesting to note that *type* dimension is discrete, while the *size* and *road quality* are real-valued, but different: one has an absolute scale (size), while the other will be close to 1.

The context space $\mathbb{C}$ is three dimensional, with one discrete dimension and two continuous ones. The next step is a definition of marginal distances d^q for each dimension. In the *type* domain, we place our products on a "sensitivity" scale: medical supplies require most attention: 5, with the food in the middle: 1 and the durables as least sensitive ones, with 0.2 value[4]. Our type distance metrics is defined as follows, using the product properties defined above:

---

[4] Inverting the scale will not change the result thanks to the distance symmetry stated in Eq. 2.

$$d^{type}(c_1, c_2) = |ln(type_1) - ln(type_2)| \tag{14}$$

In the size domain, the metric shall describe the differences between two contracts in terms of their relative size. We propose a measure

$$d^{size}(c_1, c_2) = |ln(size_1) - ln(size_2)| \tag{15}$$

The logarithmic relation captures an intuitive notion of ratio: 10 tons difference between two 20 and 30 ton transports is much more important than the same difference between two shipments of thousands of tons.

We apply the same reasoning for the road quality:

$$d^{road}(c_1, c_2) = |ln(qual_1) - ln(qual_2)| \tag{16}$$

Then we combine the above metrics using a slightly modified (weighted) "Manhattan distance":

$$d(c_1, c_2) = \alpha_1 d^{type}(c_1, c_2) + \alpha_2 d^{size}(c_1, c_2) + \alpha_3 d^{road}(c_1, c_2) \tag{17}$$

4.2 Experimental Setup

In the task allocation problem solved by the agents in our humanitarian logistics scenario, they choose one or more providers (transporters) for each contract and use their trust models to reason about their trustfulness.

In the underlying simulation model, the transporters answer the call for proposals with *bid prices* pr_b based on the nominal transportation cost and profit margins. The *real price*, that includes the cost of the cargo lost during transportation, is derived after the transport from the bid price and transporter *real trustworthiness* Θ. The Θ depends on the same parameters as those that define the $\mathbb{C}$ dimensions. Real price pr_r is determined as $pr_r = \frac{pr_b}{\Theta}$, where

$$\Theta = \Theta_{type} \cdot atan'(price) \cdot atan'(supply) \tag{18}$$

The function $atan'(x)$, used as a sigmoid approximation, is defined as a normalized *arctan*: its range is (x_{inf}, x_{sup}) (both x_{inf}, x_{sup} are in the range set) and x coordinate of its flection point is defined by parameter x_{center}. x_{slope} determines the first derivation - speed of the growth on the domain.

$$atan'(x) = \frac{1 - x_{inf}}{\pi} \cdot arctan(\frac{x_{center} - x}{x_{slope}}) \tag{19}$$

While the provider simulation is a very simple one, it is sufficiently versatile to model the performance of market actors to obtain validation scenarios for our methods.

To evaluate the performance of the evaluated trust models, we introduce the *mean loss*, defined as a difference between the real price pr_r and the bid price pr_b. In the graphs, it is aggregated per all contracts awarded in a single time step. As it is impossible to achieve the zero loss in our scenario, we introduce the optimal choice value, defining the optimal performance of the trust model.

To validate the model independence of the method presented (i.e. the fact that the restrictions placed on Θ modeling are not constraining), we have used two different trust models in our evaluation. The first model, denoted **RNT** in the graphs, we represent the trustfulness in each $\Theta(X|r_i)$ as a time-weighted average of the last N relevant observations. This means that we store N real values in each reference context r_i. The other model, denoted FNT, is a slightly simplified representation described in [10], where each $\Theta(X|r_i)$ is a triangular, asymmetrical fuzzy number.

In Figures 5, 6 and 8, we compare the performance of the trust models without the context representation (denoted **FNT** and **RNT**) with the same models enhanced with context representation, denoted **clusters RNT** and **clusters FNT** that are extended with the above mechanism.

4.3   Influence of Situation Modeling

In the first batch of experiments, we will investigate the influence of the context modeling. We will therefore compare several trust models with and without the context components in the scenarios with increasing level of situation influence on the provider performance. The changes in the performance are modeled by changes of the coefficients in the Eq. 18 and 19.

In the first scenario (Fig. 5), the performance of all the providers is flat over the whole space $\mathbb{C}$ - the outcome of the delegation/contracting is independent of the situation. We may note that the general methods perform slightly better, as their learning process is more efficient, but the differences remain minor.

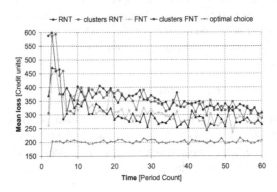

Fig. 5. Scenario with trustfulness independent of situation – all methods perform comparably

In the second scenario (Fig. 6), we have introduced a strong, but one dimensional situation dependence with one best provider per cargo type. We can see that in this case, context-based methods easily outperform the general trust and reach the optimum relatively fast. In Fig. 7, we can see that the depicted sub-market (defined by the contracts in one part of the context space) is rapidly dominated by the most trustful provider, while the others are restrained to the services where they perform better.

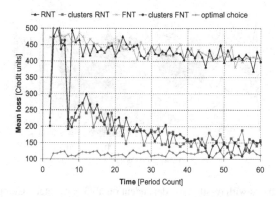

Fig. 6. Scenario with trustfulness dependent on the cargo type only

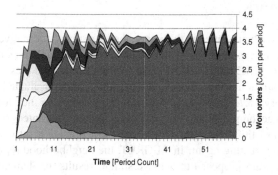

Fig. 7. Market shares example with trustfulness dependent on the cargo type only

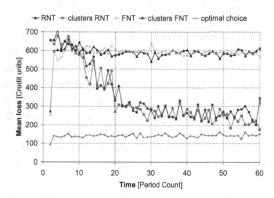

Fig. 8. Scenario with trustfulness dependent on all 3 parameters, with an adequate metrics

When we introduce a full 3D context dependence, we obtain the results shown in Fig. 8 and Fig. 9. We can see that the task is more difficult due to the increased dimensionality, but the context modeling solves the problem. The slower learning pace is clear when we compare the Fig. 9 with Fig. 7 – the market domination is slower.

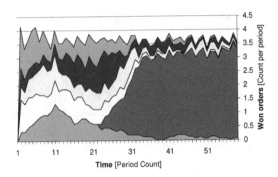

Fig. 9. Market shares with trustfulness dependent on all 3 parameters. Compare with Fig. 7.

5   Related Work

Besides the approach presented in this paper, there are many alternative approaches that attempt to relax the assumptions of the baseline trust mechanism as mentioned in the introduction. The FIRE model [9] presents two relevant techniques – role-based trust and certified reputation – that improve the reasoning about evaluated partners lacking interaction history. *Role-based trust* evaluates the partners on the basis of their social role, taking into account the endorsements of the others, guarantees or group membership. Evaluation of this trust component is performed using a set of rules applied on belief base of the trusting agent. In REGRET, the neighborhood reputation [8,6] and system reputation are proposed to achieve similar results, implemented by means of rules or aggregated reputation of the members of the evaluated agent's group.

Certified reputation, also proposed in [9], uses the ratings provided by the evaluated agent to assess its trustfulness. These ratings are created by the agents that interacted with the evaluated agent in the past, and are typically electronically signed. While this mechanism is valuable it is inherently biased – evaluated agent is free to chose only the ratings that are positive regarding its performance. Moreover, the electronic signature by itself proves only the *integrity* of the message. *Authenticity* of the message can be proved only if the evaluating agent can attach the signing key to a respectable (or at least known) information source. Otherwise, nothing can prevent agents from collusion, or creation of false evaluators and boosting of their ratings. Therefore, this technique is of limited interest if the identity of the agents is not proven.

The problem of the identity is also prominent in [15], where the agents (robots) are identified by their group membership, instead of the individual identity.

6   Conclusions and Future Work

In this paper, we have presented a mechanism that supports the use of existing trust models in the open, unmanaged and dynamic environments, where the agents can enter and leave the system at their own will, can change or disguise their identity easily and where the conditions under which the system operates are frequently changing.

We address the above particularities of the open environments by introduction of an intelligent mechanism that decouples the trustfulness from the statement of identity (e.g AID, address or name) and attaches the trustfulness to the important points in the Identity-Context metric space. It shall be noted that while their implementation is joint, the identity management and context representation parts of the module can be effectively split, using only half of the mechanism functionality. This is applicable fr example in the supply chain solutions, where the identity of the actors is certain and verifiable, but we still need to manage the context.

Besides the above-mentioned problem of securing open multiagent systems, trust models extended with the presented mechanism can be used in a completely novel contexts. In our current research [14], we experiment with use of such models for network intrusion detection and protection, in attempt to automatize the response to worm-type attacks. In this deployment, we don't evaluate individual hosts or applications, but rather the actual connections and their characteristics, taking into account the relevant statistics off the network traffic [16] as a context for the decision. The trustfulness of the connections is deduced from the status of the protected hosts and other alarms.

Such deployment changes completely the place of the trust model in the agent architecture – instead of being an accessory, protective module, it forms a core part of the business reasoning of the agent, under the assumption that rapid, automatic and intelligent response to attacks delivered by distributed security system is the best option how to address the worm threat. The underlying reasoning behind our approach is that the epidemics-like attacks in the virtual ecosystem shall be addressed by the approaches that are inspired by the real immune systems.

Acknowledgment

This material is based upon work supported by the European Research Office of the US Army under Contract No. N62558-07-C-0001. Any opinions, findings and conclusions or recommendations expressed in this material are those of the author(s) and do not necessarily reflect the views of the European Research Office of the US Army. Also supported by Czech Ministry of Education grants 1M0567 and 6840770038.

References

1. Marsh, S.: Formalising trust as a computational concept (1994)
2. Falcone, R., Castelfranchi, C.: Social trust: a cognitive approach, pp. 55–90 (2001)
3. Sabater, J., Sierra, C.: Review on computational trust and reputation models. Artif. Intell. Rev. 24, 33–60 (2005)
4. Ramchurn, S.D., Huynh, D., Jennings, N.R.: Trust in multiagent systems. The Knowledge Engineering Review 19 (2004)
5. Stallings, W.: Cryptography and network security: principles and practice, 2nd edn. Prentice-Hall, Inc., Englewood Cliffs (1999)
6. Sabater, J., Sierra, C.: Reputation and social network analysis in multi-agent systems. In: Alonso, E., Kudenko, D., Kazakov, D. (eds.) Adaptive Agents and Multi-Agent Systems. LNCS (LNAI), vol. 2636, pp. 475–482. Springer, Heidelberg (2003)

7. Pechoucek, M., Rehak, M., Marik, V.: Expectations and deployment of agent technology in manufacturing and defence: case studies. In: Pechoucek, M., Steiner, D., Thompson, S.G. (eds.) AAMAS Industrial Applications, pp. 100–106. ACM Press, New York (2005)

8. Sabater, J., Sierra, C.: Social regret, a reputation model based on social relations. SIGecom Exch. 3, 44–56 (2002)

9. Huynh, T.D., Jennings, N.R., Shadbolt, N.R.: An integrated trust and reputation model for open multi-agent systems. Journal of Autonomous Agents and Multi-Agent Systems 13, 119–154 (2006)

10. Rehák, M., Foltýn, L., Pĕchouček, M., Benda, P.: Trust model for open ubiquitous agent systems. In: Intelligent Agent Technology, 2005 IEEE/WIC/ACM International Conference. Number PR2416 in IEEE (2005)

11. Josang, A., Gray, E., Kinateder, M.: Simplification and analysis of transitive trust networks. Web Intelligence and Agent Systems 4, 139–162 (2006)

12. Yu, B., Singh, M.P.: Detecting deception in reputation management. In: AAMAS '03, pp. 73–80. ACM Press, New York (2003)

13. Duda, R.O., Hart, P.E., Stork, D.G.: Pattern Classification, 2nd edn., p. 654. John Wiley & Sons, New York (2001)

14. Rehák, M., Tožička, J, Pĕchouček, M., prokopová, M., Foltýn, L.: Autonomous protection mechanism for joint networks in coalition operations. In: Knowledge Systems for Coalition Operations 2007, Proceedings of KIMAS'07 (2007)

15. Birk, A.: Boosting cooperation by evolving trust. Applied Artificial Intelligence 14, 769–784 (2000)

16. Ertoz, L., Eilertson, E., Lazarevic, A., Tan, P.N., Kumar, V., Srivastava, J., Dokas, P.: Minds - minnesota intrusion detection system. In: Next Generation Data Mining, MIT Press, Cambridge (2004)

Learning Initial Trust Among Interacting Agents

Achim Rettinger[1], Matthias Nickles[1], and Volker Tresp[2]

[1] AI/Cognition Group, Technical University of Munich,
D-85748 Garching bei München, Germany
{achim.rettinger,matthias.nickles}@cs.tum.edu
[2] Corporate Technology, Siemens AG,
Information and Communications, Munich, Germany
volker.tresp@siemens.com

Abstract. Trust learning is a crucial aspect of information exchange, negotiation, and any other kind of social interaction among autonomous agents in open systems. But most current probabilistic models for computational trust learning lack the ability to take context into account when trying to predict future behavior of interacting agents. Moreover, they are not able to transfer knowledge gained in a specific context to a related context. Humans, by contrast, have proven to be especially skilled in perceiving traits like trustworthiness in such so-called *initial trust situations*. The same restriction applies to most multiagent learning problems. In complex scenarios most algorithms do not scale well to large state-spaces and need numerous interactions to learn. We argue that trust related scenarios are best represented in a system of relations to capture semantic knowledge. Following recent work on nonparametric Bayesian models we propose a flexible and context sensitive way to model and learn multidimensional trust values which is particularly well suited to establish trust among strangers without prior relationship. To evaluate our approach we extend a multiagent framework by allowing agents to break an agreed interaction outcome retrospectively. The results suggest that the inherent ability to discover clusters and relationships between clusters that are best supported by the data allows to make predictions about future behavior of agents especially when initial trust is involved.

Keywords: Trust in Multiagent Systems, Information Agents, Agent Negotiation, Initial Trust, Relational Learning.

1 Introduction

The assessment of trust values is getting increasingly important in distributed information systems since contemporary developments such as the Semantic Web, Service Oriented Architecture, Information Markets, Social Software, Pervasive and Ubiquitous Computing and Grid Computing are targeted mainly at open and dynamic systems with interacting autonomous entities. Such entities possibly show a highly contingent behavior, and it is often not feasible to implement effective mechanisms to enforce socially fair behavior as pursued in mechanism

M. Klusch et al. (Eds.): CIA 2007, LNAI 4676, pp. 313–327, 2007.

design or preference aggregation. Although computational trust has been focussed by research in Artificial Intelligence for several years (for an overview see [1]), current approaches still lack certain features of human trustability assessment which we consider to be of high importance for the computational determination of trust values in open systems. E.g., recent studies in psychology [2] have shown that people can robustly draw trait inferences like trustworthiness from the mere facial appearance of unknown people after a split second. Although seemingly neither the time span nor the available information allow to make a well-founded judgement, the derived trust (or distrust) provides after all a foundation for immediate decision making, and a significant reduction of social complexity especially under time pressure. Whereas the "quality" of such so-called *initial trust* (i.e., trusting someone without having accumulated enough experiences from relevant past behavior of the trustee) might be limited in the described scenario, this example shows that humans are able to estimate the trustability of others using information which are at a first glance unrelated to the derived expectation. (e.g., the facial appearance, or any contextual information in general). In contrast, the vast majority of approaches to empirical trust value learning in Artificial Intelligence lack this ability, as these approaches strongly rest on well-defined past experiences with the trustee, from which it is directly concluded that the trustee will behave in the future as he did in the past, regardless of the concrete context (cf. Section 6 for related work). These approaches come to their limits in cases where the trustor could not make such experiences and thus has to rely on "second order" information such as the context of the respective encounter instead. In order to make such initial trust computationally feasible, we not only need to relate trust values to a specific context, but we also need to provide a mechanism in order to take over contextualized trust to a new, possibly somewhat different context.

In particular, the general requirements that we wish to meet are:

Context sensitivity and trust transfer: Contextual information that might be related to the trust decision to be made needs to be incorporated. This shall include attributes of the person one needs to trust, attributes of the external circumstances under which the trust decision is made, and actions and promises the person has given to seek one's confidence. Furthermore, specific trust values gained in a certain context need to be *transferrable* to new, unknown "trigger" situations.

Multi-dimensionality: Most trust models assign a single trust value per agent. This ignores the fact that human trust decisions are made in relation to a whole spectrum of aspects (e.g., what a person is likely to do, such as the expected outcome of some information trading, even in the same context. For instance a certain information supplier agent might be trustworthy in terms of delivery date, but not in terms of information quality (e.g., precision, topicality, credibility...). Combining several trust related measures as in our approach is considerably much more flexible. In contrast, most existing approaches to trust still relate trust to "whole persons" only instead of their contextualized behavior.

At this, we focus on *interaction-trust* (i.e., (dis-)trust formed by agents during the course of an interaction regarding their opponents' behavior) in order to tailor our model to the specifics of the probably most relevant application field for empirical trustability assessment.

The remainder of this work is organized as follows: The next two Sections describes the basic scenario underlying our approach. Section 4 introduces our model for relational learning of initial trust, and Section 5 explores the general capabilities of our model with example data. Section 6 presents an application of initial trust learning in the context of simulated social interaction in order to provide a concrete evaluation of our approach. Section 7 discusses related work, and Section 8 outlines future research directions and concludes.

2 Modeling Interactions

Our scenario can be based on one of the most general frameworks for learning interactions in multiagent systems namely general-sum stochastic games (see [3]). A stochastic game can be represented as a tuple (A, C, Ac, R, T)[1]. A is the set of agents, C is the set of *stage games* (sometimes denoted as *states*), Ac is the set of actions available to each agent, R is the immediate reward function and T is a stochastic transition function, specifying the probability of the next stage game to be played.

It is in the nature of trust that we are dealing with incomplete and partially observable information. We neither assume the knowledge of the reward function R of the opponent nor their current state C. In fully observable games with perfect monitoring, incentives to betray can be estimated and trust becomes irrelevant because agents can be punished effectively [4]. Furthermore trust decisions require general sum games where joined gains can be exploited. Both zero-sum (e.g., [5]) and common-payoff (e.g., [6]) games are not relevant because either there are no joint gains or the agents' interests do not conflict.

Building on that formal setting our goal is to predict trust values O^e associated with the expectation of the next actions Ac given agent A and state C. We neither are trying to learn a strategy or policy nor are we interested in finding equilibria or proofing convergence. But we make contributions on how to scale MAL to more complex scenarios and show how an opponent model can be learned efficiently:

Predicting the next action of an opponent is an essential part of any model-based approaches to MAL [7]. The best-known instance of a model-based approach is fictitious play [8] where the opponent is assumed to be playing a stationary strategy. The opponent's past actions are observed, a mixed strategy is calculated according to the frequency of each action and then the best response is played, accordingly. This technique does not scale well to a large state-space $|C|$ as we experienced in our second experiment (Section 6): The same stage game is on average not observed before 400 interactions. Thus, this kind of naive

---

[1] Our notation differs slightly from the commonly used ones, where A denotes actions and S states. Our notation should become clear in the next section.

approach does not allow to make an informed decision before 400 interactions and is obviously not suited for initial trust scenarios.

In our approach we make use of two techniques to face this issue. First, we allow to model any context related to the next trust-decision in a rich relational representation. This includes non-binding arrangements among agents also known as "cheap talk" [4] which take place before the actual interaction O^e is carried out and which are denoted as O^p. Second, we make use of techniques from the mature field of *Transfer Learning* [9] to reuse knowledge from previous interactions for potentially unknown future actions.

3   Modeling Interaction-Trust

The basic precondition for the emergence of trust are entities and social interactions between those entities. Hence, we chose a scenario that is interaction-centered as seen from the perspective of one agent who needs to trust (trustor) in someone/something (trustee). As usual in agent trust scenarios, (dis-)trust is related to the expected occurrence of some promised outcome (e.g., the communication of correct and precise information as negotiated before with some information trading agent, or the delivery of some other kind of product at the agreed price). The basic interaction-trust scenario then consists of:

1. A set of agents A (trustees) that are willing to interact with the trustor, each characterized by a set of observable attributes Att^A. An agent can be considered as a person or more general any instance that can be trusted, like an information source, a company, a brand, or an authority.
2. A set of external conditions or state C with corresponding attributes Att^C. An apparent condition would be the type of service provided by the trustee, for instance a specific merchandize or an information supply in case of information trading agents. Moreover this implies all external facts comprising this particular state like the trustor's own resources or the current market value of the merchandize in question.
3. A relation $interacts(a, c)$ with a set of relationship attributes Att^O capturing all negotiable interaction issues depending on a specific agent $a \in A$ and specific conditions $c \in C$. In general those attributes can be directly manipulated by the interacting agents and separated into two different sets:
   (a) Promised outcome O^p: Attributes Att^{O^p} of this set are (in general) observable before the trust-act is carried out.

       A typical attribute of this category is for example the price for the merchandize or the scope of the services offered, such as the amount and precision of information in case of a negotiation among agents regarding the delivery of information. A promised outcome $o^p \in O^p$ is an assignment of values to the corresponding attribute vector Att^{O^p}, which can be negotiated by the trustor and trustee. In game theory this kind of non-binding negotiations among agents before the actual interaction takes place is known as "cheap talk" [4].

(b) Effective outcome O^e: The set of attributes Att^{O^e} are not observable until the trust-act has been carried out. Those attributes act as a feedback or judgment for the trustee in respect to his expectations. Att^{O^e} can be thought of as quality aspects of the merchandize, like the delivery time. From a decision theoretic point of view those attributes are the objectives or interests of the trustor and need to be optimized in a multi-criteria optimization problem. From a MAL perspective Att^{O^e} depends on the actions Ac carried out by the opponent.

This way of modeling interaction-trust scenarios allows us to capture almost any context relevant for trust-based decision making.

Our goal is to learn the value function $o^p \to o^e$ that allows to predict o^e from a given o^p offered by agent a under external conditions c. Moreover, it might be possible to calculate the utility of the trustor for a given o^e. Hence, the ultimate objective is to find the utility function $o^p \to [0, 1]$. If this function is known the trustor knows what assignment to O^p he should try to achieve (e.g., in a negotiation) to maximize its payoff / reward.

4   Infinite Relational Trust Model

Relational models are an obvious formalization of requirements arising from the relational nature of entities in social, biological, physical and many other fields. The benefits of the relational model for multiagent learning include amongst others:

1. Relational models exhibit flexible and sophisticated modeling capabilities. For typical interaction scenarios in the real world there are more than two types of players and the number of players in each single interaction is flexible or unknown beforehand. In this case propositional models can hardly describe the data and its behavior. Relational models represent correlations both, between the features of an entity and between features of related entities.
2. By transforming the data into a flat representation, also known as propositionalization, the structural information can get lost. Moreover, there is no standard procedure for propositionalization. In [10] manifold propositionalization approaches and their disadvantages are analyzed. In general propositionalization causes high computational costs , since the complexity increases exponentially in parameters, attributes and relations. Another problem of the propositionalization process is the generation of too many irrelevant features. The low quality of propositional features becomes increasingly problematic in more complex interaction models. In contrast, relational models do not need this preprocessing phase at all and do not generate redundant information.
3. Relational models and its dynamics can be intuitively visualized by graphical model. This makes complex models more comprehensible and easier to analyze.

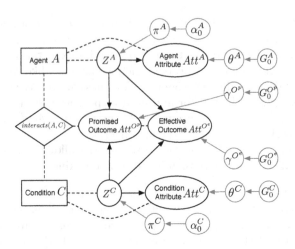

Fig. 1. Infinite Relational Trust Model

Recently [11] and [12] independently introduced infinite relational models (IRM), which express interactions via a potentially infinite number of hidden variables associated with entities instead of difficult structure learning in PRM. Those latent variables play a key role and comprise the inherent structure of the data. As additional features of entities they can improve the accuracy of the learned model.

Considering the properties of nonparametric probabilistic relational models our approach intuitively follows from the interaction scenario that we want to model:

Entity and relationship classes are the two basic building blocks of such a model. In our scenario agents A and states C are both modeled as entities with a corresponding relation $interacts(A, C)$. As a visual representation we make use of the DAPER model (cf. [13]). Figure 1 illustrates the DAPER model for the interaction scenario. Entity classes A and C are depicted as rectangles and the relationship class as a rhombus. Actual evidence Att is modeled as attribute classes of entities and relationships (oval). Local distribution classes denoting the parameters and hyperparameters of the probability distributions are shown in small gray circles. The direction of arrows shows the statistical dependency or the sampling process.

The most distinctive feature of our approach are the hidden variables Z (circles). They provide clustering capabilities of entities with a *potentially infinite* number of clusters. Assuming for every entity class one hidden variable our model contains Z^A and Z^C with r^a and r^c clusters, respectively.

4.1   Sampling and Inference

Given the model the essential goal is to infer the conditional distribution

$$P(Z^C, Z^C | Att^A, Att^C, Att^O)$$

of cluster assignments Z^A and Z^C given evidence about relationship attributes Att^{O^p} and Att^{O^e}. This posterior distribution can be formed from the generative models by

$$P(Att_1^{O^e}, ..., Att_k^{O^e}, z_1^A, ..., z_m^A, z_1^C, ..., z_n^C) =$$

$$\prod_{l=1}^{k} P(Att_l^{O^e} | z_1^A, ..., z_m^A, z_1^C, ..., z_n^C) \prod_{i=1}^{m} P(z_i^A) \prod_{j=1}^{n} P(z_j^C)$$

where we have k actions carried out by m agents and n states. Similar formulas hold for the joint distributions of $P(Att^{O^p}, Z^A, Z^C)$, $P(Att^A, Z^A)$ and $P(Att^C, Z^C)$.

The prior on cluster assignments π^A and π^C is a Dirichlet distribution with hyperparameters α_0^A and α_0^C respectively, where sampling of both Z^A and Z^C can be induced by a Chinese Restaurant Process: $Z|\alpha_0 \sim CRP(\alpha_0)$. By the use of the Chinese Restaurant Process the number of clusters can be determined in an unsupervised fashion. Entities are assigned to (potentially new) clusters corresponding to the size of the existing clusters. Entity attributes Att^A and Att^C are samples from multinomial distributions with parameters $\theta^A \sim G_0^A = Dir(\cdot|\beta^A)$, $\theta^C \sim G_0^C = Dir(\cdot|\beta^C)$) and are generated for each cluster in Z^A and Z^C. The same applies for the relationship attributes Att^{O^p} and Att^{O^e} which can be induced by a multinomial distribution with parameters $\gamma^{O^p} \sim G_0^{O^p}$, and $\gamma^{O^e} \sim G_0^{O^e}$. However, γ needs to be generated for every combination of entity attribute clusters, resulting in $r^A \times r^C$ parameter vectors.

Now inference can be carried out based on Gibbs sampling by estimating $P(Z|Att) \propto P(Att|Z)P(Z)$. For instance the probability of agent i being assigned to cluster k is proportional to $P(z_i^A = k|Z_{j \neq i}^A, Att_i^A, \theta^A, \gamma^{O^p}, \gamma^{O^e}, Z^C) \propto N_k P(Att_i^A | \theta_k^A, \gamma_{k,*}^{O^p}, \gamma_{k,*}^{O^e})$ where N_k is the number of agents already assigned to cluster k and $\gamma_{k,*}$ notes the relation parameters of agent cluster k and all state clusters. Finally, standard statistical parameter estimation techniques can be used for estimating $\gamma_{k^A, k^C}^{O^e}$ from given cluster assignments.

The parameters α_0 and β affect the number of clusters and the certainty of priors and can be tuned. However, we experienced that results were quite robust without extensive tuning. Moreover, our experiments are rather targeted at feasibility than absolute performance, so we fixed $\alpha_0^A, \alpha_0^C = 10$ and $\beta^A, \beta^C = 20$ in all our experiments.

For a detailed description of the algorithm we refer to [11]. We extended the algorithm, as just described, to enable the handling of more than one relationship attribute. Using an arbitrary number of relationships is essential to enable a rich representation of the interaction context and multidimensional trust values.

4.2   Implications

The ultimate goal of the model is to group entities into clusters Z. A good set of partitions allows to predict the value of attributes Att^{O^p} and Att^{O^e} by their mere cluster assignments. Hereby, our model assumes that each entity belongs

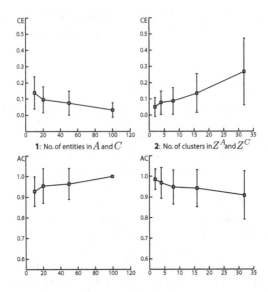

Fig. 2. Results on experiment 1: Synthetic data, setup 1 and 2. Top row graphs show the classification error metric (CE), subjacent graphs show the related accuracy (AC).

to exactly one cluster. It simultaneously discovers clusters and the relationships in-between clusters that are best supported by the data, ignoring irrelevant attributes.

Although the value of attributes is determined entirely by the cluster assignment of associated entities, there is no need for direct dependencies between attributes or extensive structural learning. The cluster assessment of a entity is influenced by all corresponding attributes and cluster assessments of related entities. This way information can propagate through the whole network while the infinite hidden variables Z act as "hubs". As shown in [11] this allows for a collaborative filtering effect. Cross-attribute and cross-entity dependencies can be learned which is not possible with a "flat" propositional approach that assumes independent and identical distributed (i.i.d.) data.

At the same time the number of clusters needs *not* to be fixed in advance. Thus, it can be guaranteed that the representational power is unrestricted.

5   Experiment 1: Synthetic Data

To explore the learning and modeling capabilities of our IRTM we generated synthetic data and evaluated its ability to find clusters in this data. For this purpose we constructed an interaction-trust scenario with the fixed number of 2 entity attributes per entity and 2 relationship attributes, one for O^p and one for O^e. The number of entities $|A|$ and $|C|$ was prespecified but varied in different runs, as well as the underlying clustersize r^a and r^c for Z^a and Z^c. Each entity was randomly assigned to a cluster and its attributes were sampled from a multinomial distribution with 4 possible outcomes and parameter vector θ each. θ in turn, was once randomly generated for each cluster. Accordingly, $r^a \times r^c$ Bernoulli-parameters γ for relationship attribute att^{O^p} and att^{O^e} were constructed.

In Figure 2 and 3 two different error metrics measuring the performance of IRTM averaged over 10 runs are shown. The top row graphs visualize the classification error metric (CE) for clusterings while the bottom row depicts the accuracy (AC) of classifying att^{O^e} correctly. Both are supplemented by a 95% confidence interval. CE reflects the correspondences between the estimated cluster labels and the underlying cluster labels measuring the difference of both (cf. [14]). A value of 0 relates to an exact match, 1 to maximum difference. In this experiment AC is a binary classification task and denotes the ratio of classifying att^{O^e} correctly. Results are averaged over both hidden variables Z^a and Z^c.

5.1 Evaluation

We considered three different experimental setups:

1. We analyzed the performance for different numbers of entities with fixed cluster sizes $r^a = r^c = 4$. The performance shown in Figure 2-1 expectedly suffers for small numbers of entities $|A| = |C| < 20$. Nonetheless, this result suggests that the IRTM is quite robust even with few training samples. This makes it especially interesting for initial trust problems as discussed in the next Section.
2. Correctly recovering different cluster sizes r^a and r^c while the number of entities was fixed to $|A| = |C| = 50$ was the goal of setup 2. In Figure 2-2 we see that the IRTM underestimates the cluster sizes if $r^a = r^c > 16$. This suggests that the number of combinations in such a simple scenario is not enough and entities from different clusters tend to become alike. Still, the AC is almost perfect. Besides that the number of entites per cluster ($|A|/r^a$ and $|C|/r^c$, respectively) gets so small that not all clusters are represented in the training set.
3. Finally, missing and noisy data sets were used in two different ways for training:
   (a) Half of the relationship attribute O^e data was omitted while missing values for O^p was varied. The variance of all measures in figure 3-3a increases with the increase of missing values. Still, the AC is good although cluster correspondences deviate. This clearly shows that dependencies across relationship-attributes have a significant effect on the performance and can be exploited by IRTM. As mentioned before, standard techniques working with a "flat" vector-based attribute-value representations cannot use such information. In contrast IRTM can propagate information through the network.
   (b) First, evidence for O^e was partially omitted. The AC in Figure 3-3b expectedly drops because less training samples of the effective outcome that is to be predicted are available. Still, clustering abilities are hardly affected because other attributes can replace the missing information.

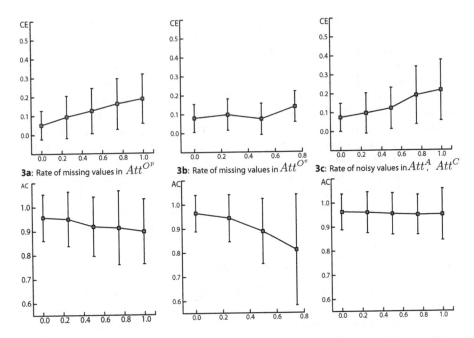

Fig. 3. Results on experiment 1: Synthetic data, setup 3a-c. Top row graphs show the classification error metric (CE), subjacent graphs show the related accuracy (AC).

(c) Second, in order to measure the influence of the entity attributes we added noise to Att^A and Att^C. With the used parameter settings IRTM did obviously (see Figure 3-3c) not suffer in predicting AC. However the ability to infer the correct clusters was slightly hindered.

6   Experiment 2: Negotiation Data

Finding an agreement amongst a group of conflicting interests is one of the core issues of distributed Artificial Intelligence. For instance auctions, information markets, preference aggregation and judgement aggregation, game theory and automated negotiations are all research areas that deal with those kind of problems. However most of the approaches neglect the fact that finding the best agreeable solution is not sufficient if the execution of the negotiated outcome can not be enforced by the interaction mechanism. Especially in open systems where agents can enter and leave or change their identity at will, initial trust plays an important role in this regard. The purpose of the IRTM is to make predictions about Att^{O^e} which can be utilized by the agent to adjust its negotiation strategy or trading decisions.

In order to investigate this issue we extended the implementation of a multiagent negotiation framework by an additional trading step. As defined before, let O^p be the promised outcome the agents are negotiating over (e.g., the punctual delivery of information some information agents requested or offered to supply, respectively). This outcome is without loss of generality specified by a set of discrete attributes Att^{O^p}. Now given an assignment of values O^p that two agents have agreed on and promised to fulfill the agents enter an additional trading step where each of them is free to change the assignments of values related to their commitments. Doing so, the agent can decide whether to stick to a bargain or break it at will. One interaction round in this negotiation framework consists of three phases:

1. Negotiation: A strategy that calculates a possible outcome O^p both parties can agree on (e.g., an exchange of goods).
2. Trading: The decision made by every agent whether to stick to a bargain or break it (possibly only partially). The outcomes regarding the agent's obligations are executed according to the agent's decision.
3. Evaluation: The agents can review the effective actions Att^{O^e} of the opponent by observing the received goods and draw conclusions for future interactions

This procedure is repeated over a specified number of rounds with different types of agents.

6.1   Evaluation

Four different agent types were used as opponents in the negotiation game. Every round the negotiation outcome O^p and the effective outcome O^e was recorded. To keep it simple, all agent types follow the same static negotiation strategy but each one acts differently in the trading phase. The agent denoted *Greedy* always maximizes its utility regardless of O^p. *Sneaky*-agent only deviates from O^p if it increases its utility by a large margin, while *Honest*-agent always sticks to O^p. Finally, the agent named *Unstable* deviates only slightly from O^p (by giving away $+/-1$ amount) if its utility is increased hereby.

As the negotiation strategies were the same for all agent types the negotiation outcome was modeled as attributes of C and not of O^p. Furthermore no specific attributes for A were available except for its identity. Besides the raw negotiation outcome and the state of the own resources, features describing the risk of losing utility and the chance of gaining utility were extracted and added to Att^C. Att^{O^e} was set to be the binary classification task whether the utility would increase less than negotiated or not. This way about 120 interactions were carried out per agent type containing a total of 165 different negotiation outcomes alltogether. $1/3$ of the data was randomly withhold and used for testing. Again, all results are averaged over 10 runs.

The predictive performance was measured by calculating the area under the ROC curve (AUC). We compare the results of IRTM to two content based approaches, namely a support vector machine (SVM) using a PolyKernel and a Decision Tree (DecTree, ID3). The SVM and DecTree got an additional input by assigning each agent in A an unique ID number. This way the relational model did not have more information than a "flat" model. We also evaluated

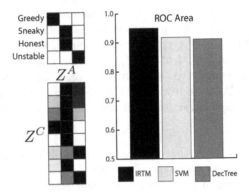

Fig. 4. Results on experiment 2: Negotiation data. Top left shows final clustering of agent types. Bottom left visualizes $P(O^e)$ for each pair of clusters (the darker the more probable). Bar graph shows AUC for classifying $P(O^e)$.

the clustering abilities by plotting the most frequent assignment of cluster by the IRTM.

In the top left of Figure 4 one can see that in the end the four agent types (rows) were clustered into three groups in Z^A (columns). Interestingly, the assignment of *Sneaky*- and *Honest*-agent to the same cluster suggests that it is a good strategy to act reliable and provide confidence most of the time in order to convince an opponent of the own trustworthiness. But if it is clear that the gain is really worth it one should betray the opponent's trust.

The rectangles in the lower left corner of Figure 4 visualize $P(O^e|Z^A, Z^C)$. From the 165 different negotiation outcomes and external conditions 8 clusters emerged in Z^C. Each row indicates one condition-cluster Z_i^C, each column an agent-cluster Z_i^A. Thus, each element stands for $P(O^e)$ given the cluster assignments. Brighter rectangles indicate a lower probability for a utility increase as negotiated. As expected the first column (*Greedy*-agent cluster) is on average brighter than the third column (*Unstable*-agent cluster) which in turn is brighter than the middle column (*Sneaky*- and *Honest*-agent).

The overall performance, shown in the bar graph on the right of Figure 4, demonstrates that IRTM has a slightly better performance in classifying $P(O^e)$ than the SVM and the DecTree.

The inherent clustering of the IRTM suggests that it is especially well suited for initial trust situation when unknown but related agents and conditions are observed. Actually, entities can be correctly assigned to a cluster without having seen a single effective outcome related to this entity just by their attributes. To check this assumption we gathered data from interactions with another *Unstable* type agent and evaluated the performance for different numbers of training samples. In the top graph of Figure 5 the AUC is plotted for different numbers of training samples. Especially for a small sample size ≤ 10 the performance of IRTM is clearly better than those of the content based approach.

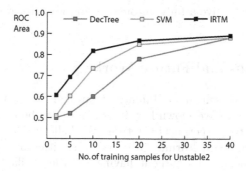

Fig. 5. Results on experiment 2: Negotiation data. Graph shows AUC for different number of samples in an initial trust setup.

7 Related Work

As already pointed out, connecting trust to the trusted agent alone without considering contextual and different aspects (dimensions) of trust is not sufficient in many scenarios. Whereas much research on trust concede the importance of context information, most of them do not actually use such information for the calculation of trust degrees in a general and automatic way [15]. To our knowledge using contextual information for initial trust assessment and the transfer of trust between contexts is completely new.

Regarding its dimensionality, most work represent trust as a single discrete or continuous variable associated with one specific agent. Modeling trust in multiple dimensions is only considered by a few elaborate approaches such as [16]. We leave it to the actual scenario how trust needs to be modeled in this respect. In principle, IRTM can handle an arbitrary number of trust variables, each associated with one aspect of the trustor's expectations and represented with any probability distribution needed.

Analogously, we argue that a fine grained modeling of relations between agents and their environment is essential to capture the essence of trust, especially in initial trust situations. There exist a few approaches that can take relationships into account when modeling trust. But in most of this research such relationships are either only considered as reputation or recommendations [17], or as interactions between a group of agents (e.g., [18]). The manifold different kinds of relations that exist between two agents in a specific situational context are not modeled in detail. In addition, most learning techniques are "improvised" for one specific scenario only.

Assessing initial trust values for unknown agents based on pre-specified membership to a certain group has been addressed by [19]. A group-based reputation architecture is proposed here where new agents are assessed according to their pre-specified membership to a certain group of agents. Likewise, the *TRAVOS-C* system proposed by [15] includes rudimentary ideas from hierarchical Bayes modeling by assigning parameter distributions to groups of agents but doesn't come to the point to give a fully automated and intuitive way of how to build clusters.

8   Conclusions and Future Work

In this work, we presented an Infinite Relational Trust Model (IRTM) for interaction-trust and have shown how interactions can be modeled and learned in theory and in two experimental setups. We believe that our model will be especially useful for trust learning in initial trust situations, where the trustor interacts with other agents without having recorded sufficiently enough relevant past experiences in order to judge trustability using traditional methods. E.g., this would typically be the case in short-lived communities of practice, where information agents gather in a kind of ad-hoc manner in order to exchange knowledge, or in open information markets, where mutually more or less unknown information sellers and buyers interact with each other.

IRTM is more powerful and flexible in representing intial trust and fine grained contextual relations, adding a new level of semantics to trust learning. The experimental results suggest that IRTM shows a performance comparable to a "flat" feature-based machine learning approach if trained with independent and identical distributes (i.i.d.) data. We expect to see superior performance of IRTM if no i.i.d. assumption is made and cross-attribute and cross-entity dependencies can be exploited. However, our second experiment shows that in initial trust situations the IRTM can outperform a traditional feature-based approach even if the i.i.d. assumption is made. Besides that, IRTM can handle missing attribute values and enables a clustering analysis which is not possible in existing feature-based trust learning approaches.

Furthermore the experiments deliver preliminary insights into the effect of different strategies on trustworthiness in negotiations. We plan on continuing our work in this direction. Furthermore we intend to address issues like reputation and recommendations which should naturally fit in our relational model.

References

1. Ramchurn, S.D., Hunyh, D., Jennings, N.R.: Trust in multi-agent systems. Knowledge Engineering Review (2004)
2. Willis, J., Todorov, A.: First impressions: Making up your mind after a 100-ms exposure to a face. Psychological Science 17(7), 592–598 (2006)
3. Hu, J., Wellman, M.P.: Multiagent reinforcement learning: Theoretical framework and an algorithm. In: Shavlik, J.W. (ed.) ICML, pp. 242–250. Morgan Kaufmann, San Francisco (1998)
4. Murray, C., Gordon, G.: Multi-robot negotiation: Approximating the set of subgame perfect equilibria in general sum stochastic games. In: Schölkopf, B., Platt, J., Hoffman, T. (eds.) Advances in Neural Information Processing Systems 19, MIT Press, Cambridge, MA (2007)
5. Littman, M.L.: Markov games as a framework for multi-agent reinforcement learning. In: ICML, pp. 157–163 (1994)

6. Wang, X., Sandholm, T.: Reinforcement learning to play an optimal nash equilibrium in team markov games. In: Becker, S., Obermayer, S.T. (eds.) Advances in Neural Information Processing Systems 15, pp. 1571–1578. MIT Press, Cambridge, MA (2003)

7. Shoham, Y., Powers, R., Grenager, T.: If multi-agent learning is the answer, what is the question? Technical Report 122247000000001156, UCLA Department of Economics (2006)

8. Brown, G.: Iterative solution of games by fictitious play. In: Activity Analysis of Production and Allocation, John Wiley and Sons, New York (1951)

9. Caruana, R.: Multitask learning. Mach. Learn. 28(1), 41–75 (1997)

10. Krogel, M.A.: On Propositionalization for Knowledge Discovery in Relational Databases. PhD thesis, die Fakultät für Informatik der Otto-von-Guericke-Universität Magdeburg (2005)

11. Xu, Z., Tresp, V., Yu, K., Kriegel, H.-P.: Infinite hidden relational models. In: Proceedings of the 22nd International Conference on Uncertainity in Artificial Intelligence (UAI 2006) (2006)

12. Kemp, C., Tenenbaum, J.B., Griffiths, T.L., Yamada, T., Ueda, N.: Learning systems of concepts with an infinite relational model. In: AAAI (2006)

13. Heckerman, D., Meek, C., Koller, D.: Probabilistic models for relational data. Technical Report MSR-TR-2004-30, Microsoft Research (2004)

14. Meilă, M.: Comparing clusterings: an axiomatic view. In: ICML '05: Proceedings of the 22nd international conference on Machine learning, pp. 577–584. ACM Press, New York, USA (2005)

15. Teacy, W.T.L.: Agent-Based Trust and Reputation in the Context of Inaccurate Information Sources. PhD thesis, Electronics and Computer Science, University of Southampton (2006)

16. Maximilien, E.M., Singh, M.P.: Agent-based trust model involving multiple qualities. In: Proceedings of the 4th International Joint Conference on Autonomous Agents and Multiagent Systems (AAMAS) (2005)

17. Sabater, J., Sierra, C.: REGRET: reputation in gregarious societies. In: Müller, J.P., Andre, E., Sen, S., Frasson, C. (eds.) Proceedings of the Fifth International Conference on Autonomous Agents, pp. 194–195. ACM Press, New York (2001)

18. Ashri, R., Ramchurn, S.D., Sabater, J., Luck, M., Jennings, N.R.: Trust evaluation through relationship analysis. In: AAMAS '05: Proceedings of the fourth international joint conference on Autonomous agents and multiagent systems, pp. 1005–1011. ACM Press, New York, USA (2005)

19. Sun, L., Jiao, L., Wang, Y., Cheng, S., Wang, W.: An adaptive group-based reputation system in peer-to-peer networks. In: Deng, X., Ye, Y. (eds.) WINE 2005. LNCS, vol. 3828, pp. 651–659. Springer, Heidelberg (2005)

A Probabilistic Framework for Decentralized Management of Trust and Quality*

Le-Hung Vu and Karl Aberer

School of Computer and Communication Sciences
Ecole Polytechnique Fédérale de Lausanne (EPFL)
CH-1015 Lausanne, Switzerland
{lehung.vu,karl.aberer}@epfl.ch

Abstract. In this paper, we propose a probabilistic framework targeting three important issues in the computation of quality and trust in decentralized systems. Specifically, our approach addresses the multi-dimensionality of quality and trust, taking into account credibility of the collected data sources for more reliable estimates, while also enabling the personalization of the computation. We use graphical models to represent peers' qualitative behaviors and exploit appropriate probabilistic learning and inference algorithms to evaluate their quality and trustworthiness based on related reports. Our implementation of the framework introduces the most typical quality models, uses the Expectation-Maximization algorithm to learn their parameters, and applies the Junction Tree algorithm to inference on them for the estimation of quality and trust. The experimental results validate the advantages of our approach: first, using an appropriate personalized quality model, our computational framework can produce good estimates, even with a sparse and incomplete recommendation data set; second, the output of our solution has well-defined semantics and useful meanings for many purposes; third, the framework is scalable in terms of performance, computation, and communication cost. Furthermore, our solution can be shown as a generalization or serve as the theoretical basis of many existing trust computational approaches.

1 Introduction

Quality and trust have become increasingly important factors in both our social life and online commerce environments. In many e-business scenarios where competitive providers offer various functionally equivalent services, quality is often the most decisive criterion that helps to build trust among participants in the system and influences their selection of most prospective partners. For example, between two file hosting service providers, a user would aim for the one allowing the storage of larger files in longer periods, offering higher download and upload speed under better pricing conditions. Similarly, there are several other types of services that are highly differentiated

* The work presented in this paper was (partly) supported by the Swiss National Science Foundation as part of the project: "Computational Reputation Mechanisms for Enabling Peer-to-Peer Commerce in Decentralized Networks", contract No. 205121-105287. We also thank Radu Jurca, Wojciech Galuba, Vasilios Darlagiannis and the reviewers for their useful comments on this paper.

M. Klusch et al. (Eds.): CIA 2007, LNAI 4676, pp. 328–342, 2007.

by their quality of service (QoS) features such as Internet TV/radio stations, online music stores or teleconferencing services. Therefore, appropriate mechanisms for accurate evaluation of service quality have become highly necessary.

In large scale decentralized systems with no central trusted authorities, past performance statistics are usually considered as indicative of future quality behaviors [19]. An autonomous agent (or peer) acting on behalf of its users often uses ratings or recommendations from other peers to estimate QoS and trustworthiness of its potential partners before involving into costly transactions. Such scenarios necessitate appropriate solutions to the following important issues:

The credibility of observation data: collected ratings from various sources can either be trustworthy or biased depending on the inherent behaviors and motivation of the ones sharing the feedback. Thus the understanding and evaluation of malicious reporting behaviors need to be taken into consideration adequately [5, 7, 11].

The multi-dimensionality of quality and trust: QoS and trust are multi-faceted, inter-dependent, and subject to many other factors. This issue further complicates our estimations of service quality and peer behaviors, because we have to take into account various dependencies among quality parameters and related factors, whose values can only be observed indirectly via (manipulated) ratings. The incompleteness of the observation data set is an additional issue to be considered.

The subjectivity of quality and trust: the evaluation and perception of quality and trust information strongly depend on the view or execution context of the evaluating user. For example, according to its personalized preferences, a peer may estimate the trustworthiness of another based on certain quality dimensions of the latter. Generally peers have different interpretations on the meaning of a trust value, therefore the recommendation and/or propagation of such formulated quantity in large scale systems are inappropriate. A more suitable approach is to enable a peer to do its own evaluation of well-defined quality attributes of the others based on collected experience, from which to personally evaluate the trustworthiness of its prospective partners.

The three above issues are *inter-related* and even *inseparable* for many application scenarios. In this perspective, we believe that more generalized results are still missing as most research efforts are either ad-hoc in nature or only focus on specialized aspects. Many trust computational models in the literature either rely on various heuristics [1, 25] or produce trust values with ambiguous meanings based on the transitivity of trust relationships [13, 26]. Other probabilistic-based trust evaluation approaches, e.g., [2, 6, 15, 17, 24], are still of limited applications since they do not appropriately take into account the effects of contextual factors, the multi-dimensionality of trust, quality, and the relationships among participating agents.

In this paper, we propose a probabilistic framework to support the subjective computation and personalized use of trust and quality information. Our approach addresses the above questions adequately by offering several advantages:

- **Support personalized modeling of quality and trust:** using graphical models, our framework enables a peer to describe the quality and behaviors of the others flexibly according to personalized preferences. To the best of our knowledge, this work is among the first ones taking into account natural dependencies among QoS parameters, related contextual factors, and inherent behaviors of peers for the reliable estimation of trust and quality.

- **Offer a general decentralized approach to evaluate quality and trust:** our solution facilitates the use of various probabilistic learning and inference algorithms to evaluate quality and behaviors of peers. In our current implementation, we introduce the Expectation-Maximization (EM) and the Junction Tree (JTA) algorithms enabling a peer to estimate quality and behaviors of others locally and independently. These algorithms produce reliable estimates without using ad hoc heuristics as in many other approaches. As a side effect, our framework can be seen as the *generalization* or *theoretical basis* of many representative trust computational models in the literature, e.g., [25, 6, 24, 17, 2, 23, 22, 15].

- **Provide informative results with well-defined semantics:** since we use probabilistic graphical models to represent quality and trust, our computation produces outputs with clear and useful meanings, for example, it evaluates the probability that a peer is honest when reporting or the probability that a file hosting provider offers its service with high download speed for a client with a certain type of Internet connection and willing to pay a certain price. This output can be used by the evaluating peer in many ways: (1) for its subjective trust evaluation, i.e., compute its trust on another according to various preferences, given the estimated values of different quality dimensions of the latter; (2) to choose the most appropriate service for execution given many functionally equivalent ones offered by the different peers in the system; and (3) to decide to go for interactions with other peers knowing their reporting behaviors, e.g., sharing and asking for experiences from them.

- **Have good performance and scalability:** our implementation yields good performance even with high level of malicious rating users in the system. It is also applicable and works well in case the collected multi-dimensional recommendations data set is incomplete. Additionally, our analysis and empirical experimentations have shown that the implemented framework is *scalable* in terms of performance, computation and communication cost.

In the next section we will describe our general framework for the personalized modeling and evaluation of trust and quality. Section 3 presents our implementation of this framework for the learning and inference of quality and trust. Section 4 presents our analytical and experimental results clarifying the advantages of our proposed approach. Section 5 summarizes the related work and Section 6 concludes the paper.

2   General Framework Architecture

Our proposed framework (Fig. 1) has the following components to support the computation and personalized use of trust and quality information:

Modeling: since trust and quality are multi-faceted and subject to the personalized view of a user, the modeling of trust and quality is of high importance. Such modeling enables the personalization and reusability of our computational approach in many different application scenarios, given that the user can obtain necessary domain knowledge to

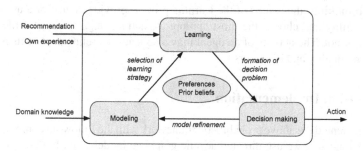

Fig. 1. The probabilistic quality and trust computational framework

formalize their problems appropriately. We propose the use of probabilistic graphical models, e.g., Bayesian networks, for the following important reasons:

- **Reality and generality:** such an approach models the reality elegantly: on one hand, the nature of quality and trust is probabilistic and dependent on various environmental settings; on the other hand, those dependencies can be easily obtained in most application domains, e.g., as causal relationships among the factors, which increase the reusability of our solution.

- **Personalized support:** user preferences and prior beliefs can be easily incorporated into a model, e.g., via the introduction of observed variables representing the behavior of trusted friends and hidden variables representing the quality and trustworthiness of unknown partners.

- **Tool availability:** algorithms to learn and inference on graphical models are widely available for our implementation and analysis of the obtained results.

Let us take the file hosting scenario as an example. With certain helps from the user, a peer can identify most relevant QoS properties of such service (download, upload speed, maximal number of concurrent downloads, etc.) as well as related environmental factors (Internet connection speed, subscription price) to build an appropriate quality/trust model. These variables and dependencies among them can be easily obtained as causal relationships among corresponding factors in the domain, or even from the service description published by the provider peer.

Learning: given the quality model of the target peer, the evaluating peer selects a suitable learning algorithm to estimate the quality and trustworthiness of its target based on collected recommendations and own experience. Specifically, recommendations are ratings by other peers/users on certain quality dimensions of the peer being evaluated, e.g., the download and upload speed of its service. The selection of a learning algorithm usually depends on the constructed model, properties of the observation data set, and user preferences. For example, given an incomplete observation data set on a model with many hidden variables, the EM algorithm is a potential candidate. If the model is less complex and if the user prefer a full posterior distribution of the quality and behavior of the target peer, a Bayesian-based learning approach is more suitable. Subsequently, the computation of quality and trustworthiness of the target peer can be done by (probabilistically) inferencing on this learned model.

Decision making: the results of the learning process provide inputs for a user to evaluate his utility and choose the most appropriate actions, e.g., to select a file hosting provider or not. The actions of an agent may also lead to a refinement of its model on the others' quality and behaviors.

3   Solution Implementation

We implemented the above probabilistic framework with the following components. For the modeling step, we use directed acyclic graphical models, i.e., Bayesian networks, since we believe that they are sufficiently expressive to represent the causal relationships among various factors in our scenario, e.g., quality parameters, contextual variables, and peer behaviors. We apply the EM algorithm to learn the conditional probabilities of the constructed models and use the Junction Tree Algorithm (JTA) [10] to inference on them for the estimation of peer quality and behaviors. The trust-based decision-making step is not included in the current implementation, yet it can be incorporated easily.

3.1   Personalized Quality and Behavior Modeling

Depending on its personalized preferences and prior beliefs on the outside world, an evaluating peer P_0 can model its view on the quality of a target peer P accordingly. We identify the following most typical models that a peer may use for various scenarios:

- **The basic QoS model** (Fig. 2a) is built on the assumption of the evaluating peer that all recommending peers are honest in giving feedback. Thus, those values reported by a peer are also its actual observations on the service quality. The nodes e_l, $1 \leq l \leq t$ represent those execution environmental (or contextual) factors influencing those quality attributes q_i, $1 \leq i \leq m$ of the service provided by peer P. This model is most useful in case the judging peer only collects rating from its trusted friends in the network.
- **The simplified QoS model** (Fig. 2b) extends the basic one to the case where P_0 believes that the collected recommendations are possibly biased. The node b denotes the reporting behavior of all recommenders, and v_i denotes their reported values on a quality attribute q_i. This model only considers the overall distribution of all peers' behaviors and is useful in case there are few reports on quality of the target peer P for the trust and quality evaluation step.
- **The extended QoS model** (Fig. 2c) extends the previous model to the case where P_0 believes that each reporter P_j may exhibit a different behavior b_j, $1 \leq j \leq n$. The variable v_{ij} denotes a rating value by a peer P_j on a quality attribute q_i of P. This extensive model is used if the judging peer P_0 also wants to evaluate the individual trustworthiness of each peer P_j in terms of giving recommendations.

A rounded square surrounding a node in Fig. 2 represents similar variables with the same dependencies with the others. Shaded nodes are variables whose values are observable and blank nodes are hidden (latent) variables to be learned from the collected reports. The numbers t, n, m are respectively the number of contextual factors, reporting peers, and QoS attributes that will be used in later analysis. For brevity reason, from now on we simply use the notation x to denote any node in a QoS model, π_x to

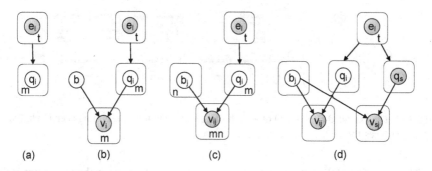

Fig. 2. a. The basic QoS model of a service/peer; b. The simplified QoS model; c. The extended QoS model; d. Example personalized QoS model with some observed QoS properties q_s

denote the list of all parent nodes of x, and y to denote a state value of x. The set of conditional probability table (CPT) entries $p(x|\pi_x)$ of a QoS model is then referred to as its set of parameters θ henceforth. Using such probabilistic graphical models enable us to compute the probabilistic quality of a peer only by analyzing and traversing on corresponding graphs, thus widening the applicability of our approach.

A peer can include other prior beliefs and preferences to build its personalized QoS model based on the three typical models above. For example, let us assume that P_0 can collect correct observations on certain quality parameters q_s of the target peer P, either via its own experience or via the ratings of its trusted friends in the network. In that case P_0 can model those nodes q_s in the extended QoS model of P as visible to reduce the number of latent variables (Fig. 2d). There may also be dependencies among the different quality attributes q_i and among the nodes e_l, which are not shown for the clarity of presentation. The dynamic of various quality factors can be taken into account by considering time as a contextual variable in the models. Furthermore, it is possible to model the collusive groups in P_j by introducing further latent variables in the simplified and extended QoS models, but the learning will then be much more complicated. Consequently, in this paper we limit our analysis to the case where the evaluating peer collects random feedback from the network such that the behaviors of P_j can be assumed as approximately indepedent to each other.

A concrete modeling example for a file hosing service in our previous scenario is given in Fig. 3a. In this figure, M, D, U respectively represent the maximal number of concurrent downloads, the download and upload speed offered by the file hosting provider being evaluated. The variables P, I denote the subscription price and the Internet connection type of an observing peer having used this file hosting service. Actually P, I define the context in which this observing peer perceives those quality attributes M, D, U. Other variables M_1, D_1, U_1 are correspondingly ratings of a specific peer P_1 on different quality dimensions M, D, U, assuming that the behavior of P_1 is b_1. The state space of each quality variable in a QoS graphical model can be modeled as binary (for good or bad quality conformance) or as discrete values representing different ranges of values of a QoS variable or an environmental factor, depending on the nature of the node and the viewpoint of the judging peer (see Fig. 3b).

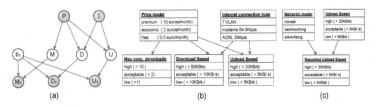

Fig. 3. a. The extended QoS model of a file hosting service with one reporting peer P_1; b. Example state spaces for quality and environmental nodes

The modeling of the behavior of a reporting peer also depends on the viewpoint of the evaluating peer. For example, in Fig. 3a, a reasonable classification of behaviors is the following. At the time of reporting, a peer can exhibit one of three possible behaviors: *honest*, *badmouthing*, or *advertising*. An honest peer experiencing an upload speed $U = y$ reports the same value in its feedback, thus $p(U_1 = y|b_1 = honest, U = y) = 1.0$. An advertising peer mainly increases its observed quality values and a badmouthing one decreases the quality it perceives, therefore $p(U_1 = low|b_1 = badmouthing, U = y) = 1.0$, and $p(U_1 = high|b_1 = advertising, U = y) = 1.0$. Note that via this modeling, a reporting peer can alternatively appear as honest, badmouthing, or advertising with certain probability, e.g., the variable b_1 may follow different probabilistic distributions. Consequently, certain parameters of a quality model can be defined according to prior beliefs of the evaluating peer or based on domain-dependent constraints as part of this modeling step. Other unknown parameters of a QoS model shall be learned from the observation data set as will be explained later on.

3.2 Subjective Learning of Quality and Trust

Learning the model parameters. The first learning step in our framework is the estimation of unknown parameters θ of the QoS model of the peer P being evaluated. Given a set of reports on different QoS properties of P, we formulate an observation data set $R_p = \{r^\mu, 1 \leq \mu \leq N\}$ on the Bayesian network of the model. Each r^μ has a visible part v^μ including values of those variables appearing in related reports and a hidden part h^μ representing unknown values. Variables in h^μ may include hidden nodes and unreported values on QoS variables or contextual factors.

Given an observation data set R_p on a model, there are two well-known approaches for estimating the model parameters θ:

- **Frequentist-based approach:** methods of this category estimate the model parameters such that they approximately maximize the likelihood of the observation data set R_p, using different techniques like Maximum Likelihood Estimation, gradient methods, or Expectation-Maximization (EM) algorithm, etc. In this solution class, the EM algorithm appears to be a potential candidate in our case since it works well on a general QoS model whose log likelihood function is too complex to be optimized directly. In addition, the EM algorithm can deal with incomplete data and is shown to converge relatively rapidly to a (local) maximum of the log likelihood. However, the disadvantages of an EM-based approach are its possibility to reach to a sub-optimal result and its sensibility to the sparseness of observation data set.

- **Bayesian method:** this approach considers parameters of the model as additional unobserved variables and computes a full posterior distribution over all nodes conditional upon observed data. The summation (or integration) over the unobserved variables then gives us estimated posterior distributions of the model parameters, which are more informative than the results of frequentist-based methods. Unfortunately, this approach is expensive and may lead to large and intractable Bayesian networks, especially if the original QoS model is complex and the observation data set is incomplete.

In this paper, we study the use of the EM algorithm in our framework to learn the quality and behavior of peers encoded as unknown variables in a QoS model. This application of an EM algorithm in our implementation is mainly due to its genericity and promising performance. The use of other learning methods in our framework, e.g., an approximate Bayesian learning algorithm, to compare with an EM-based approach is part of our future work and thus beyond the scope of this paper.

An outline of the EM parameter learning on a general QoS graphical model with discrete variables, i.e., any model in Fig. 2 or a derivation, is given in Algorithm 1. This algorithm is run locally and independently by a peer in the system to evaluate the quality and behaviors of other peers, after the evaluating peer has constructed an appropriate QoS model of its targets.

Algorithm 1. $QoSGraphicalModelEM(R_p = \{r^\mu = \langle h^\mu, v^\mu \rangle, 1 \leq \mu \leq N\})$

1: Initialize the unknown parameters $p(x = y | \pi_x = z)$;
2: /* y, z are possible state values of variables x and π_x */
3: **repeat**
4:   **for** each observation case r^μ **do**
5:     **for** each node x **do**
6:       Compute $p(x = y, \pi_x = z | v^\mu, \theta)$ and $p(\pi_x = z | v^\mu, \theta)$;
7:     **end for**
8:   **end for**
9:   Compute $E_{xyz} = \sum_\mu p(x = y, \pi_x = z | v^\mu, \theta)$ and $E_{xz} = \sum_\mu p(\pi_x = z | v^\mu, \theta)$;
10:   **for** each node x **do**
11:     $p(x = y | \pi_x = z) = E_{xyz}/E_{xz}$;
12:   **end for**
13:   Compute the log likelihood $LL(\theta) = \sum_\mu log p(v^\mu | \theta)$ with current θ;
14: **until** convergence in $LL(\theta)$ or after a maximal number of iterations;

The first line of Algorithm 1 initializes the unknown model parameters, i.e., the unknown conditional probability table (CPT) entries of the corresponding Bayesian network. Depending on its preferences, the evaluating peer initializes the conditional probability of each quality node accordingly, e.g., either randomly or as its prior beliefs on the service advertisement of the evaluated peer. Specifically, in case of the extended and the simplified models, the evaluating peer can define the corresponding CPT entries $p(b_j)$ (or $p(b)$) appropriately depending on its prior beliefs on the trustworthiness of P_j (or on the overall hostility of the environment).

Lines $4 - 9$ of Algorithm 1 implement the Expectation step of the EM algorithm, where we compute the expected counts E_{xyz} and E_{xz} of the events $(x = y, \pi_x = z)$ and $(\pi_x = z)$, given the observation data set R_p and current parameters θ. We can use any exact or approximate probabilistic inference algorithm in this step to compute the posterior probabilities $p(x = y, \pi_x = z|v^\mu, \theta)$ and $p(\pi_x = z|v^\mu, \theta)$. Our current implementation uses the Junction Tree Algorithm (JTA) [10] since it produces exact results, works for all QoS models, and is still scalable in our scenario (see Section 4). The Maximization step of the EM algorithm is implemented in lines $10 - 12$ of Algorithm 1, therein we update the model parameters $p(x|\pi_x)$ such that they maximize the observation data's likelihood, assuming that the expected counts computed in lines $4 - 9$ are correct. The two Expectation and Maximization steps are iterated till the convergence of the log likelihood of the observation data, which gives us an estimation of all model parameters $p(x = y|\pi_x = z)$.

Subjective Computation of Quality and Trust. After the parameter learning step, the evaluating peer can compute the probability that a target provides a quality level $q_i = y$ under its own execution context $\phi_{q_i}^*$ as follows: first, it sets the value of each variable in the target's QoS model with the values corresponding to its environmental setting $\phi_{q_i}^*$; second, it runs the JTA inference algorithm on the learned model to compute the conditional distribution $P(q_i = y|\phi_{q_i}^*)$. For brevity reason, we refer the interested readers to [10] for a detailed description of the JTA algorithm and to the long version of this paper [21] for an example use of this algorithm to inference on a QoS model.

Additionally, the evaluating peer can estimate its personalized trust on a potential partner depending on its preferences and objectives conveniently. For example, the trust level that the evaluating peer has on a potential partner can be modeled as the joint distribution of some quality attributes of the latter, which can also be computed using the JTA algorithm. More generally, the jugding peer P_0 can formulate any decision problem based on the learned QoS model according to its personalized objectives and use inference techniques to derive a best decision appropriately. All these probabilistic learning and inference procedures can be done automatically by analyzing the QoS model, which falicitates the application of our framework in a wide range of scenarios.

4   Experimental Results

Computation cost. Since the most expensive step in our computational framework is the learning of parameters of a QoS model, we focus on the evaluation of its scalability. We can show that the complexity of each EM iteration of Algorithm 1 for the extended model in Fig. 2c is $O(Kn^2)$, where K is the average number of reports by each of n reporting peers P_j. For the simplified and the basic models, the computation cost of each EM iteration is $O(nK)$ (see [21], Section 4.1). Given the fact that the EM algorithm converges fast in practice (one can also bound the number of EM iterations appropriately) and we consider a limited number of recent reports K per peer, our EM learning algorithm (Algorithm 1) is scalable in terms of computation cost.

Communication cost. Suppose that we use a logarithmic routing overlay, e.g., Chord [20], for searching the feedback in the network, then the communication cost required by the evaluating peer P_0 for its learning process is $O(Knlog|V|)$, where $|V|$ denotes the number of all peers in the system [21]. This cost mainly depends on the number of direct reporting peers n. Thus one important question is to find out the number of direct reporting peers n that P_0 needs to ask for feedback and the number of reports K required for an acceptably accurate estimation of the quality of a peer P.

Experimental setting. We implement the presented computational framework using the BNT toolbox [16] and test it extensively under various conditions. We perform our experiments with the aforementioned file hosting scenario to study the accuracy of the estimated values under various practical settings: (1) the evaluating peer uses different ways to model the service quality of its target, e.g., using the basic, simplified, and extended QoS models; (2) there are different reporting behaviors; and (3) the recommendation set is incomplete and sparse. On one hand, we set the conditional probabilities of the QoS parameters of the service according to some (small) deviations from advertised QoS values to simulate possible cheating behavior of a provider peer. On the other hand, we generate the observation data set as samples on the visible variables of the QoS model according to various reporting behaviors of the service users, i.e., the reported quality values have different probabilities. We further hide a fraction of reported values to create an incomplete recommendation data set. The EM algorithm is initialized as follows: the conditional probability of each QoS parameter is set randomly and behaviors of all reporting peers are unknown. From the model with parameters learned by the EM algorithm, we estimate the distributions of the reporting behaviors of the peers and the QoS features of the service. We deliberately use the above setting since it is of the worst case scenario where the evaluating peer has no information of any other peer and no trusted friends. The experiments in other cases where the evaluating peer has different preferences and certain trusted friends are subject to our future work. However, we believe that we can obtain even better results in those cases, since we will have more evidence to learn on models with fewer hidden variables.

The accuracy of the approach is measured as the normalized square root error of the quality and behavior estimations. The experimental settings and corresponding results are summarized as in Table 1, where f_c, f_u represent the percentage of cheating and uncertain reporting peers with varied behaviors. The column f_i denotes the percentage of missing data in the observation data set. Each experiment type consists of 20 experiments, each of which is averaged over 20 run times. These experimentations give us the following conclusions:

- **Performance scalability:** generally the number of direct reporting peers n does not have any major influence in the performance of the algorithm (Fig. 4), but the number of observation cases does (Fig. 5). A peer only needs to get the reports from a certain number of other peers (chosen randomly) in the system to reduce the total learning time and communication cost without sacrificing much in the accuracy. Practically, in some other experiments, we only choose a set of $n = 15$ reporting peers randomly chosen in the network to reduce the total running time of the learning algorithm.

Table 1. Summary of experimental settings and results

Experiment goal	n	K	$f_c(\%)$	$f_u(\%)$	$f_i(\%)$	Results
Performance vs. number of raters	1 to 100	5	45	5	0	Fig. 4
Performance vs. number of reports	15	1 to 20	45	5	0	Fig. 5
Performance vs. effective attacks	15	5	0 to 95	0	0	Fig. 6
Performance vs. incomplete data	15	5	50	5	0 to 95	Fig. 7

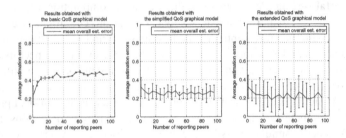

Fig. 4. Estimation errors vs. the number of raters

- **The importance of modeling** (Fig. 4, 5, 6, 7): the simplified QoS model yields best estimation results given only a few ($K = 3$ to 5) reports per peer (Fig. 5). Using the extended QoS model requires more reports to be collected for the learning step, otherwise its results are less accurate and more uncertain. This is reasonable since the extended model is much more complex and has many hidden variables to be learned. The result of the basic QoS model is the least accurate and more vulnerable to malicious reports, since in this case the modeling peer trusts all raters completely. This observation confirms the necessity of an approach which enables the subjective modeling and evaluation of quality and trust. Depending on the availability of the reports from others, its prior beliefs on other peers and personalized preferences, a peer can choose an appropriate model for its personal estimations of quality and behaviors of prospective partners.

- **The effectiveness of the EM algorithm:** the performance of the EM algorithm is tested extensively with respect to the following dimensions: (1) the average number of reports K submitted by each peer P_j; (2) the incompleteness of the observation data set; (3) and the percentage of honest reporting peers among P_js. Given enough observation data, generally using the EM learning algorithm on the simplified and the extended models can yield respectively good or acceptable estimation, even with a reasonably high number of malicious users (Fig. 6) and with an incomplete observation data set (Fig. 7).

Note that the above obtained performance is more useful in reality since many applications are not sensitive to the exact quality or behavior distributions but only their relative values for making decisions.

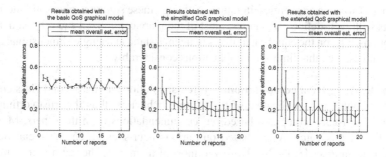

Fig. 5. Estimation errors vs. the number of reports per peer

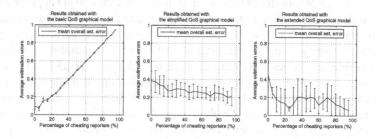

Fig. 6. Estimation errors vs. cheating behaviors

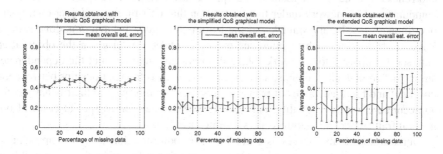

Fig. 7. Estimation errors vs. percentage of missing data

5 Related Work

Our work is mostly related to current research efforts in building reputation-based trust management systems, which can be mainly classified into two categories [4, 7, 11, 19]:

Social-control trust management systems: these approaches use reputation information as a sanctioning tool to enforce trustful behaviors in the system, e.g., applications are designed s.t. participants are motivated to fully cooperate for maximizing their utilities. Representative contributions in this field include [3, 4, 12, 14], as well as a huge number of work in economics literature. Although these solutions give solid theoretical results with well-defined semantics, most of them are centralized and rely on many assumptions, e.g., on the equal capabilities of agents, static and single-dimensional quality of resources, etc., that limit their applications considerably.

Social-learning trust computational systems: solutions in this second class consider agents' reputation as indicative of their future behaviors, thus the trustworthiness of an agent is usually computed based on recommendations from the others. Many approaches in this category are based on various heuristics [1, 25] or social network analysis techniques [8, 13, 26]. However, most of them are either ad hoc in nature [1], have high implementation overhead [13] or do not produce rigorous results with well-defined semantics [25, 26]. Other probabilistic-based trust evaluation approaches, such as [2,6,15,17,23,24] are still of limited applications since they do not appropriately take into considerations the effects of contextual factors and preferences of users, the multi-dimensionality of trust and quality, and the relationships among participating agents.

Our work belongs to the second category and can be shown as the generalization of many existing approaches of this class, e.g., [2,6,9,15,17,18,22,23,24]. For example, [6] is a special case of our solution if we suppose the reports and quality to be one-dimensional and apply the EM algorithm on the simplified QoS model to find those model parameters best fitting to the observation data set. [25] evaluates the quality and trust value of a peer given its ratings by other peers and credibility of individual raters, therefore it is similar to the learning on an extended QoS model in our framework. Yet, our solution is well-supported by the probabilistic learning and inferencing techniques and relies on no ad hoc heuristics. The approaches in [2, 15, 17, 24] are similar to our computational framework under the assumption that user ratings and peer quality are one-dimensional, binary-valued, and the learning of peers' behaviors is done using a Bayesian-based approach to inference on the extended model given the observation data set. Comparing to our work, [22, 23] do not take into account the dependencies among different quality dimensions of the peer as well as the relationships between the observed and reported values of peers in the system with respect to their behaviors for an accurate estimate of quality and trust. [9, 18] can be seen as special cases of our general framework and therefore are subsumed by ours.

Our framework also serves as a good starting point to address some important issues in decentralized reputation-based trust management. For example, it is possible to model the social relationships among agents and agents' communities. As a result, the learning and reasoning on the model can give us an estimation of the quality and trust of a target peer considering the relationship among peers in their communities. Such approach would combine the advantages of those trust computational solutions based on social network analysis and on probabilistic estimation techniques, as suggested by [7].

6   Conclusion

We have proposed a general probabilistic framework to support the personalized computation of quality and trust in decentralized systems, which can be shown as the generalization of many related approaches. Our first implementation of the framework using Bayesian network modeling, the EM and JTA algorithms have been shown to be scalable in terms of performance, run-time, and communication cost. Additionally, it exhibits several advantages: it works well given few and incomplete feedback data from the service users; it considers the dependencies among QoS attributes and various related factors to produce more accurate estimations; the computed results have well-defined semantics and are useful for many application purposes. Currently, we are interested in the modeling of peer communities and in studying the performance of our approach under the sharing of the estimated quality and behaviors among members of such social groups. We also plan to use the implemented framework to study the emergence of trustful behaviors in decentralized networks where peers use different models and learning algorithms to evaluate each other. Another issue is the unavailability of real-life experimental reports constrains us in using only synthetic data in our experimentation. Therefore, an important part of our future work is to find credible information sources to empirically test our approach in various real-life scenarios.

References

[1] Aberer, K., Despotovic, Z.: Managing trust in a peer-2-peer information system. In: CIKM '01: Proceedings of the tenth international conference on Information and knowledge management, pp. 310–317. ACM Press, New York (2001)

[2] Buchegger, S., Le Boudec, J.-Y.: A robust reputation system for P2P and mobile ad-hoc networks. In: Proceedings of the Second Workshop on the Economics of Peer-to-Peer Systems (2004)

[3] Dellarocas, C.: The digitization of word-of-mouth: Promise and challenges of online reputation systems. Management Science 49(10), 1407–1424 (2003)

[4] Dellarocas, C.: Reputation mechanism design in online trading environments with pure moral hazard. Information Systems Research 16(2), 209–230 (2005)

[5] Dellarocas, C.: Reputation Mechanisms. In: Hendershott, T. (ed.) Handbook on Economics and Information Systems, Elsevier Publishing, Amsterdam, Netherlands (2005)

[6] Despotovic, Z., Aberer, K.: A probabilistic approach to predict peers performance in P2P networks. In: Klusch, M., Ossowski, S., Kashyap, V., Unland, R. (eds.) CIA 2004. LNCS (LNAI), vol. 3191, Springer, Heidelberg (2004)

[7] Despotovic, Z., Aberer, K.: P2P reputation management: Probabilistic estimation vs. Journal of Computer Networks, Special Issue on Management in Peer-to-Peer Systems: Trust, Reputation and Security 50(4), 485–500 (2006)

[8] Guha, R., Kumar, R., Raghavan, P., Tomkins, A.: Propagation of trust and distrust. In: International World Wide Web Conference (WWW2004) (2004)

[9] Hsu, J.Y.-J., Lin, K.-J., Chang, T.-H., Ho, C.-J., Huang, H.-S., Jih, W.-R.: Parameter learning of personalized trust models in broker-based distributed trust management. Information Systems Frontiers 8(4), 321–333 (2006)

[10] Huang, C., Darwiche, A.: Inference in belief networks: A procedural guide. International Journal of Approximate Reasoning 15(3), 225–263 (1996)

[11] Jøsang, A., Ismail, R., Boyd, C.: A survey of trust and reputation systems for online service provision. Decision Support Systems (2005)

[12] Jurca, R., Faltings, B.: An incentive compatible reputation mechanism. In: Proceedings of the IEEE Conference on E-Commerce, Newport Beach, CA, USA (June 24-27), (2003)

[13] Kamvar, S.D., Schlosser, M.T., Molina, H.G.: The EigenTrust algorithm for reputation management in P2P networks. In: Proceedings of the Twelfth International World Wide Web Conference (2003)

[14] Miller, N., Resnick, P., Zeckhauser, R.: Eliciting informative feedback: The peer-prediction method. Management Science 51(9), 1359–1373 (2005)

[15] Mui, L., Mohtashemi, M., Halberstadt, A.: A computational model of trust and reputation. In: Proceedings of the 35th Hawaii International Conference on System Science (HICSS), Hawaii, USA (2002)

[16] Murphy, K.: Bayes Net Toolbox for Matlab (2006), http://bnt.sourceforge.net/

[17] Patel, J., Teacy, W.T.L., Jennings, N.R., Luck, M.: A probabilistic trust model for handling inaccurate reputation sources. In: Herrmann, P., Issarny, V., Shiu, S. (eds.) Proceedings of iTrust'05, France, pp. 193–209. Springer, Heidelberg (2005)

[18] Regan, K., Cohen, R., Poupart, P.: Bayesian reputation modeling in e-marketplaces sensitive to subjectivity, deception and change (2006)

[19] Resnick, P., Kuwabara, K., Zeckhauser, R., Friedman, E.: Reputation systems. Commun. ACM 43(12), 45–48 (2000)

[20] Stoica, I., Morris, R., Karger, D., Kaashoek, M.F., Balakrishnan, H.: Chord: A scalable peer-to-peer lookup service for internet applications. In: SIGCOMM '01, pp. 149–160. ACM Press, New York (2001)

[21] Vu, L.-H., Aberer, K.: Probabilistic estimation of peers quality and behaviors for subjective trust evaluation. Technical Report LSIR-REPORT-, -013, 2006 (2006), available at http://infoscience.epfl.ch/search.py?recid=89611

[22] Wang, Y., Cahill, V., Gray, E., Harris, C., Liao, L.: Bayesian network based trust management. In: Internation Conference in Autonomic and Trusted Computing, pp. 246–257 (2006)

[23] Wang, Y., Vassileva, J.: Bayesian network-based trust model. In: WI '03: Proceedings of the IEEE/WIC International Conference on Web Intelligence, p. 372. IEEE Computer Society Press, Los Alamitos (2003)

[24] Whitby, A., Jøsang, A., Indulska, J.: Filtering out unfair ratings in Bayesian reputation systems. The Icfain Journal of Management Research 4(2), 48–64 (2005)

[25] Xiong, L., Liu, L.: PeerTrust: Supporting reputation-based trust for peer-to-peer electronic communities. IEEE Trans. Knowl. Data Eng. 16(7), 843–857 (2004)

[26] Yu, B., Singh, M.P., Sycara, K.: Developing trust in large-scale peer-to-peer systems. In: Proceedings of First IEEE Symposium on Multi-Agent Security and Survivability, pp. 1–10. IEEE Computer Society Press, Los Alamitos (2004)

Formal Analysis of Trust Dynamics in
Human and Software Agent Experiments

Tibor Bosse[1], Catholijn M. Jonker[2], Jan Treur[1], and Dmytro Tykhonov[2]

[1] Department of Artificial Intelligence, Vrije Universiteit of Amsterdam,
1081HV, Amsterdam, The Netherlands
{tbosse,treur}@few.vu.nl
[2] Man-Machine Interaction Group, Delft University of Technology,
Mekelweg 4, Delft, The Netherlands
{catholijn,dmytro}@mmi.tudelft.nl

Abstract. Recognizing that trust states are mental states, this paper presents a formal analysis of the dynamics of trust in terms of the functional roles and representation relations for trust states. This formal analysis is done both in a logical framework and in a mathematical framework based on integral and differential equations. Furthermore, the paper presents formal specifications of a number of relevant dynamic properties of trust. The specifications provided were used to perform automated formal analysis of empirical and simulated data from two case studies, one involving two experiments with humans, and one involving simulation experiments in the context of an economic game.

Keywords: Trust, dynamics, intelligent agents, human experiments.

1 Introduction

In the literature, a variety of definitions can be found for the notion of trust. The common factor in these definitions is that trust is a complex issue relating to belief in honesty, faithfulness, competence, and reliability of the trusted system actors, see e.g., [5, 6, 7, 13, 14, 19, 22, 23, 27]. Furthermore, the definitions indicate that trust depends on the context in which interaction occurs or on the observer's point of view. The agent might for example trust that the statements made by another agent are true. Likewise, the agent might trust the commitment of another agent with respect to certain (joint) goals, or the agent might trust that another agent is capable of performing certain tasks.

In [21] trust is analysed referring to observations which in turn lead to expectations: 'observations that indicate that members of a system act according to and are secure in the expected futures constituted by the presence of each other for their symbolic representations.' In [8] it is agreed that observations are important for trust, and trust is defined as: 'trust is the outcome of observations leading to the belief that the actions of another may be relied upon, without explicit guarantee, to achieve a goal in a risky situation.' Elofson [8] notes that trust can be developed over time as the outcome of a series of confirming observations. The evolution of trust over time, also called the dynamics of trust, is addressed in this paper.

We conceive trust as an internal (mental) state property of an agent that may help him to conduct various kinds of complex behaviour. The cognitive concept trust

M. Klusch et al. (Eds.): CIA 2007, LNAI 4676, pp. 343–359, 2007.
© Springer-Verlag Berlin Heidelberg 2007

enables the agent to effectively cope with complex environments that are populated by self-interested agents. Trust is based on a number of factors, an important one being the agent's own experiences with the subject of trust; e.g., another agent. Each event that can influence the degree of trust is interpreted by the agent to be either a trust-negative experience or a trust-positive experience. If the event is interpreted to be a trust-negative experience, the agent will lose its trust to some degree, if it is interpreted to be trust-positive, the agent will gain trust to some degree. The degree to which trust is changed depends on the characteristics of the agent. Agents equipped with a concept of trust perform a form of continual verification and validation of the subjects of trust over time. For example, a car is trusted, based on a multitude of experiences with that specific car, and with other cars in general.

In this paper the dependence of trust on experiences obtained over time is the main focus, abstracting from other possible influences, which (as a form of idealisation) are assumed not present or constant. The dynamics of trust are formalised by sequences of mental states. The functional roles and the representation relations of trust states are formalised both in a logical and in a numerical mathematical way based on integral and differential equations. Properties of trust dynamics are formalised accordingly, incorporating logical and numerical aspects. Empirical and simulated data from two experiments with trust are formally analysed using an automated checker for such properties against traces [2]. The first experiment, called the scenario case study, studies dynamics of trust of humans when confronted with two different scenario's. The other experiment, called the Trust & Tracing case study, is an agent-based simulation of an economic game.

Section 2 sketches the preliminaries for modelling trust dynamics using both mathematical and logical means. To properly catch the dynamics of trust as a cognitive phenomenon, Section 3 describes trust states as mental states for which the functional or causal roles and representation relations can be analysed and formally specified in logical terms. In Section 4 a continuous approach to trust dynamics is presented showing how functional role and representation relation specifications for trust states can be defined within Dynamical Systems Theory [25]. Section 5 presents a number of formally specified properties of trust used to analyse the case studies. The scenario case study is presented in Section 6; its formal analysis, based on properties formalised in Section 5, is presented in Section 7. The Trust & Tracing case study is presented in Section 8 and analysed in Section 9, using the properties from Section 5. Section 10 is a discussion.

2 Preliminaries

In this paper, trust is considered a mental agent concept that depends on experiences. This is modelled by a mathematical function that relates sequences of experiences to trust representations: a *trust evolution* function. As a computational alternative the iterative *trust update* function relates a current trust representation and a current experience to the next trust representation. To obtain a formal mathematical framework, the following four sets are introduced. A partially ordered set EV of *experience values*, a linearly ordered time frame TIME with initial time point 0, the set ES of *experience sequences*, i.e., functions from TIME to EV, and a partially ordered set TV of *trust values*. Examples of such sets EV and TV are the (closed) interval of real numbers between -1 and 1, the set {-1, 0, 1}, or sets of qualitative labels, such as {very high, high, neutral, low, very low}. Within the sets EV and TV, subsets of positive and negative values are distinguished. A *trust evolution function* is a function te : ES x TIME → TV. A *trust update function* is a function tu : E x TV → TV. Throughout this paper we assume that the value of a trust evolution function te at time t only depends on the experiences in the past of t (*future independence*).

A logical formalisation is based on the language TTL [2]; this is a language in the class of reified predicate logic-based temporal languages as distinguished in [11, 12]. In TTL, *state atoms* are atoms over the state ontology Ont, such as experience(v) and trust(w), which express the experience value v from sort EV and trust value w from sort TV in a state, and for relations for these sorts such as <, pos and neg. A *state* is a truth assignment (of truth values true and false) to the set of (ground) state atoms; STATES(Ont) denotes the set of all states over state ontology Ont. A *trace* for state ontology Ont is a function γ: TIME → STATES(Ont); they form the set TRACES(Ont). The expression state(γ, t) |= p denotes that state property holds in the state of γ at time t. Here |= is an infix predicate of the language. Based on such atoms formulae can be formed using the predicate logic connectives, including quantifiers (e.g., over time and traces).

3 Trust States and Mental States

Following literature on Philosophy of Mind, such as [20], a mental state can be characterised in two manners:

- by the functional role it plays, defining the immediate predecessor and successor states in causal chains in which it is involved
- by its representation relations, defining how the mental state relates to other states more distant in locality and time

For each of these two aspects, trust states will be analysed by a simple example.

The Shop Example
Consider the following example, concerning agent A and a specific shop. The behaviour of agent A considered is as follows:

- agent A can go to the shop or avoid it
- when meeting somebody, agent A can tell that it is a bad shop or that it is a good shop

The following types of events determine the behaviour of agent A

negative events:

- an experience that a product bought in this shop was of bad quality
- somebody else tells A that it is a bad shop
- passing the shop, A observes that there are no customers in the shop

positive events:

- an experience that a product bought in this shop was of good quality
- somebody else tells A that it is a good shop
- passing the shop, A observes that there are customers in the shop

Assume for the sake of simplicity that only the last two experiences count for the behaviour of A, and that the past pattern a considered are histories in which the last two experiences are negative events. The future pattern b considered are the futures in which the agent avoids the shop and, when meeting somebody tells that it is a bad shop. It is assumed that, viewed from an external perspective, past pattern a leads to future pattern b.

Functional Role Specifications for a Trust State
Functional roles are described from a backward perspective (relating the trust state to states that lead to it) and a forward perspective (relating the trust state to states to which it leads). The trust state property 'very negative' is used as an illustration. For the forward perspective a relatively simple specification can be made; for example:

Functional role specification: forward
If the agent has very negative trust about the shop, then it will avoid the shop and when meeting somebody, (s)he will speak negatively about the shop.

state(γ, t) $\models$ trust(very_negative) $\Rightarrow$ $\exists$t1$\geq$ t [t$\leq$t1$\leq$t+d & $\forall$t2 [t1$\leq$t2$\leq$t1+d $\Rightarrow$
[state(γ, t2) $\models$ preparation_for(avoiding_shop) &
     state(γ, t2) $\models$ conditional_preparation_for(meets(A, B), speaks_bad_about_shop_to(A, B))]]

Here preparation for an action leads to performing the action (unless it is blocked), and conditional preparation for an action leads to preparation for the action as soon as the condition is observed. The backward perspective is inherently less simple, since trust is a type of mental state property that accumulates over longer time periods. This means that trust at a next point in time depends on present experiences but also on the present trust state. This gives it a recursive character. For example,

Functional role specification: backward
If the agent has negative or very negative trust about the shop, and it has a negative experience, then it will have very negative trust.

[state(γ, t) $\models$ trust(negative) $\vee$ trust(very_negative) & state(γ, t) $\models$ observes(e) & neg(e)] $\Rightarrow$
     $\exists$t1$\geq$t [t$\leq$t1$\leq$t+d & $\forall$t2 [t1$\leq$t2$\leq$t1+d $\Rightarrow$ state(γ, t1) $\models$ trust(very_negative)]

A numerical example with trust decay rate r is as follows:
If the agent has trust level w about the shop, and it has an experience of level v, then it will have trust of level $rw+(1-r)v$.

[state(γ, t) $\models$ trust(w) & state(γ, t) $\models$ observes(experience(v)) $\Rightarrow$
     $\exists$t1$\geq$t [t$\leq$t1$\leq$t+d & $\forall$t2 [t1$\leq$t2$\leq$t1+d $\Rightarrow$ state(γ, t1) $\models$ trust(rw+(1-r)v)]

Note that in mathematical terms a backward functional role specification corresponds to a trust update function, and can be used directly in a computational manner for simulation.

Representation Relations for a Trust State
Trust is an example of a mental state property that heavily relies on histories of experiences, as also is found in empirical work; e.g. [17]. By abstracting from these histories in the form of a trust state that accumulates the history of experiences, the future dynamics can be described on the basis of the present mental state in a simple manner. Here the past pattern can be characterised by a formula $\varphi(\gamma, t)$:

$\exists$t1<t2$\leq$t [state(γ, t1) $\models$ observes(e1) & state(γ, t2) $\models$ observes(e2) &
     neg(e1) & neg(e2) & $\forall$t3 [t1$\leq$t3$\leq$t $\Rightarrow$ $\neg$ $\exists$e3 [pos(e3) & state(γ, t3) $\models$ observes(e3)]]

Moreover, the future pattern can be characterised by a ψ(γ, t):

∃t1≥ t state(γ, t1) |= performs(avoiding_shop) & ∀ t2 ≥ t [state(γ, t2) |= observes(meeting(A, B)) ⇒
∃t3 ≥ t2 state(γ, t3) |= performs(speaking_bad_about_shop_to(A, B))]

This obtains the following temporal relational specifications for the representational content of the trust state

Representation relation: forward

∀γ, t [state(γ, t) |= trust(very_negative) ⇔ ψ(γ, t)]

Representation relation: backward

∀γ, t [φ(γ, t) ⇔ state(γ, t) |= trust(very_negative)]

A numerical example with trust decay rate r involves summation over time as follows:

If t0 is a time point and d a duration and t1 = t0+d and the agent has at time point t0 trust level w(t0) and at time points t between t0 and t1 it has experiences of level v(t), then at t1 it will have trust of level $w(t1) = r^d w(t0) + (1-r)\Sigma_{0\leq d1\leq d-1}\ r^{d1} v(t0+d-d1)$.

This property can be expressed in TTL as follows (here for any formula φ, the expression case(φ, v1, v2) indicates the value v1 if φ is true, and v2 otherwise):

[state(γ, t0) |= trust(w0) & $w1 = r^d w0 + (1-r)\Sigma_{k=0}^d \Sigma_{v:EV}$ case(state(γ, t0+k) |= experience(v), $r^{d-k}v$, 0)
⇒ state(γ, t0+d) |= trust(w1)]

In Section 4 an analysis of representation relations and functional role specifications for continuous trust values over continuous time in terms of the Dynamical Systems Theory [25] (based on integral and differential equations) will be presented.

4   A Continuous Approach

The notions functional role and representation relation can be analysed in a continuous form as well. This section considers a continuous mental state property for trust over continuous time. As an example, it is assumed that trust in the weather forecast depends on one's experiences based on continuously monitoring the actual weather and comparing the observed weather with the predicted weather. For some of the patterns of behaviour, decisions may depend on your trust in the weather forecast. In particular, the decision to take an umbrella depends not only on the weather forecast, but also on your trust in the weather forecast. For example, when the weather forecast is not bad, but

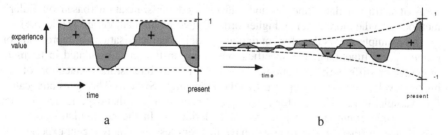

Fig. 1. Trust for continuous experiences (a) without decay and (b) with decay

your trust is low, you still will take an umbrella with you. It is assumed that for each point in time your experience with the weather forecast is a modelled by value (real number) between -1 (negative experience) and 1 (positive experience).

As a first approach (for reasons of presentation leaving decay aside for a moment) the accumulation of experiences in trust may be described by averaging the accumulation of the shaded area of the graph of experiences values over time, shown in Figure 1. So, a trust state represents a kind of average of the experiences over time. More specifically, the trust value is taken to be the real number indicating the shaded area divided by the length of the time interval, where the parts below the time axis count as negative. Within an overall trace γ the relation between trust value $tv_\gamma(t)$ at a certain point in time $t > 0$ and the experience history can be modelled by the following backward representation relation for the trust state, expressed by an integral over time until t of the experience value $ev_\gamma(t)$, i.e.,

$$tv_\gamma(t) = \int_0^t ev_\gamma(u) \, du \, / \, t$$

where for a trace γ the functions tv, ev are defined by:

$tv_\gamma(t) = v$ iff state(γ, t, internal) |= has_value(trust, v)
$ev_\gamma(t) = w$ iff state(γ, t, input) |= has_value(experience, w)

This shows the cumulative character of a (backward) representation relation. A functional role specification has a more local character. Such a local relationship can be modelled by a differential equation, which can be found from the integral relation

$$t \, tv_\gamma(t) = \int_0^t ev_\gamma(u) \, du$$

by differentiation and application of the product rule as follows:

$$d/dt \; t. \, tv_\gamma(t) = d/dt \int_0^t ev_\gamma(u) \, du$$
$$(d/dt \; t) . \, tv_\gamma(t) + t. \, d/dt \; tv_\gamma(t) = ev_\gamma(t)$$
$$tv_\gamma(t) + t. \, d/dt \; tv_\gamma(t) = ev_\gamma(t)$$

Therefore the following differential equation is obtained:

$$dtv_\gamma(t) \, / \, dt = [\, ev_\gamma(t) - tv_\gamma(t) \,] \, / \, t$$

In discretised form this provides:

$$tv_\gamma(t+ \Delta t) = tv_\gamma(t) + [[\, ev_\gamma(t) - tv_\gamma(t) \,] \, / \, t \,] \; \Delta t$$

This shows the backward functional role specification for a trust state, which has the form of a trust update function and can be used for simulation (based on Euler's method; also the more efficient higher-order Runge-Kutta methods are applicable).

The example shows that, for continuous models, characterisations of representation relations and functional role specifications show up that are formulated in terms of integrals or differential equations. This provides an interesting connection of the higher-level cognitive concept trust to the Dynamical Systems Theory as advocated, for example, in [25]. In the above example, it is not realistic that experiences very far back in time count the same as recent experiences. In the accumulation of trust, experiences further back in time will have to count less than more recent experiences, based on a kind of inflation rate, or increasing memory vagueness. Therefore a more realistic model is obtained if it is assumed that the graph of experience values against time is modified to fit it between two curves that are closer to zero further back in time. Trust then is the accumulation of the areas in the graph below, where the parts below the time axis count negative.

The limiting curves can be based, for example, on an exponential function e^{at} with $a > 0$ a real number related to the strength of the decay. The graph in Figure 1b depicts the resulting function $ev_\gamma(t)\,e^{at}$. Under these assumptions the representation relation between trust and experiences can be modelled by the integral

$$tv_\gamma(t) = \int_0^t ev_\gamma(u)\,e^{au}\,du \,.\, a / (e^{at} - 1)$$

where the factor $a / (e^{at} - 1)$ is a normalisation factor to normalise trust in the interval $[-1, 1]$. As before, a functional role specification for the trust state in the form of the following differential equation can be obtained:

$$dtv_\gamma(t)/dt \quad = [\ ev_\gamma(t) - tv_\gamma(t)\ ]\,.\, a\,e^{at} / (e^{at} - 1)$$
$$= [\ ev_\gamma(t) - tv_\gamma(t)\ ]\,.\, a\ / (1 - e^{-at})$$

5 Properties of Trust

In [18] a number of possible properties of trust evolution and trust update functions are defined in mathematical terms. For motivation and further explanation we refer to the reference mentioned. This paper contributes formalisation of the properties in logical terms. The following properties from [18] are considered (here $e, f \in ES$, $\gamma \in TRACES$, $s, t, u \in TIME$, $v \in EV$, $w \in TV$, and $e|\leq t$ denotes the restriction of sequence e to time points $\leq t$).

Future independence. Future independence expresses that trust only depends on past experiences, not on future experiences. This is a quite natural assumption that is assumed to hold for all trust evolution functions. In mathematical terms:

$e|\leq t = f|\leq t \ \& \ te(e, 0) = te(f, 0) \quad \Rightarrow \quad te(e, t) = te(f, t)$

This property can be expressed logically in TTL [2] as follows:

$\forall \gamma 1, \gamma 2, t\ [\ \forall w\ [\ state(\gamma 1, 0) \models trust(w) \Leftrightarrow state(\gamma 2, 0) \models trust(w)\ ]\ \&$
$\forall t1 \leq t\ \forall v\ [\ state(\gamma 1, t1) \models experience(v) \Leftrightarrow state(\gamma 2, t1) \models experience(v)\ ]\ ] \Rightarrow$
$\quad \forall w\ [\ state(\gamma 1, t) \models trust(w) \Leftrightarrow state(\gamma 2, t) \models trust(w)\ ]$

Note that as the property refers to two different histories (and compares them), it cannot be expressed in modal temporal logic: then only reference can be made to one history.

Limited memory d. Limited memory expresses that trust only depends on past experiences in a certain time interval of duration d back in time from the present. In mathematical terms:

$e|\geq t-d = f|\geq t-d \Rightarrow \quad te(e, t) = te(f, t)$

This property can be expressed logically in TTL as follows:

$\forall \gamma 1, \gamma 2, t\ [\ \forall t1 \leq t\ [\ t1 \geq t-d \Rightarrow$
$\quad \forall v\ [\ state(\gamma 1, t1) \models experience(v) \Leftrightarrow state(\gamma 2, t1) \models experience(v)\ ]\ ] \Rightarrow$
$\quad \forall w\ [\ state(\gamma 1, t) \models trust(w) \Leftrightarrow state(\gamma 2, t) \models trust(w)\ ]$

Trust Monotonicity. Monotonicity expresses that the more positive experiences are, the higher the trust. Mathematically:

$e \leq f \ \& \ te(e, 0) \leq te(f, 0) \quad \Rightarrow \quad te(e, t) \leq te(f, t)$

Note that this property again refers to two different histories; it can be expressed logically as follows:

$\forall\gamma1, \gamma2, t$
[$\forall w1,w2$ [state($\gamma1$, 0) |= trust(w1) & state($\gamma2$, 0) |= trust(w2) $\Rightarrow$ w1$\leq$w2] & $\forall t1\leq t$ $\forall v1,v2$
[state($\gamma1$, t1) |= experience(v1) & state($\gamma2$, t1) |= experience(v2) $\Rightarrow$ v1$\leq$v2]] $\Rightarrow$ $\forall w1,w2$ [state($\gamma1$, t) |= trust(w1) & state($\gamma2$, t) |= trust(w2) $\Rightarrow$ w1$\leq$w2]

The two different histories again imply that this property cannot be expressed in modal temporal logic.

Positive trust extension. Positive (or negative) trust extension expresses that trust is increasing if only positive (negative) experiences are encountered, i.e., after a positive (negative) experience, trust will become at least as high (low) as it was. In mathematical terms:

$\forall s,t$ [$\forall u \in$ TIME [s $\leq$ u < t $\Rightarrow$ e_u positive] $\Rightarrow$ te(e, s) $\leq$ te(e, t)

This property can be expressed logically as follows

$\forall s,t$ [$\forall u \forall v$ [s $\leq$ u < t & state(γ, u) |= experience(v) $\Rightarrow$ pos(v)]]
$\Rightarrow$ $\forall w1,w2$ [state(γ, s) |= trust(w1) & state(γ, t) |= trust(w2) $\Rightarrow$ w1$\leq$w2]

Trust Flexibility: degree of trust gaining d. The property degree of trust gaining (or dropping) expresses, independent of the trust state, after how many positive (or negative) experiences trust will be positive (or negative). In mathematical terms:

$\forall t$ [$\forall k \in$ TIME [t-d < k $\leq$ t $\Rightarrow$ e_k negative] $\Rightarrow$ te(e, t) negative

This property can be expressed logically as follows

$\forall t, d$ [state(γ, t) |= trust(w) & $\forall t1$ [[t-d$\leq$t1$\leq$t & state(γ, t1) |= experience(v)] $\Rightarrow$ pos(v)] $\Rightarrow$ pos(w)]

Positive limit approximation. Positive limit approximation expresses that it is always possible to reach maximal trust, if a sufficiently long period with only positive experiences is encountered (the same for the negative case). Mathematically:

If a t exists such that for all s > t it holds that e_s is maximal (in EV),
then a t' exists such that te(e, t) is maximal (in TV) for all t > t'.

This property can be expressed logically as follows:

$\exists t$ $\forall t1\geq t$ [state(γ, t1) |= experience(v) $\Rightarrow$ $\neg$ $\exists v1$ v1>v] $\Rightarrow$ $\exists t'$ $\forall t1\geq t'$ [state(γ, t1) |= trust(w) $\Rightarrow$ $\neg$ $\exists w1$ w1>w]

This property can be relaxed a bit by specifying a margin such that trust should become that close to the maximal value. In [18] also the following properties of trust update functions are defined and related to properties of trust evolution functions. They are formalised logically as follows.

Trust Update Monotonicity. A trust update function tu is monotonic if higher experience values and higher trust values lead to higher trust update values. In mathematical terms:

ev1 $\leq$ ev2 & tv1 $\leq$ tv2 $\Rightarrow$ tu(ev1, tv1) $\leq$ tu(ev2, tv2)

This property can be expressed logically as follows

$\forall\gamma1, \gamma2, t$ $\forall v1,v2, w1, w2$ [state($\gamma1$, t) |= experience(v1) $\wedge$ trust(w1)&
state($\gamma2$, t) |= experience(v2) $\wedge$ trust(w2) $\Rightarrow$ v1$\leq$v2 & w1$\leq$w2] $\Rightarrow$
$\forall w1,w2$ [state($\gamma1$, t+1) |= trust(w1) & state($\gamma2$, t+1) |= trust(w2) $\Rightarrow$ w1$\leq$w2]

Positive and negative trust extension. This property states that positive (negative) experiences lead to higher (lower) trust values. In mathematical terms:

ev positive $\Rightarrow$ tu(ev, tv) $\geq$ tv

This property can be expressed in logical terms as follows:

∀v1,w1,w2 [[state(γ, t) |= experience(v1) ∧ trust(w1) & state(γ, t+1) |= trust(w2)] ⇒ w1≤w2]

Strict positive (negative) monotonic progression. This property is a stronger version of the previous one and states that positive (negative) experiences lead to strictly higher (lower) trust values, as long as this is possible. In mathematical terms:

ev positive and tv not maximal (in T) ⇒ tu(ev, tv) > tv

This property can be expressed logically as follows

∀v1,w1,w2 [[state(γ, t) |= experience(v1) ∧ trust(w1) & ∃w3 w3>w1 & state(γ, t+1) |= trust(w2)]⇒ w1<w2]

Negative or positive trust fixation of degree d. After d negative events the agent will never trust anymore and its trust will remain the least possible. After d positive events the agent will forever trust (even when faced with negative events) and its trust will remain maximal.

∀t, d [state(γ, t) |= trust(w) & ∀t1 [[t-d≤t1≤t & state(γ, t1) |= experience(v)] ⇒ pos(v)] ⇒
∀t2≥t [state(γ, t) |= trust(w2) ⇒ ¬∃w w>w2]]

6 The Scenario Case Study

This section describes an experiment based on 294 subjects, taken from [17]. From the 294 subjects, 238 subjects (81%) completed the full questionnaire. The other 19% of the subjects were either not able to complete the questionnaire because of technical problems, decided to stop during the experiment, or did not respond to a question within a given time limit of 15 minutes between each two questions. Only the data obtained from subjects that fully completed the questionnaire have been used. In the

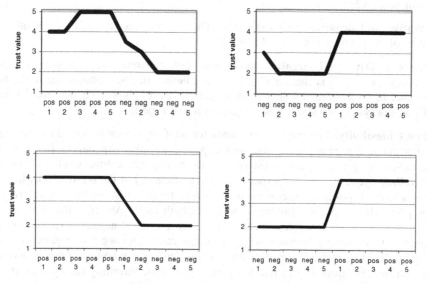

Fig. 2. Dynamics of trust in Photocopier (upper two), resp. Travel Agency (lower two) experiments, both for positive experiences first and negative experiences first

test the effect of experiences with an organisation or an object on the trust in that organisation or object is measured. The effect of the experiences on trust is measured by describing various experiences in small stories and instructing the subject to express his or her trust in the object or organisation after having gone through such experiences, using a five-points trust rating scale [17]. In this scale, trust value 3 represents neutral trust, value 4 and 5 represent positive trust, and value 1 and 2 represent negative trust.

Each scenario consisted of an introduction and ten distinctive stories, five of which were positive (written and validated as to induce trust) and five of which were negative (written and validated as to induce distrust). For more details, see [17]. In Figure 2 the median of the trust values of the population for both the photocopier and the travel agency are plotted. It shows that indeed trust increases when the subject had a positive experience and that trust decreases upon a negative experience. There is a clear difference between plots starting with negative experiences and plots that start with positive experiences. To determine the significance of this difference, a 2-way between subjects ANOVA test was performed on the means of the positive and negative experiences within a single scenario. The ANOVA test takes both the experience (positive or negative) and the order in which experiences are presented (positive-first or negative-first) into account. The results show that both factors have an effect on trust in the object or organisation at a significance level beyond 0,001.

7 Analysis of the Scenario Case Study

The outcomes of the experiments as described in Section 6 and (in particular) the traces shown in Figure 2 have been compared to the most relevant dynamic properties from Section 5:

Positive and negative trust extension. These properties are satisfied for all four types of human traces shown in Figure 2.

Strict positive and negative monotonic progression. These properties failed according to the TTL checker [2] for all four human traces, probably due to the low number of possible trust values (if there are few trust values to choose from, people do not always choose a new trust value after a new experience).

Trust Flexibility. The property succeeds for all four human traces, given the right value for d. For example, for the photocopier, 3 negative experiences in a row are sufficient to get a negative trust (no matter how positive trust was), and for the travelling agency 2 negative experiences are sufficient. For the positive side, in both cases only 1 positive experience is sufficient to get trust positive again, so the property 'degree of trust gaining 1' holds for both cases. An effect that does occur, however, in the photocopier context, is that after a series of negative experiences, the level of trust does not become as high as in the case of no negative experiences (see Figure 2). More refined properties than the ones above can be formulated to account for this relative form of trust fixation. Notice that in the travelling agency context this effect does not occur.

Positive limit approximation. This property assumes that the traces under investigation are of infinite length. Therefore, it makes no sense to check it against the traces of this case study. (For example, if for t one chooses the last time point of the trace, then the property is not very informative). However, analytically it can be shown that for the photocopier an upper limit of 5 and a lower limit of 2 are reached, whilst for the travel agency an upper limit of 4 and a lower limit of 2 are reached.

Limited memory d. According to the TTL checker, this property succeeds for all four human traces, given the right value for d. This can be illustrated by considering the same example as for Trust Flexibility above: for the photocopier, 3 negative experiences in a row are sufficient to get a negative trust (no matter what the previous trust value was), and for the travelling agency 2 negative experiences are sufficient.

8 Trust and Tracing Case Study

The Trust and Tracing game [24] is a research tool designed to study human behaviour with respect to trust in commodity supply chains and networks in different institutional and cultural settings. The game played by human participants is used both as a tool for data gathering and as a tool to make participants feed back on their daily experiences.

The focus of study is on trust in a business partner when acquiring or selling commodities with invisible quality. There are five roles: traders (producers, middlemen and retailers), consumers and a tracing agency. The real quality of a commodity is known by producers only. Sellers may deceive buyers with respect to quality, to gain profits. Buyers have either to rely on information provided by sellers (Trust) or to request a formal quality assessment at the Tracing Agency (Trace). This costs a tracing fee for the buyer if the product is what the seller stated (honest). The agency will punish untruthful sellers by a fine. Middleman and Retailers have an added value for the network by their ability to trace a product cheaper than a consumer can.

Commodities usually flow from producers to middlemen, from middlemen to retailers and from retailers to consumers. Players receive 'monopoly' money upfront. Producers receive sealed envelopes representing lots of commodities. Each lot is of a certain commodity type (represented by the colour of the envelope) and of either low or high quality (represented by a ticket covered in the envelope). The envelopes may only be opened by the tracing agency, or at the end of the game to count points collected by the consumers. The player who has collected most points is the winner in the consumer category. In the other categories the player with maximal profit wins.

Sessions played until 2005, see [16], provided insights, such as: participants who know and trust each other beforehand tend to start trading faster and trace less. The afterwards indignation about deceits that had not been found out during the game is higher in these groups than it is when participants do not know each other. The objective of [16] was to model the behaviour of sellers and buyers using the concept of trust. The model is inspired by [19] who define reliability trust as "trusting party's probability estimate of success of the transaction". This choice allows for considering economic aspects; agents may decide to trade with low-trust partners if loss in case of deceit is low. In the paper trust is defined as a subjective probability. This choice allows trust to be related with risk. In the agent-based simulations of [16], the agent's *trust model* is based on tracing results and the trust update schema proposed in [18]:

$$\text{trust}_{t+1}(\delta^+)=(1-\delta^+)\,\text{trust}_t+\delta^+ \quad \text{if the experience is positive}$$
$$\text{trust}_{t+1}(\delta^-)=(1-\delta^-)\,\text{trust}_t \quad \text{if the experience is negative}$$

where trust_t represents trust after the t transactions. The value of $\text{trust}=1$ represents complete trust, $\text{trust}=0$ represents complete distrust, and $\text{trust}=0.5$ represents complete uncertainty. This model is an asymmetric trust update function, with either completely positive or completely negative experience. A negative experience, or losing something, may have stronger impact than a positive experience, or gaining the same thing. This is known as the endowment effect [1, 26]. δ^+ and δ^- are impact factors of positive and negative experiences respectively. They are related by an endowment coefficient e.

$$\delta^+ = e\,\delta^-, \quad 0<e\leq 1$$

This trust update function has the following properties: monotonicity, positive and negative trust extension, and strict positive and negative progression.

For buyers, *trading* entails the trust-or-trace decision. In human interaction, this decision depends on human factors that are not sufficiently well understood to incorporate in a multi-agent system. To not completely disregard these intractable factors, the trust-or-trace decision is modelled as a random process instead of as a deterministic process. The agglomerate of all these intractable factors is called the confidence factor. The distribution involves experience-based trust in the seller, the value ratio of high versus low quality, the cost of tracing, and the buyer's confidence factor.

Tracing reveals the real quality of a commodity. The tracing agent executes the tracing and punishes cheaters as well as traders reselling bad commodities in good faith. The tracing agent only operates on request and requires some tracing fee. Several factors influence the tracing decision to be made after buying a commodity. The most important factors for the current research are the buyer's *trust* in seller and his confidence. Trust is modelled as a subjective evaluation of the probability that the seller would not cheat on the buyer. Confidence reflects the preference of a particular player to trust rather than trace, represented as a value on the interval [0,1]. The other factors are: the satisfaction ratio of the commodity (tracing makes more sense for valuable products than for products with small satisfaction ratio), and the tracing costs which depend on the depth to be traced.

Agent models were validated against the game sessions played by the humans (see [16] for details). Thus, computer simulations were performed for the same game setups with populations of 15 agents: 3 producers, 3 middlemen, 3 retailers, 6 consumers. A set of validation game setups (combinations of free parameters of the agent models) was built to be able to compare output of the computer simulations and the results of the human games. The values of free parameters were selected uniformly from their definition intervals to confirm the models capability to reproduce desired input-output relationships and explore their sensitivities. Some of the highlights of the outcomes are as follows.

Effects of confidence on tracing. The difference in output variables with respect to high and low levels of confidence is not significant for neutral risk-taking agents (i.e., agents that evaluate the potential profit equally high as the risk of deception in their buying/selling decision function). For high-risk-taking agents (i.e., agents that prefer potential profit over the risk of deception) the amount of traces decreases for highly confident players and increases for lowly confident players.

Effects of confidence on cheating. The number of cheats is high for highly confident players in games with dominating number of risk-neutral players. Surprisingly, the results show the opposite for risk-taking players: some dishonest sellers do not get traced and punished in highly risk-taking game configurations, so they are encouraged to continue their fraudulent practices.

Effects of confidence on certification and guaranteeing. The number of guarantees is linked with agent's confidence through the concept of the tracing trust. High confidence leads to a lower number of traces, meaning fewer deceptions are discovered and consequently a higher average tracing trust. High tracing trust decreases the seller's risk and thus increases number of guarantees provided by the seller (a guarantee increase seller's costs in case of deception).

In all experiments, effects of risk-taking attitude are consistent: high risk-taking leads to more cheating, less certificates and increased willingness to give guarantees and to rely on them. Differences in risk-taking attitude outweigh changes in other parameters. This result corresponds with observations from human games.

9 Analysis of Trust and Tracing Case Study

The model as introduced in the previous section has been tested in a case study. According to [15], initial trust and honesty of the agents have the strongest influence on the evolution of trust throughout the game. Thus, the following 4 games with homogeneous agents were chosen from the validation set of the game setups to be played using the multi-agent simulation system:

- Distrusting buyers and dishonest sellers
- Distrusting buyers and honest sellers
- Trusting buyers and dishonest sellers
- Trusting buyers and honest sellers

Distrusting buyers have a low initial value of trust (0.1) and develop trust when confronted with honest sellers. Vice versa, trusting buyers have a high initial trust value (1.0) and decrease it through tracing and discovering cheaters in games with dishonest sellers. One additional game was played having only one dishonest agent and in which the rest of the agents are all honest and trusting. Each game played is logged as set of traces generated by each of the agents. Each trace logs the trust and honesty values, and all experiences influencing these. In the logs the trust of an agent is represented by the predicate

    has_cheating_trust(AgentA,AgentB,*trust_value*).

This expression means that AgentA has trust value of *trust_value* w.r.t. AgentB. The positive experience influencing the AgentA's trust is recorded using predicate successful_trade(AgentA, AgentB). The predicate has_cheated(AgentA, AgentB) represents the negative experience (a cheat has been discovered by the tracing agency) that AgentB cheated on AgentA. In a similar way agents log their honesty level that is updated in the same way as their level of trust. The predicate has_honesty(AgentA, honesty_value) represents the current value of honesty of AgentA. If the predicate potential_cheat(AgentA) holds, then the value of the honesty of AgentA decreases. If the predicate punishment(AgentA) holds, then the honesty value of AgentA increases.

The game traces were checked against a number of trust properties proposed in Section 5. To this end, the properties of Section 5 have been slightly modified, to make them compatible with the predicates as mentioned above. For example, the properties have been parameterised with a variable for the trusting agent and a variable for the trusted agent, and have been checked for all combinations of agents. The results were as follows:

Positive and negative trust extension. The TTL checker [2] showed that these properties hold for the trust model of the software agents in the Trust and Tracing game. This fact can be also confirmed analytically by the trust update function (see Section 8).

Strict positive and negative monotonic progression. These properties were also confirmed by the TTL checker for all software agents. This is a consequence of the fact that the software agents update their trust using only strictly positive or negative experiences.

Trust flexibility. The automated checks pointed out that this property holds only for a part of the traces for a given value of d. This effect can be explained by the differences in the initial value of trust in the agents across the different games. Furthermore, due to the randomness of the agent's partner selection and the endowment effect (see Section 8) model and the time limits of the games the agents can have insufficient number of trades (experiences) to establish positive trust level .

Positive limit approximation. As explained in Section 7, it makes no sense to check this property against traces of finite length. However, analytically it can be shown that the trust update function converges to the maximum level of trust (1.0) while the agent perceives a continuous sequence of positive experiences.

Limited memory d. According to the TTL checker this property does not hold for the software agents. After each new experience value and performed trust update, the agents still remember all the previous experiences. The history of the experiences is weighted, meaning that the older an experience gets the lower its weight in the current value of trust. In addition, the weight drops exponentially causing the agent to ignore old experiences rapidly.

10 Discussion

This paper focused on the dependence of trust on experiences over time, abstracting from other possible influences, which are assumed as an idealisation to be not present or constant. For trust states as mental states, following literature in Philosophy of Mind such as [20], functional role and representation relation specifications were formalised both in a logical way and in a numerical mathematical way according to a

Dynamical Systems Theory approach (based on integral and differential equations; cf. [25]). It is shown how a backward functional role specification in mathematical terms corresponds to a trust update function in the discrete case and a differential equation in the continuous case and is directly usable in a computational manner for simulation. Moreover, it is shown how a backward representation relation can be obtained (as a repeated application of the functional role specification) by a summation (discrete case) or integration (continuous case) of the (inflating) experiences over time.

Trust may be influenced by experiences of different types. This paper models differences between experiences by mapping them into one overall set of distinct experience 'values'. In addition, more explicit distinctions between different dimensions of experience could be made. Also other cognitive or emotional factors could be integrated, such as the concept of expectation. The work presented in [5, 8, 13, 14] addresses some of these other aspects of trust, which could be integrated. This is left for future work.

In our work we focus on the dynamic properties of the trust models where trust is represented by a single variable. However, we acknowledge the fact that trust is a complex cognitive phenomenon [5] and might be represented by multiple factors or have a more complex relationship with decision making in humans and software agents. To apply our analytical method in such complex cases break down the trust model into single aspect. For each single aspect the proposed dynamic properties can be studied. As an example, this method was applied to the Trust and Tracing case study [14, 15], where different types of trust are used depending on the context of the decision making. The trust-or-trace decision is based on the cheating trust that estimates the subjective probability of truthful behaviour of the opponent while the partner selection decision is based on the negotiation trust, revealing the subjective probability of reaching an agreement with a given partner.

The requirements imposed on models for trust dynamics may depend on individual characteristics of agents; therefore, a variety of models that capture these characteristics may be needed. The approach put forward here enables the explication of these characteristics. Formal specification of both qualitative and quantitative models is supported, based on trust evolution functions and trust update functions. Validation has taken place on the basis of extensive empirical and simulated data in different contexts.

To formalise and analyse dynamic phenomena, often it is implicitly or explicitly claimed that temporal logic, e.g., linear time or branching time temporal logic, is useful; e.g., [3, 9, 10] Other literature claims that the Dynamical Systems Theory, based on differential equations is a suitable approach to dynamics of cognitive phenomena; cf. [25]. In this study it has been found that a number of basic properties for trust dynamics cannot be expressed in standard temporal logic, nor in the form of integral and differential equations. For example, the quite elementary property 'trust monotonicity' that expresses that better experiences lead to more trust is not expressible. The reason for this lack of expressivity is the impossibility to refer to and compare different histories. In the logical language TTL used here, traces are first class citizens; for example, variables and quantifiers can be used over them. In this way explicit reference can be made to histories, and they can be compared.

References

1. Amamor-Boadu, V., Starbird, S.A.: The value of anonimity in supply chain relationships. In: Bremmers, H.J., Omta, S.W.F., Trienekens, J.H., Wubben, E.F.M. (eds.) Dynamics in Chains and Networks, pp. 238–244. Wageningen Academic Publishers, Holland (2004)
2. Bosse, T., Jonker, C.M., Meij, L., van der Sharpanskykh, A., Treur, J.: Specification and Verification of Dynamics in Cognitive Agent Models. In: Proceedings of the Sixth International Conference on Intelligent Agent Technology, IAT'06., pp. 247–254. IEEE Computer Society Press, Los Alamitos (2006)
3. Burgess, J.P.: Temporal logic. In: Gabbay, D.M., Guenther, F. (eds.) Handbook of Philosophical Logic, vol. 2, Reidel, Dordrecht (1984)
4. Castelfranchi, C., Falcone, R.: Principles of Trust for MAS: Cognitive Anatomy, Social Importance, and Quantification. In: Demazeau, Y. (ed.) Proceedings of the Third International Conference on Multi-Agent Systems, pp. 72–79. IEEE Computer Society, Los Alamitos (1998)
5. Castelfranchi, C., Falcone, R., Social Trust, A.: Social Trust: A Cognitive Approach. In: Castelfranchi, C., Tan, Y.H. (eds.) Trust and Deception in Virtual Societies, pp. 55–90. Kluwer Academic Publishers, Dordrecht (2001)
6. Demolombe, R.: To trust information sources: a proposal for a modal logical framework. In: Proceedings of the First International Workshop on Trust, pp. 9–19 (1998)
7. Elofson, G.: Developing Trust with Intelligent Agents: An Exploratory Study. In: Proceedings of the First International Workshop on Trust, pp. 125–139 (1998)
8. Fisher, M.: Temporal Development Methods for Agent-Based Systems. Journal of Autonomous Agents and Multi-Agent Systems 10, 41–66 (2005)
9. Fisher, M., Wooldridge, M.: On the formal specification and verification of multi-agent systems. International Journal of Co-operative Information Systems, IJCIS. In: Huhns, M., Singh, M. (eds.) special issue on Formal Methods in Co-operative Information Systems: Multi-Agent Systems, 6 (1), 37–65 (1997)
10. Galton, A.: Temporal Logic. Stanford Encyclopedia of Philosophy, URL (2003), http://plato.stanford.edu/entries/logic-temporal/#2
11. Galton, A.: Operators vs Arguments: The Ins and Outs of Reification. Synthese 150, 415–441 (2006)
12. Gambetta, D.: Trust. Basil Blackwell, Oxford (1990)
13. Grandison, T., Sloman, M.: A Survey of Trust in Internet Applications. IEEE Communications Surveys (2000)
14. Jonker, C.M., Meijer, S., Tykhonov, D., Verwaart, T.: Agent-based Simulation of the Trust and Tracing Game for Supply Chains and Networks. Technical Report, Delft University of Technology
15. Jonker, C.M., Meijer, S., Tykhonov, D., Verwaart, D.: Modelling and Simulation of Selling and Deceit for the Trust and Tracing Game. In: Castelfranchi, C., Barber, S., Sabater, S., Singh, M. (eds.) Proceedings of the Trust in Agent Societies Workshop, pp. 78–90. Springer, Heidelberg (2005)
16. Jonker, C.M., Schalken, J.J.P., Theeuwes, J., Treur, J.: Human Experiments in Trust Dynamics. In: Jensen, C., Poslad, S., Dimitrakos, T. (eds.) iTrust 2004. LNCS, vol. 2995, pp. 206–220. Springer, Heidelberg (2004)
17. Jonker, C.M., Treur, J.: Formal analysis of models for the dynamics of trust based on experiences. In: Garijo, F.J., Boman, M. (eds.) MAAMAW 1999. LNCS, vol. 1647, pp. 221–232. Springer, Heidelberg (1999)

18. Jøsang, A., Presti, S.: Analysing the Relationship between Risk and Trust. In: Jensen, C., Poslad, S., Dimitrakos, T. (eds.) Trust 2004. LNCS, vol. 2995, pp. 135–145. Springer, Heidelberg (2004)
19. Kim, J.: Philosophy of Mind. Westview Press, Boulder (1996)
20. Lewis, D., Weigert, A.: Social Atomism, Holism, and Trust. In: Sociological Quarterly, pp. 455–471 (1985)
21. Marsh, S.: Trust and Reliance in Multi-Agent Systems: a Preliminary Report. In: Proc. of the Fourth European Workshop on Modelling Autonomous Agents in a Multi-Agent World, MAAMAW'92, Rome (1992)
22. Marsh, S.: Trust in Distributed Artificial Intelligence. In: Castelfranchi, C., Werner, E. (eds.) MAAMAW 1992. LNCS, vol. 830, pp. 94–112. Springer, Heidelberg (1994)
23. Meijer, S., Hofstede, G.J.: The Trust and Tracing game. In: Proceedings of 7th Int. workshop on experiential learning. IFIP WG 5.7 SIG conference, Aalborg, Denmark (2003)
24. Port, R.F., van Gelder, T. (eds.): Mind as Motion: Explorations in the Dynamics of Cognition. MIT Press, Cambridge, Mass (1995)
25. Rabin, M.: Psychology and Economics. Journal of Economic Literature 36, 11–46 (1998)
26. Ramchurn, S.D., Hunyh, D., Jennings, N.R.: Trust in Multi-Agent Systems, Knowledge Engineering Review (2004)

Author Index

Lecture Notes in Artificial Intelligence (LNAI)